C0-BXP-060

Submanifolds and holonomy

CHAPMAN & HALL/CRC
Research Notes in Mathematics Series

Submission of proposals for consideration
Suggestions for publication, in the form of outlines and representative samples, are invited by the Editorial Board for assessment. Intending authors should approach one of the main editors or another member of the Editorial Board, citing the relevant AMS subject classifications. Alternatively, outlines may be sent directly to the publisher's offices. Refereeing is by members of the board and other mathematical authorities in the topic concerned, throughout the world.

Preparation of accepted manuscripts
On acceptance of a proposal, the publisher will supply full instructions for the preparation of manuscripts in a form suitable for direct photo-lithographic reproduction. Specially printed grid sheets can be provided. Word processor output, subject to the publisher's approval, is also acceptable.

Illustrations should be prepared by the authors, ready for direct reproduction without further improvement. The use of hand-drawn symbols should be avoided wherever possible, in order to obtain maximum clarity of the text.

The publisher will be pleased to give guidance necessary during the preparation of a typescript and will be happy to answer any queries.

Important note
In order to avoid later retyping, intending authors are strongly urged not to begin final preparation of a typescript before receiving the publisher's guidelines. In this way we hope to preserve the uniform appearance of the series.

CRC Press UK
Chapman & Hall/CRC Statistics and Mathematics
23 Blades Court
Deodar Road
London SW15 2NU
Tel: 020 8875 4370

Jürgen Berndt, Sergio Console,
and Carlos Olmos

Submanifolds and holonomy

CHAPMAN & HALL/CRC

A CRC Press Company

Boca Raton London New York Washington, D.C.

Library of Congress Cataloging-in-Publication Data

Berndt, Jürgen
Submanifolds and holonomy / Jürgen Berndt, Sergio Console, Carlos Olmos.
p. cm. -- (Research notes in mathematics ; 434)
Includes bibliographical references and index.
ISBN 1-58488-371-5 (alk. paper)
1. Submanifolds. 2. Holonomy groups. I. Console, Sergio. II. Olmos, Carlos. III. Title.
IV. Chapman & Hall/CRC research notes in mathematics series ; 434.
QA649.B467 2003
516,3′62—dc21 2003041924
CIP

Direct all inquiries to CRC Press LLC, 2000 N.W. Corporate Blvd., Boca Raton, Florida 33431.

Visit the CRC Press Web site at www.crcpress.com

International Standard Book Number 1-58488-371-5
Library of Congress Card Number 2003041924
Printed in the United States of America 1 2 3 4 5 6 7 8 9 0
Printed on acid-free paper

Für meine Eltern Marianne und Erhard
Ai miei genitori Franca e Ettore
A mis padres Rosa y Enrique

Preface

The geometry of curves and surfaces has attracted mathematicians, physicists and other scientists for many centuries. Starting from simple geometric observations, mathematicians produce highly sophisticated theories that often lead not just to a deeper understanding of the observations made at the beginning, but also to further questions. Curves are one-dimensional objects and surfaces have two dimensions. One question that often arises is: What happens in higher dimensions? This is a natural question, since experience tells us that, in many instances, more than two dimensions are relevant. The generalizations of curves and surfaces to higher dimensions are submanifolds.

In this book we deal with particular questions about the geometry of submanifolds.

For Jürgen Berndt, the gateway to this area has been the classification by Elie Cartan of isoparametric hypersurfaces in real hyperbolic spaces. In his doctorate thesis he investigated the analogous problem for complex hyperbolic spaces. Surprisingly, a full classification is still not known, and recent results show that this problem is much more difficult than expected. These recent results stem from the author's investigations about isometric actions with an orbit of codimension one, so-called cohomogeneity one actions. Cohomogeneity one actions are currently of interest in Riemannian geometry for the construction of metrics with special properties, for instance, Einstein metrics, metrics with special holonomy and metrics of positive sectional curvature. The investigation of actions on manifolds and the geometry of their orbits is a central theme in his research.

Submanifold geometry is the primary research topic of Sergio Console. He has been particularly interested in the interaction of algebraic and geometric methods for the study of the Riemannian geometry and the topology of submanifolds of space forms with simple geometric invariants, for example, isoparametric or homogeneous submanifolds. In particular, he learned from the third author how to use holonomy methods in submanifold geometry, a theme he discussed much with the first author in 1995 when they both worked at the University of Cologne. This was the beginning of the plan to write the present monograph, and collaboration on this project started when all the authors met in Turin in 1997.

Carlos Olmos is mainly interested in local and global submanifold geometry in space forms, in particular in problems related to the so-called normal holonomy that combines local and global methods. He is also interested in Riemannian and Lorentzian geometry. The subjects of his doctoral thesis, directed by Cristian Sánchez, motivated most of his research.

Many available textbooks deal with the geometry of curves and surfaces, the classical topic for introductory courses to differential geometry at universities. In contrast, only few books deal with submanifolds of higher dimensions. Although many books on differential geometry contain chapters about submanifolds, these chapters are often quite short and contain only basic material. A standard reference for submanifold geometry has been *Geometry of Submanifolds* by Bang-yen Chen, but this book was written in 1973 and concerned research problems that were of interest at that time. Books dealing with more recent problems from submanifold geometry are *Critical Point Theory and Submanifold Geometry* (1988) by Richard Palais and Chuu-lian Terng, *Submanifolds and Isometric Immersions* (1990) by Marcos Dajczer et al., *Tubes* (1990) by Alfred Gray, and *Lie Sphere Geometry with Applications to Submanifolds* (1992) by Thomas Cecil. To some extent, these books deal with topics that also appear in our book, but, for these problems, our approach is different and relies on methods involving the holonomy group of the normal connection of a submanifold. These methods originated from the Normal Holonomy Theorem that was proved by the third author in 1990. The Normal Holonomy Theorem is the analogue for submanifold geometry in space forms of Marcel Berger's classification of holonomy groups of Riemannian connections on manifolds. Since 1990, normal holonomy has developed as a powerful tool in submanifold geometry. The purpose of this book is to present a modern and self-contained introduction to submanifold geometry with emphasis on topics where the tool of normal holonomy had great impact. This book is aimed at researchers and graduate students in mathematics, in particular in geometry, and could be used as a textbook for an advanced graduate course.

We briefly describe the contents of this book. Until now, the main applications of normal holonomy concern submanifolds of space forms, that is, manifolds of constant sectional curvature. For this reason, we first present an introduction to submanifolds in space forms and discuss in detail the fundamental results about such submanifolds. Important examples of submanifolds of Euclidean spaces are orbits of linear Lie group actions, and, for this reason, we investigate in great detail the geometry of such orbits. Then we introduce the concept of normal holonomy and present the Normal Holonomy Theorem together with its proof and some applications. In great detail, we apply the tool of normal holonomy to study isoparametric submanifolds and their focal manifolds, orbits of linear Lie group actions and homogeneous submanifolds, and homogeneous structures on submanifolds. At the end of the book we discuss generalizations to submanifolds of Riemannian manifolds, in particular of Riemannian symmetric spaces. In an appendix, we summarize the necessary facts about Riemannian manifolds, Lie groups and Lie algebras, homogeneous spaces, symmetric spaces and flag manifolds, which the reader might find helpful.

Several proofs presented in the book have never appeared in the literature. For instance, we present a new proof of Cartan's theorem about the existence of totally geodesic submanifolds of Riemannian manifolds, a result that is hard to find in the literature. An advantage of this book is that it contains much material that is currently accessible only in a large number of published articles in various journals. The book also contains a number of open problems that might attract the reader.

Of course, there are many interesting and fascinating problems in submanifold geometry that are not touched on in this book. The reason is simply that there are too many of these problems. Our selection of topics for this book has been motivated by normal holonomy and, naturally, also by personal taste and interest.

To produce most of the illustrations we used the software SUPERFICIES by Angel Montesinos Amilibia of Universidad de Valencia. SUPERFICIES is freely distributed, with source code, under GNU General Public License and is available at ftp://topologia.geomet.uv.es/pub/montesin/

We would like to thank Simon Chiossi, Antonio Di Scala, Anna Fino, Sergio Garbiero and Simon Salamon for their careful reading of parts of the manuscript and for their suggestions for improvements.

Contents

Chapter 1

Introduction

These notes are motivated by recent progress in submanifold geometry in space forms, using new methods based on the holonomy of the normal connection. Particular progress has been made in the framework of homogeneous submanifolds, isoparametric submanifolds and their generalizations. In this monograph we present an introduction to this topic and a thorough survey of all main results in this area. The proofs presented here are, to some extent, new, resulting in a more unified treatment of this topic. At the end of the book, we discuss generalizations of some problems to more general manifolds, in particular symmetric spaces.

The study of submanifolds in Euclidean space has a long tradition, and many beautiful results and theories emerged from it. The first objects of interest were surfaces in 3-dimensional Euclidean space on which certain geometric or analytic properties were imposed. For example, Weingarten surfaces were defined by a functional relationship between their principal curvatures. This class of surfaces contains minimal surfaces and surfaces of constant mean curvature, both of which still attract much interest. The immediate generalization to higher dimension is the study of hypersurfaces in Euclidean spaces of arbitrary finite dimensions. A fundamental result of this theory states that principal curvatures, together with the integrability conditions given by the equations of Gauss and Codazzi, determine uniquely the hypersurface up to a rigid motion of the Euclidean space. Note that higher dimensional hypersurfaces are more rigid than surfaces. Indeed, by the Beez-Killing Theorem, the second fundamental form of an m-dimensional hypersurface with $m > 2$ is generically determined by the first, see [12] and Exercise 2.8.1.

In higher codimension the situation is slightly more complicated, since one can choose among infinitely many normal directions. Each normal direction induces a set of principal curvatures, and the information regarding all these sets of principal curvatures is encoded in the second fundamental form, or shape operator, of the submanifold. The canonical derivative of Euclidean space induces in a natural way a metric connection on the normal bundle of the submanifold, the so-called normal connection. The second fundamental form, the normal connection, and the integrability conditions given by the equations of Gauss, Codazzi and Ricci determine the submanifold locally, up to a rigid motion of the Euclidean space. Such higher complexity is the reason that research on the local geometry of submanifolds of Euclidean space with simple geometric invariants is relatively recent compared with surface geometry.

A very influential paper by Chern, do Carmo and Kobayashi [55] on this topic was published in 1971. In this paper the authors studied submanifolds with second fundamental form of constant length and this was the starting point for quite a few investigations in submanifold geometry. For instance, parallelism of the second fundamental form was first mentioned in [55], a geometric feature widely studied in the 1980s. Submanifolds with parallel second fundamental form in Euclidean spaces are locally orbits of a distinguished class of representations, namely the isotropy representations of semisimple symmetric spaces, or s-representations for short. These s-representations have a fundamental role in the context of our investigations. For many reasons, the orbits of s-representations play a role in submanifold geometry similar to that of symmetric spaces in Riemannian geometry.

A simple geometric condition for submanifolds of higher codimension is flatness of the normal bundle (the first thorough account on the subject is maybe the book by B.Y. Chen [48] published in 1973). In this case, the normal connection is trivial and the geometric data are all encoded in the shape operator, consisting in this case of a commuting family of selfadjoint operators that can be simultaneously diagonalized. When the principal curvatures (with respect to parallel normal fields) are constant, one gets a very important class of submanifolds: isoparametric submanifolds. These submanifolds are generalizations of isoparametric hypersurfaces, which were introduced at the beginning of the 20th century in the context of geometrical optics and studied by Segre, Levi Civita and É. Cartan, among others.

There is a strong link between isoparametric submanifolds and s-representations. In fact, as a consequence of a result by Thorbergsson [219], the orbits of s-representations are almost all submanifolds with constant principal curvatures, that is, principal curvatures with respect to parallel normal fields along curves are constant.

In this book we will explore the central position of s-representations in the framework of submanifold geometry in space forms. At the same time, we will illustrate a method for investigating the local geometry of submanifolds of space forms. The main tools are the following:

- *Reduction of codimension* (Theorem 2.5.1): allows one to ignore the part of the normal bundle on which the shape operator vanishes.
- *Moore's Lemma for local splitting* [139]: permits splitting a submanifold locally into irreducible components.
- *Normal Holonomy Theorem* [173]: this result yields the decomposition of the representation of the normal holonomy group on the normal spaces into the sum of irreducible representations, all of which are s-representations. It can be regarded as a kind of extrinsic analogue to Rham's decomposition theorem and Berger's classification of Riemannian holonomy groups. The Normal Holonomy Theorem involves geometric constructions such as focal manifolds and holonomy tubes.

We will also present some recent results on the geometry of homogeneous submanifolds of space forms. In the case of hyperbolic spaces, the strategy is to regard them as hypersurfaces of a Lorentz space and Lorentzian holonomy plays an important rôle.

We now summarize the contents of this book.

In Chapter 2 we explain the basics of submanifold geometry in space forms. We introduce the main local invariants and derive the fundamental equations for submanifolds. Then we investigate some simple conditions on local invariants. For example, the vanishing of the second fundamental form characterizes *totally geodesic submanifolds*, which we shall consider in Section 2.4, where we classify totally geodesic submanifolds of space forms. If a submanifold M of a Riemannian manifold $\bar{M}$ is contained in a totally geodesic submanifold of $\bar{M}$ of dimension less than the dimension of $\bar{M}$, one says that there is a *reduction of the codimension of* M.

In Section 2.5 we explain Theorem 2.5.1 (reduction of the codimension), the first of our three basic tools. A natural generalization of totally geodesic submanifolds is that of *totally umbilical submanifolds*, which means that the second fundamental form is proportional to the metric.

In Section 2.6 we discuss the classification of totally umbilical submanifolds of space forms. The second of our main tools, Moore's Lemma for the local splitting of submanifolds, is explained in Section 2.7.

Chapter 3 is devoted to the study of an important class of submanifolds, namely those arising as orbits of isometric actions of Lie groups on Riemannian manifolds. These submanifolds, which we shall call *(extrinsically) homogeneous*, have a strong regularity, since their geometric invariants are the same at each point (modulo some suitable identification).

In Section 3.1 we present the general setup, introducing some basic concepts such as orbit types, principal orbits, isotropy and slice representations. We will investigate orbits from two different viewpoints: the geometry of a single orbit, and the geometry of the whole set of orbits. Indeed, an action on a Riemannian manifold $\bar{M}$ determines a foliation (often singular) whose leaves are the orbits of the action. For some distinguished types of representations, like s-representations and polar actions, introduced in Section 3.2, it is more interesting to investigate the whole orbit foliation, since the subgroup of isometries of the ambient space that leaves some orbit invariant does not depend on the orbit.

Polar actions on $\mathbb{R}^n$ are characterized by the existence of a linear subspace, called a *section*, that intersects every orbit and lies perpendicular to the orbits at intersection points. Now, s-representations are polar (the tangent space to any flat in the symmetric space is a section) and, by Dadok's Theorem 3.2.15 [63], have the same orbits as polar actions. The existence of a section implies that the orbit foliation has remarkable geometric properties. The orbits are all equidistant and their tangent spaces are parallel. Moreover, if N is a principal orbit, the normal bundle to N is flat with trivial holonomy, and the principal curvatures of N with respect to any parallel normal field are constant. This leads to the study of isoparametric submanifolds of $\mathbb{R}^n$, which will be carried out later in Chapter 5.

In Section 3.3 we will reverse our approach, starting from a homogeneous Riemannian manifold and examining whether it can be viewed as an orbit in some Riemannian manifold or, in other terms, if it admits an equivariant embedding. The study of a single orbit is carried out in Sections 3.4 and 3.5, where we investigate homogeneous submanifolds of space forms.

In Section 3.5 we show how the theory of homogeneous submanifolds of the hyperbolic space H^n can be used to obtain general results on the action of a connected Lie subgroup of $O(n, 1)$ on the Lorentzian space $\mathbb{R}^{n,1}$, [69].

In Section 3.6 we describe the extrinsic geometry of orbits and give, among other things, a description of the second fundamental form of the orbit of a representation of a Lie group G in terms of the corresponding Lie algebra representation. As already mentioned, symmetric submanifolds of $\mathbb{R}^n$ (and their generalizations to spaces of constant curvature) are historically one of the first class of homogeneous submanifolds that have been studied and classified. Section 3.7 is devoted to this topic.

In Sections 3.8 and 3.9, we consider classes of submanifolds sharing properties with homogeneous ones. The most classical "homogeneous-like" property is the constancy of principal curvatures, which characterizes isoparametric hypersurfaces of space forms. Historically, these hypersurfaces are defined as regular level sets of isoparametric functions, so that they determine an orbit-like foliation of the ambient space. Thus, isoparametric hypersurfaces are very close to being homogeneous (and actually, in many cases, they are). In higher codimension, a natural "homogeneous-like" property is that the algebraic type of the second fundamental form does not depend on the point. This is an extrinsic version of curvature homogeneous manifolds [225].

In Chapter 4 we explore holonomy methods for studying submanifold geometry. In Section 4.1 we recall some important results about the holonomy of a Riemannian manifold, which will allow us to make a comparison with results on normal holonomy. Important in the extrinsic context is the Normal Holonomy Theorem 4.2.1 [173], which asserts that the nontrivial part of the action of the normal holonomy group on a normal space is an s-representation. The Normal Holonomy Theorem is an extrinsic analogue of Berger's Theorem on Riemannian holonomy, and one of its main consequences is the recognition that orbits of s-representations play a similar role in submanifold geometry as Riemannian symmetric spaces in Riemannian geometry. This is illustrated in Section 4.4, where we define some important tools for the study of submanifolds with some regularity (e.g., submanifolds with parallel normal fields whose shape operator has constant eigenvalues, isoparametric submanifolds): focalizations, building holonomy tubes. These tools will be very important in the subsequent chapters.

Chapter 5 is devoted to the study of certain generalizations of isoparametric hypersurface to higher codimensions.

In Section 5.2 we will discuss some geometric properties of isoparametric submanifolds. Among them is the important fact, due to Terng [216] (and to Carter and West [38] in the particular case of codimension three), that one can associate a finite reflection group to isoparametric submanifolds, the *Coxeter group*. The singular levels of isoparametric maps are actually focal manifolds of the isoparametric submanifolds. Thus, isoparametric maps determine a singular foliation of the ambient space.

If M is a fixed isoparametric submanifold of $\mathbb{R}^n$, the leaves are the parallel manifolds $M_\xi = \{p+\xi(p) \mid p \in M\}$, where ξ is an arbitrary parallel normal vector field. Suppose that one drops the assumption that the normal bundle is flat in the definition of isoparametric submanifold, and requires only that the shape operator A_ξ have constant eigenvalues for any parallel normal vector field $\xi(t)$ along any piecewise differentiable curve. Then one defines a submanifold of a space form with constant principal curvatures. Strübing studied these submanifolds in [205] (even though he called them isoparametric) and noticed that the focal manifolds of an isoparametric hypersurface are submanifolds with constant principal curvatures. This result was generalized by Heintze, Olmos and Thorbergsson in [96] to isoparametric submanifolds. Indeed, in [96] the converse is proved as well, namely that a submanifold with constant principal curvatures is either isoparametric or a focal manifold of an isoparametric submanifold (Theorem 5.3.3 here). The paper [205] of Strübing is actually of great importance for the methods adopted by him: he constructed tubes around isoparametric submanifolds and used normal holonomy for the study of submanifolds with constant principal curvatures. These are the same methods we make use of extensively.

In Section 5.5 we examine a slightly more general situation than that of an isoparametric submanifold. We suppose there exists a (locally defined) parallel normal section that is not umbilical and isoparametric, i.e., the eigenvalues of the shape operator A_ξ in the ξ direction are constant (and A_ξ is not a multiple of the identity). Our aim is to study the geometric consequences of this property. What we will show is that this imposes severe restrictions on the geometry of the submanifold. Namely, if a submanifold of the sphere with this property does not (locally) split, then it is a submanifold with constant principal curvatures [61], (Theorems 5.5.2 and 5.5.8). This result also has a global version for complete simply connected submanifolds [70], (Theorem 5.5.8).

In Chapter 6 we continue the study of geometric invariants by distinguishing submanifolds with constant principal curvatures from other submanifolds. We weaken the above condition on the existence of a (nontrivial) parallel isoparametric normal field. We require only that the submanifold admits "enough" parallel normal fields or, in other words, that the normal holonomy group has a nontrivial pointwise fixed subspace whose dimension is called the rank of the immersion. In the case of a homogeneous submanifold M of Euclidean space it was proved in [175] that if the rank is larger than or equal to 2 then M is an orbit of an s-representation. A crucial fact in the original proof is the following: the curvature normals of a homogeneous submanifold (which can be defined as in the isoparametric case, taking into consideration only directions in the flat part of the normal bundle) have constant length. In [70] it is actually shown that this property alone, together with the same higher rank assumption, yields a generalization (Theorem 6.1.7) of the above higher rank rigidity result. Unlike the theorems on higher isoparametric rank rigidity (Theorem 5.5.2 and 5.5.8), Theorem 6.1.7 is global and is, in fact, false without the completeness assumption. As a consequence, one can derive a global characterization of an isoparametric submanifold: a complete immersed and irreducible submanifold $f : M^m \to \mathbb{R}^n$, $m \geq 2$

with flat normal bundle is isoparametric if and only if the distances to its focal hyperplanes are constant on M.

Moreover, we apply these higher rank rigidity results to pursue a study of normal holonomy (and, more generally, $\nabla^\perp$-parallel transport) of a homogeneous submanifold. In the more general setting of homogeneous (pseudo)-Riemannian vector bundles, the holonomy algebra can be described in terms of projection of Killing vector fields on the homogeneous bundle (see [60], for more details). In the case of Riemannian manifolds, this yields Kostant's method for computing the Lie algebra of the holonomy group of a homogeneous Riemannian manifold. One can compute normal holonomy of homogeneous submanifolds by projecting on the normal spaces the Killing vector fields determined by the action (Theorem 6.2.7).

In Chapter 7, we give a differential characterization of homogeneous submanifolds. Using this framework, we characterize orbits of s-representations and we study isoparametric submanifolds, giving a proof of Thorbergsson's theorem, which asserts that isoparametric submanifolds of higher codimension are homogeneous and actually orbits of s-representations.

In Chapters 8 and 9, we generalize certain topics to a broader class of manifolds. In Chapter 8, we first discuss the fundamental equations for submanifolds in Riemannian manifolds. One of the basic methods for studying submanifolds in general Riemannian manifolds is to investigate tubes around them and their focal sets. The tool for this is Jacobi field theory, which will be explained in detail. We then continue with a discussion of totally geodesic, totally umbilical and symmetric submanifolds in Riemannian manifolds.

We present a proof of Cartan's theorem on local existence of totally geodesic submanifolds in Riemannian manifolds, and Hermann's theorem about the existence of complete totally geodesic submanifolds in Riemannian manifolds. We discuss how totally umbilical submanifolds are related to extrinsic spheres and present a characterization of extrinsic spheres by circles. We finally discuss the relation between symmetric submanifolds and submanifolds with parallel second fundamental form in general Riemannian manifolds.

In Chapter 9, we keep to submanifold theory within symmetric spaces. Symmetric spaces are natural generalizations of space forms possessing a beautiful geometric structure. We discuss the classification problems of totally geodesic submanifolds, of totally umbilical submanifolds and extrinsic spheres, of symmetric submanifolds, of submanifolds with parallel second fundamental form and of homogeneous hypersurfaces.

In the Appendix, we briefly recall basic material needed for this book: Riemannian manifolds, Lie groups and Lie algebras, homogeneous spaces, symmetric spaces and flag manifolds.

Chapter 2

Basics of submanifold theory in space forms

In this chapter, we present the basics of submanifold theory in spaces of constant curvature, or briefly, in space forms. In the literature there are mainly three different definitions for a submanifold of a Riemannian manifold. Let M and $\bar{M}$ be Riemannian manifolds. When we have an isometric immersion from M into $\bar{M}$ we say that M is an *immersed submanifold* of $\bar{M}$. When M is a subset of $\bar{M}$ and the inclusion $M \hookrightarrow \bar{M}$ is an isometric immersion, then M is said to be a *submanifold* of $\bar{M}$. If, in addition, the inclusion is an embedding, then M is said to be an *embedded submanifold* of $\bar{M}$. Note that a submanifold is embedded if and only if its topology is the one that is induced from the ambient space. The immersion of a real line as a figure eight in a plane is an example of an immersed submanifold that is not a submanifold. And a dense geodesic on a torus is an example of a submanifold that is not embedded. The local theories for these three kinds of submanifolds are the same, the only difference arises when dealing with global questions. Therefore, when we deal with local properties of submanifolds, we make no distinction and just say submanifold.

The Riemannian metric on a manifold induces a Riemannian metric on a submanifold in a natural way. More precisely, let M be a submanifold of a Riemannian manifold $\bar{M}$. At each point $p \in M$, the inner product $\langle \cdot , \cdot \rangle_p$ on $T_p\bar{M}$ induces an inner product on T_pM that we denote by the same symbol. This family of inner products on the tangent spaces of M forms a Riemannian metric on M, the so-called *induced Riemannian metric*. Note that this is a local notion and has to be interpreted for an isometric immersion $f : M \to \bar{M}$ by means of the formula $\langle X, Y \rangle_p = \langle f_{*p}X, f_{*p}Y \rangle_{f(p)}$ for all $p \in M$ and $X, Y \in T_pM$. We will always view a submanifold of a Riemannian manifold with the metric that is induced in this way.

We now give a more detailed description of the contents of this chapter. In Section 2.1, we start with the fundamental equations of submanifold theory. The equations of first order, the so-called Gauss and Weingarten formulae, define the basic objects for the study of submanifolds: the second fundamental form, the shape operator and the normal connection. The second fundamental form and the shape operator contain the same information and just provide different viewpoints of the same aspects. The fundamental equations of second order, the so-called equations by Gauss, Codazzi and Ricci, represent higher-dimensional generalizations of the Frenet equations that are familiar to us from the differential geometry of curves. The Gauss-Codazzi-Ricci equations determine locally a submanifold of a space form in a unique way up to isometric congruence of the space form. This is the content of the fundamental theorem of local submanifold geometry in space forms, Theorem 2.1.2.

As an application of the fundamental equations we present the standard models for the three different types of spaces forms in Section 2.2: the Euclidean space, the sphere and the real hyperbolic space.

If ξ is a normal vector of a submanifold M at a point p, the shape operator A_ξ of M in direction ξ is a self-adjoint endomorphism of the tangent space T_pM of M at p, and hence is diagonalizable. Its eigenvalues are the so-called principal curvatures of M at p in direction ξ. Almost all geometric properties of a submanifold involve the shape operator, or equivalently, the second fundamental form, and in particular the principal curvatures. For this reason, we investigate principal curvatures more thoroughly in Section 2.3.

The simplest condition one can impose on the second fundamental form is that it vanishes. This characterizes *totally geodesic submanifolds*, which we consider in Section 2.4. The main result of that section is the classification of totally geodesic submanifolds of space forms.

If a submanifold M of $\bar{M}$ is contained in a totally geodesic submanifold of $\bar{M}$ of dimension less than $\dim \bar{M}$, one says there is a *reduction of the codimension of* M. In Section 2.5, we derive a sufficient condition for reduction of codimension in space forms, Theorem 2.5.1.

A natural generalization of totally geodesic submanifolds is that of *totally umbilical* submanifolds, which means that in each normal direction the shape operator is a multiple of the identity. A basic example is a sphere in Euclidean space. In Section 2.6, we derive the classification of totally umbilical submanifolds of space forms.

Another reduction process for submanifolds is that of splitting as an extrinsic product. This so-called *reducibility of submanifolds* is discussed in Section 2.7. The main result is Moore's Lemma. Both Moore's Lemma and the theorem on the reduction of codimension are fundamental tools for the study of submanifolds of space forms.

2.1 The fundamental equations for submanifolds of space forms

In this section, we present the fundamental equations for submanifolds of space forms and discuss one of their major applications, the Fundamental Theorem of Local Submanifold Geometry in Space Forms. A few details about space forms, or spaces of constant sectional curvature, can be found in the next section. In this section, we denote by $\bar{M}$ an n-dimensional space of constant curvature κ.

a) The fundamental equations of first order

We first want to derive the fundamental equations of first order, which then induce the main objects for the study of submanifolds: second fundamental form, shape operator and normal connection. These equations can be generalized without any problems to any Riemannian manifold as ambient space, a case that will be studied later.

The Riemannian metric on $\bar{M}$ induces along M an orthogonal splitting of $T\bar{M}$

$$T\bar{M}|M = TM \oplus \nu M \ .$$

The vector bundle νM is called the *normal bundle* of M. The fibre at $p \in M$ of νM is the *normal space at* p and is denoted by $\nu_p M$ (or $\nu_p(M)$). Let X, Y be vector fields on M. In order to differentiate them with respect to the Levi Civita connection $\bar{\nabla}$ of $\bar{M}$ we have to extend them to vector fields on $\bar{M}$. But it turns out that, for our purposes, it does not matter how the extension is done, therefore, we introduce no new symbols. We decompose $\bar{\nabla}_X Y$ into its tangent part $(\bar{\nabla}_X Y)^\top$ and its normal part $(\bar{\nabla}_X Y)^\perp$. Then the *Levi Civita connection* ∇ of M is given by

$$(\bar{\nabla}_X Y)^\top = \nabla_X Y \ ,$$

and one defines the *second fundamental form* of M by

$$\alpha(X, Y) := (\bar{\nabla}_X Y)^\perp \ .$$

This gives the orthogonal decomposition

$$\bar{\nabla}_X Y = \nabla_X Y + \alpha(X, Y),$$

which is called the *Gauss formula*. The Gauss formula and the vanishing of the torsion of $\bar{\nabla}$ and ∇ imply that the second fundamental form is a symmetric tensor field with values in the normal bundle of M.

A section of νM is called a *normal vector field* of M. Let ξ be a normal vector field of M and decompose $\bar{\nabla}_X \xi$ into its tangent and normal component. The normal part induces a connection $\nabla^\perp$ on νM, the so-called *normal connection* on M. We now define

$$A_\xi X := -(\bar{\nabla}_X \xi)^\top .$$

The tensor field A_ξ is called the *shape operator of* M *in direction* ξ and is related to the second fundamental form α by the equation

$$\langle \alpha(X, Y), \xi \rangle = \langle A_\xi X, Y \rangle \ .$$

The symmetry of α implies that A_ξ is a selfadjoint tensor field on M. The previous equation also shows that for each $p \in M$ the endomorphism $A_\xi(p)$ does not depend on the extension of ξ_p as a normal vector field. Thus, we can define the shape operator with respect to any normal vector of M. The collection of all these endomorphisms is called the *shape operator* of M and is denoted by A. The orthogonal decomposition

$$\bar{\nabla}_X \xi = -A_\xi X + \nabla^\perp_X \xi$$

is known as the *Weingarten formula*.

The formulas of Gauss and Weingarten are first order equations.

b) The fundamental equations of second order

We will now derive three equations of second order, namely the equations of Gauss, Codazzi and Ricci. For this we first recall that the covariant derivatives of the second fundamental form and of the shape operator are given by the formulas

$$(\nabla^{\perp}_X \alpha)(Y,Z) = \nabla^{\perp}_X \alpha(Y,Z) - \alpha(\nabla_X Y, Z) - \alpha(Y, \nabla_X Z)\,,$$
$$(\nabla_X A)_\xi Y = (\nabla_X A_\xi)Y - A_{\nabla^{\perp}_X \xi} Y \;=\; \nabla_X(A_\xi Y) - A_\xi(\nabla_X Y) - A_{\nabla^{\perp}_X \xi} Y\,.$$

These two covariant derivatives are related by

$$\langle (\nabla^{\perp}_X \alpha)(Y,Z), \xi\rangle = \langle (\nabla_X A)_\xi Y, Z\rangle = \langle (\nabla_X A_\xi)Y, Z\rangle - \langle A_{\nabla^{\perp}_X \xi} Y, Z\rangle\,.$$

Let R and $\bar{R}$ be the Riemannian curvature tensor of M and $\bar{M}$, respectively. Recall that if $\bar{M}$ is a space of constant curvature κ, its Riemannian curvature tensor is of the form

$$\bar{R}(X,Y)Z = \kappa(\langle Y,Z\rangle X - \langle X,Z\rangle Y)\,.$$

We now relate R and $\bar{R}$ with the extrinsic invariants α, A and $\nabla^{\perp}$. Let X, Y, Z be vector fields on M. Using the formulas of Gauss and Weingarten we obtain

$$\begin{aligned}
\bar{R}(X,Y)Z &= \bar{\nabla}_X \bar{\nabla}_Y Z - \bar{\nabla}_Y \bar{\nabla}_X Z - \bar{\nabla}_{[X,Y]} Z \\
&= \bar{\nabla}_X(\nabla_Y Z + \alpha(Y,Z)) - \bar{\nabla}_Y(\nabla_X Z + \alpha(X,Z)) \\
&\quad -(\nabla_{[X,Y]} Z + \alpha([X,Y],Z)) \\
&= \nabla_X \nabla_Y Z + \alpha(X, \nabla_Y Z) - A_{\alpha(Y,Z)} X + \nabla^{\perp}_X \alpha(Y,Z) \\
&\quad -\nabla_Y \nabla_X Z - \alpha(Y, \nabla_X Z) + A_{\alpha(X,Z)} Y - \nabla^{\perp}_Y \alpha(X,Z) \\
&\quad -\nabla_{[X,Y]} Z - \alpha(\nabla_X Y, Z) + \alpha(\nabla_Y X, Z) \\
&= R(X,Y)Z - A_{\alpha(Y,Z)} X + A_{\alpha(X,Z)} Y \\
&\quad +(\nabla^{\perp}_X \alpha)(Y,Z) - (\nabla^{\perp}_Y \alpha)(X,Z)\,.
\end{aligned}$$

The tangential component of this equation gives

$$\kappa(\langle Y,Z\rangle X - \langle X,Z\rangle Y) = (\bar{R}(X,Y)Z)^{\top} = R(X,Y)Z - A_{\alpha(Y,Z)} X + A_{\alpha(X,Z)} Y\,,$$

and the normal component gives

$$0 = (\bar{R}(X,Y)Z)^{\perp} = (\nabla^{\perp}_X \alpha)(Y,Z) - (\nabla^{\perp}_Y \alpha)(X,Z)\,,$$

since $\bar{M}$ has constant curvature κ. The first equation is called the *Gauss equation*, the second one the *Codazzi equation*. If W is another vector field on M, the Gauss equation can be rewritten as

$$\begin{aligned}
&\kappa(\langle Y,Z\rangle\langle X,W\rangle - \langle X,Z\rangle\langle Y,W\rangle) \\
&= \langle R(X,Y)Z, W\rangle - \langle \alpha(Y,Z), \alpha(X,W)\rangle + \langle \alpha(X,Z), \alpha(Y,W)\rangle\,.
\end{aligned}$$

And if ξ is a normal vector field, the Codazzi equation can be rewritten as

$$\langle(\nabla_X A)_\xi Y, Z\rangle - \langle(\nabla_Y A)_\xi X, Z\rangle = 0\,.$$

Using again the formulas of Gauss and Weingarten, we obtain

$$\begin{aligned}
0 &= \bar{R}(X,Y)\xi = \bar{\nabla}_X\bar{\nabla}_Y\xi - \bar{\nabla}_Y\bar{\nabla}_X\xi - \bar{\nabla}_{[X,Y]}\xi \\
&= \bar{\nabla}_X(-A_\xi Y + \nabla^\perp_Y\xi) - \bar{\nabla}_Y(-A_\xi X + \nabla^\perp_X\xi) + A_\xi[X,Y] - \nabla^\perp_{[X,Y]}\xi \\
&= -\nabla_X(A_\xi Y) - \alpha(X, A_\xi Y) - A_{\nabla^\perp_Y\xi}X + \nabla^\perp_X\nabla^\perp_Y\xi \\
&\quad +\nabla_Y(A_\xi X) + \alpha(Y, A_\xi X) + A_{\nabla^\perp_X\xi}Y - \nabla^\perp_Y\nabla^\perp_X\xi \\
&\quad +A_\xi\nabla_X Y - A_\xi\nabla_Y X - \nabla^\perp_{[X,Y]}\xi \\
&= (\nabla_Y A)_\xi X - (\nabla_X A)_\xi Y + R^\perp(X,Y)\xi + \alpha(A_\xi X, Y) - \alpha(X, A_\xi Y)\,.
\end{aligned}$$

Here,

$$R^\perp(X,Y)\xi = \nabla^\perp_X\nabla^\perp_Y\xi - \nabla^\perp_Y\nabla^\perp_X\xi - \nabla^\perp_{[X,Y]}\xi$$

is the curvature tensor of the normal bundle with respect to the normal connection $\nabla^\perp$, the so-called *normal curvature tensor* of M. The tangential part of the latter equation yields again the Codazzi equation. The normal part gives the so-called *Ricci equation*, namely

$$0 = (\bar{R}(X,Y)\xi)^\perp = R^\perp(X,Y)\xi + \alpha(A_\xi X, Y) - \alpha(X, A_\xi Y)\,.$$

If η is another normal vector field of M, the Ricci equation can be rewritten as

$$\langle R^\perp(X,Y)\xi, \eta\rangle = \langle [A_\xi, A_\eta]X, Y\rangle\,,$$

where $[A_\xi, A_\eta] = A_\xi A_\eta - A_\eta A_\xi$. If $R^\perp$ vanishes, one says that M has *flat normal bundle*. The geometric interpretation of a flat normal bundle is that parallel translation with respect to $\nabla^\perp$ of normal vectors along curves with the same initial and end point in M depends only on the homotopy class of the curve. This will be discussed later in more detail in the context of normal holonomy.

Note that, for submanifolds of space forms, the geometric interpretation of the Ricci equation is that the normal curvature tensor measures the commutativity of the shape operators. We summarize the fundamental equations in

THEOREM 2.1.1

Let M be a submanifold of a space form $\bar{M}$ of constant curvature κ. Then the following equations hold for all vector fields X, Y, Z, W on M and all normal vector fields ξ, η of M:

Gauss equation:

$$\begin{aligned}
\langle R(X,Y)Z, W\rangle = {} & \kappa(\langle Y,Z\rangle\langle X,W\rangle - \langle X,Z\rangle\langle Y,W\rangle) \\
& +\langle\alpha(Y,Z), \alpha(X,W)\rangle - \langle\alpha(X,Z), \alpha(Y,W)\rangle\,;
\end{aligned}$$

Codazzi equation:

$$(\nabla^{\perp}_X \alpha)(Y, Z) = (\nabla^{\perp}_Y \alpha)(X, Z) \ ;$$

Ricci equation:

$$\langle R^{\perp}(X, Y)\xi, \eta\rangle = \langle [A_\xi, A_\eta]X, Y\rangle \ .$$

The fundamental equations of Gauss, Codazzi and Ricci play an analogous role in submanifold geometry of space forms as the Frenet equations in the differential geometry of curves. Namely, they suffice to determine, up to isometries of the ambient space, a submanifold of a space form. This is the conclusion of the fundamental theorem of local submanifold geometry.

THEOREM 2.1.2 (Fundamental Theorem of Local Submanifold Geometry)
Let M be an m-dimensional Riemannian manifold, ν a Riemannian vector bundle over M of rank p, ∇' a metric connection on ν and $\alpha(X, Y)$ a symmetric tensor field on M with values in ν. Define $A : \nu \to \mathrm{End}(TM)$ by $\langle A_\xi X, Y\rangle = \langle \alpha(X, Y), \xi\rangle$ for $X, Y \in T_pM$, $\xi \in \nu_p$, $p \in M$. Suppose α, A and ∇' satisfy the equations of Gauss, Codazzi and Ricci for some real number κ. Then, for each point $p \in M$, there exists an open neighbourhood U of p in M and an isometric immersion f from U into a space form $\bar{M}^n(\kappa)$ of constant curvature κ, $n = m + p$, such that α and A are the second fundamental form and shape operator of f, respectively, and ν is isomorphic to the normal bundle of f. The immersion f is unique up to an isometry of $\bar{M}^n(\kappa)$. Moreover, if two isometric immersions have the same second fundamental forms and normal connection, they locally coincide up to an isometry of the ambient space.

PROOF We give a proof for $\kappa = 0$, that is, $\bar{M}^n(\kappa) = \mathbb{R}^n$. The proof for the general case is similar and can be found, for instance, in [203].

Let $E = TM \oplus \nu$ be the Whitney sum of the Riemannian vector bundles TM and ν over M. We define a connection $\hat{\nabla}$ on E by

$$\hat{\nabla}_X Y = \nabla_X Y + \alpha(X, Y) \quad \text{and} \quad \hat{\nabla}_X \xi = -A_\xi X + \nabla'_X \xi$$

for all vector fields X, Y on M and sections ξ in ν. Then the Gauss-Codazzi-Ricci equations imply that $\hat{\nabla}$ is a flat connection, that is, the curvature of $\hat{\nabla}$ vanishes. Thus, there exists an open neighborhood V of p in M and a $\hat{\nabla}$-parallel frame field $(\xi_1, ..., \xi_n)$ of E over V. Such a frame field is unique up to a linear isometry of $\mathbb{R}^n$. We denote by η_i the metric dual one-form of ξ. Then we have

$$\begin{aligned} d\eta_i(X, Y) &= X\langle \xi_i, Y\rangle - Y\langle \xi_i, X\rangle - \langle \xi_i, [X, Y]\rangle \\ &= \langle \hat{\nabla}_X \xi_i, Y\rangle + \langle \xi_i, \nabla_X Y\rangle + \langle \xi_i, \alpha(X, Y)\rangle \\ &\quad -\langle \hat{\nabla}_Y \xi_i, X\rangle - \langle \xi_i, \nabla_Y X\rangle - \langle \xi_i, \alpha(Y, X)\rangle - \langle \xi_i, [X, Y]\rangle = 0 \ , \end{aligned}$$

since by construction the TM-part of $\hat{\nabla}$ coincides with the Levi Civita connection on M, and because α is symmetric by assumption. Thus η_i is closed, and hence, there exists a function f_i on some open neighborhood U_i of p in M such that $df_i = \eta_i$. Let U be the intersection of all U_i, $i = 1, \ldots, n$. Then $f = (f_1, ..., f_n) : U \to \mathbb{R}^n$ is the isometric immersion with the required properties. Note that, once we fix the frame field, f is unique up to translation, so f is unique up to an isometry of $\mathbb{R}^n$. Observe that there is a bundle isomorphism between E and $T\mathbb{R}^n$ restricted to U, which is the identity on TM.

Finally, assume that two isometric immersions have the same second fundamental form and normal connection. We define as above Riemannian vector bundles E and E'. Then there are bundle isomorphisms between E and $T\mathbb{R}^n|M$ and E' and $T\mathbb{R}^n|M$, which are the identity on TM and preserve both metric and connection. By the same arguments as in the first part of the proof, we can see that the immersions differ locally by a rigid motion of $\mathbb{R}^n$. ∎

c) Equations of higher order

The fundamental equations of first and second order are the basic tools for investigating the geometry of submanifolds. However, one can derive further useful equations of higher order.

We discuss here an example of a third order equation. To begin with, recall that the second covariant derivative of the second fundamental form is given by

$$\begin{aligned}(\nabla^2_{X_1 X_2}\alpha)(X_3, X_4) = {} & \nabla^\perp_{X_1}(\nabla^\perp_{X_2}\alpha)(X_3, X_4) - (\nabla^\perp_{\nabla_{X_1} X_2}\alpha)(X_3, X_4) \\ & -(\nabla^\perp_{X_2}\alpha)(\nabla_{X_1} X_3, X_4) - (\nabla^\perp_{X_2}\alpha)(X_3, \nabla_{X_1} X_4) \ .\end{aligned}$$

Then, taking the covariant derivative $\nabla^\perp_{X_1}$ of the equation

$$(\nabla^\perp_{X_2}\alpha)(X_3, X_4) = \nabla^\perp_{X_2}\alpha(X_3, X_4) - \alpha(\nabla_{X_2} X_3, X_4) - \alpha(X_3, \nabla_{X_2} X_4) \ ,$$

a straightforward computation yields the so-called *Ricci formula*

$$\nabla^2_{X_1 X_2}\alpha - \nabla^2_{X_2 X_1}\alpha = -\hat{R}(X_1, X_2) \cdot \alpha \ .$$

The curvature operator $\hat{R}(X_1, X_2)$ acts on the tangent space as the Riemannian curvature tensor and on the normal space as the normal curvature tensor. The notation $\hat{R}(X_1, X_2) \cdot \alpha$ means that $\hat{R}(X_1, X_2)$ acts on the tensor α as a derivation.

REMARK 2.1.3 By taking the trace of the operator ∇^2 one defines the *Laplace-Beltrami operator* Δ. For instance,

$$\Delta\alpha := \mathrm{tr}(\nabla^2\alpha) \ .$$

Below we will use some formulae involving $\Delta\alpha$, for instance (cf. [55, formula (3.12)])

$$\frac{1}{2}\Delta||\alpha||^2 = ||\nabla^\perp\alpha||^2 + \langle\alpha, \Delta\alpha\rangle \ ,$$

where the norms and inner product are the usual ones of tensors. The term $\langle \Delta\alpha, \alpha \rangle$ can be computed directly in terms of the second fundamental form and the normal curvature tensor. This was done, for instance, in the paper by Chern, do Carmo and Kobayashi [55] (mentioned in the introduction) for the case of a minimal submanifold of a space form of constant curvature κ. In this case, the relation is

$$\langle \Delta\alpha, \alpha \rangle = n\kappa ||\alpha||^2 - ||\alpha \circ \alpha^t||^2 - ||R^\perp||^2,$$

where α^t is the adjoint of α regarded as a homomorphism from $TM \oplus TM$ to νM. □

2.2 Models of space forms

A large part of this book deals with problems in space forms. For this reason, we now take a closer look at the standard models of these spaces. The application of the fundamental equations simplifies their description.

a) The Euclidean space $\mathbb{R}^n$

Consider $\mathbb{R}^n$ as an n-dimensional smooth manifold equipped with the standard smooth structure. At each point $p \in \mathbb{R}^n$ we identify the tangent space $T_p\mathbb{R}^n$ of $\mathbb{R}^n$ at p with $\mathbb{R}^n$ by means of the isomorphism

$$T_p\mathbb{R}^n \to \mathbb{R}^n \ , \ \dot{\gamma}_v(0) \mapsto v \ ,$$

where $\gamma_v(t) := p + tv$. Using this isomorphism we get an inner product $\langle \ , \ \rangle$ on $T_p\mathbb{R}^n$ by the usual dot product on $\mathbb{R}^n$, that is,

$$\langle v, w \rangle = \sum_{i=1}^{n} v_i w_i \ .$$

This family of inner products defines a Riemannian metric $\langle \ , \ \rangle$ on $\mathbb{R}^n$. We call $\mathbb{R}^n$ equipped with this Riemannian metric the *n-dimensional Euclidean space*, which we also denote by $\mathbb{R}^n$. By means of the above isomorphism, the Levi Civita connection ∇ of $\mathbb{R}^n$ coincides with the usual derivative D of $\mathbb{R}^n$. It is then a straightforward exercise to check that the Riemannian curvature tensor of $\mathbb{R}^n$ vanishes. The isometry group $I(\mathbb{R}^n)$ of $\mathbb{R}^n$ is the semidirect product $O(n) \rtimes \mathbb{R}^n$, where $\mathbb{R}^n$ acts on itself by left translations. Explicitly, the action of $O(n) \rtimes \mathbb{R}^n$ on $\mathbb{R}^n$ is given by $(A, a) \cdot x = Ax + a$ and the group structure of $I(\mathbb{R}^n)$ is given by the formula $(A, a) \cdot (B, b) = (AB, Ab + a)$. The identity component $I^o(\mathbb{R}^n)$ of $I(\mathbb{R}^n)$ is $SO(n) \rtimes \mathbb{R}^n$ and the quotient group $I(\mathbb{R}^n)/I^o(\mathbb{R}^n)$ is isomorphic to $\mathbb{Z}_2$.

b) The sphere $S^n(r)$

Let r be a positive real number, and consider the sphere

$$S^n(r) = \{p \in \mathbb{R}^{n+1} \mid \langle p, p \rangle = r^2\}$$

with radius r and center 0 in $\mathbb{R}^{n+1}$. It is a smooth submanifold of $\mathbb{R}^{n+1}$ with a unit normal vector field ξ defined by

$$\xi_p := \frac{1}{r} p \,,$$

where we again use the canonical isomorphism $T_p\mathbb{R}^n \cong \mathbb{R}^n$. Differentiating ξ with respect to tangent vectors of $S^n(r)$, we obtain for the shape operator A_ξ of $S^n(r)$ with respect to ξ the expression

$$A_\xi X = -\frac{1}{r} X$$

for each tangent vector X of $S^n(r)$. The Gauss equation then gives us the Riemannian curvature tensor R of $S^n(r)$, namely

$$R(X,Y)Z = \frac{1}{r^2}(\langle Y, Z \rangle X - \langle X, Z \rangle Y) \,.$$

This implies that $S^n(r)$ has constant sectional curvature r^{-2}. We usually denote the unit sphere $S^n(1)$ by S^n. The isometry group $I(S^n(r))$ of $S^n(r)$ is the orthogonal group $O(n+1)$ acting on $S^n(r)$ in the obvious way. The identity component $I^o(S^n(r))$ of $I(S^n(r))$ is $SO(n+1)$, and the quotient group $I(S^n(r))/I^o(S^n(r))$ is isomorphic to $\mathbb{Z}_2$.

c) The hyperbolic space H^n

There are various models for the hyperbolic space. One of them is constructed in a similar way to the sphere, but starting from a Lorentz space. We will refer to it as the *standard model*. Consider $\mathbb{R}^{n+1}$ equipped with the bilinear form

$$\langle v, w \rangle = \sum_{i=1}^{n} v_i w_i - v_{n+1} w_{n+1}$$

of signature $(n, 1)$. Identifying each tangent space of $\mathbb{R}^{n+1}$ with $\mathbb{R}^{n+1}$ as described above, we get a Lorentzian metric on $\mathbb{R}^{n+1}$, which we also denote by $\langle\ ,\ \rangle$. The smooth manifold $\mathbb{R}^{n+1}$ equipped with this Lorentzian metric is called Lorentz space and will be denoted by $\mathbb{R}^{n,1}$. Let r be a positive real number and

$$H^n(r) := \{p \in \mathbb{R}^{n,1} \mid \langle p, p \rangle = -r^2 \;,\; p_{n+1} > 0\} \,.$$

This is a connected smooth submanifold of $\mathbb{R}^{n,1}$ with time-like unit normal vector field

$$\xi_p := \frac{1}{r}\, p \,.$$

The tangent space $T_pH^n(r)$ consists of all vectors orthogonal to ξ_p and hence is a space-like linear subspace of $\mathbb{R}^{n,1}$. Thus, the Lorentz metric of $\mathbb{R}^{n,1}$ induces a Riemannian metric on $H^n(r)$.

An affine subspace W of $\mathbb{R}^{n,1}$ is Riemannian, Lorentzian or degenerate if the restriction of $\langle\ ,\ \rangle$ to the vector part of W is positive definite, has signature $(\dim(W) - 1, 1)$ or is degenerate, respectively.

The shape operator of $H^n(r)$ with respect to ξ is given by

$$A_\xi X = -\frac{1}{r} X$$

for all tangent vectors X of $H^n(r)$. The Gauss equation, which is valid also in the Lorentzian situation, then gives for the Riemannian curvature tensor R of $H^n(r)$ the expression

$$R(X,Y)Z = -\frac{1}{r^2}(\langle Y, Z\rangle X - \langle X, Z\rangle Y)\ .$$

It follows that $H^n(r)$ has constant sectional curvature $-r^{-2}$. We write H^n instead of $H^n(1)$. The orthogonal group $O(n,1)$ of all transformations of $\mathbb{R}^{n,1}$ preserving the Lorentzian inner product consists of four connected components, depending on whether the determinant is 1 or -1 and the transformation is time-preserving or time-reversing. The time-preserving transformations in $O(n,1)$ are those that leave $H^n(r)$ invariant and form the isometry group $I(H^n(r))$ of $H^n(r)$. The identity component $I^o(H^n(r))$ is $SO^o(n,1)$ and the quotient group $I(H^n(r))/I^o(H^n(r))$ is isomorphic to $\mathbb{Z}_2$.

Several other classical models of hyperbolic space are very useful for visualizing geometric aspects of H^n, for instance, for visualizing geodesics. We briefly mention two of them.

The first of these models is known as the *half plane model*

$$\{x \in \mathbb{R}^n \mid x = (x_1, \ldots, x_n), x_n > 0\}\ ,$$

endowed with the Riemannian metric

$$ds^2 := \langle\cdot,\cdot\rangle / x_n^2\ .$$

In this model, the geodesics are either lines orthogonal to the hyperplane $x_n = 0$ or circles intersecting the hyperplane $x_n = 0$ orthogonally.

The second model is known as the *Poincaré disk model* and is given by the open ball

$$\{x \in \mathbb{R}^n : ||x|| < 2\}$$

with the Riemannian metric

$$ds^2 := \langle\cdot,\cdot\rangle / \left(1 - \frac{||x||^2}{4}\right)^2\ .$$

In this model, the geodesics are circles orthogonal to the boundary sphere $||x|| = 2$ of the ball (including the degenerate circles given by diameters).

d) The classification problem for space forms

The Riemannian manifold

$$\bar{M}^n(\kappa) := \begin{cases} S^n(\kappa^{-1/2}) & , \text{ if } \kappa > 0 \\ \mathbb{R}^n & , \text{ if } \kappa = 0 \\ H^n((-\kappa)^{-1/2}) & , \text{ if } \kappa < 0 \end{cases}$$

is connected and simply connected and often referred to as a standard space of constant curvature κ. A connected Riemannian manifold M^n of constant curvature κ is called a *space form*, or sometimes also *real space form* to distinguish it from complex and quaternionic space forms. It is called a spherical, flat or hyperbolic space form depending on whether $\kappa > 0$, $\kappa = 0$ or $\kappa < 0$. Any space form M^n of constant curvature κ admits a Riemannian covering map $\bar{M}^n(\kappa) \to M^n$. A classical problem is to determine all compact space forms. A theorem by Bieberbach says that any compact flat space form M is covered by a flat torus, where the group of deck transformations is a free Abelian normal subgroup of the first fundamental group $\pi_1(M)$ of M with finite rank. The spherical space forms have been classified by J.A. Wolf [243]. The even-dimensional case appears to be quite simple, as one can show that any even-dimensional spherical space form is isometric either to the sphere or to the real projective space of corresponding dimension and curvature. The theory of hyperbolic space forms is more subtle and still an active research field.

2.3 Principal curvatures

The shape operator or second fundamental form is the fundamental entity in submanifold theory. Practically all geometrical problems concerning submanifolds involve them in one or another way. In the course of this book we will deal with submanifolds whose second fundamental form has a "regular" behaviour especially to what concerns its eigenvalues, called principal curvatures.

Various properties of α or A lead to interesting classes of submanifolds. For instance, the vanishing of α leads to totally geodesic submanifolds, which will be discussed later.

The *mean curvature vector field* H of an m-dimensional submanifold M of $\bar{M}$ is defined by

$$H := \frac{1}{m}\operatorname{tr}\alpha \,,$$

and $h := ||H||$ is the *mean curvature function* of M. A *minimal submanifold* is a submanifold with vanishing mean curvature function. This class of submanifolds has already attracted mathematicians for a long time. There is a great variety of literature concerning minimal submanifolds and, in particular, minimal surfaces. We refer the interested reader to [66].

A simple condition for principal curvatures on a hypersurface M is that they satisfy some functional relation, in which case, one calls M a *Weingarten hypersurface*. This is a classical topic; for a modern treatment of the subject, see [187]. For higher codimension, Terng generalized this notion by requiring that the submanifold has flat normal bundle and the principal curvatures satisfy a polynomial relation [217] (see Exercise 2.8.5).

In the course of this book we will encounter various kinds of properties of the second fundamental form or the shape operator that lead to interesting areas of mathematics. We start with discussing principal curvatures in more detail.

a) Principal curvatures and principal curvature vectors

Let M be submanifold of a space form $\bar{M}$. As usual, the shape operator of M is denoted by A and the second fundamental form by α. Recall that A and α are related by the equation

$$\langle \alpha(X,Y), \xi \rangle = \langle A_\xi X, Y \rangle ,$$

where $X, Y \in T_p M$ and $\xi \in \nu_p M$, $p \in M$. Because of the symmetry of α, the shape operator A_ξ of M is selfadjoint. Its eigenvalues are the *principal curvatures of M with respect to ξ*. An eigenvector of A_ξ is called a *principal curvature vector of M with respect to ξ*, and the eigenvectors corresponding to some principal curvature form a *principal curvature space*. The *multiplicity of a principal curvature* is the dimension of the corresponding principal curvature space. As $A_{s\xi}X = sA_\xi X$ for all $s \in \mathbb{R}$, the principal curvatures of M with respect to $s\xi$ are precisely the principal curvatures of M with respect to ξ multiplied with the factor s, and the principal curvature spaces are the same for all $0 \neq s \in \mathbb{R}$. For this reason, one is often interested only in the principal curvatures with respect to unit normal vectors. If, in particular, M is a hypersurface of $\bar{M}$, that is, if the codimension of M in $\bar{M}$ is one, and if ξ is a local or global unit normal vector field on M, one often speaks of the principal curvatures of M without referring to ξ. Note that the principal curvature spaces with respect to linearly independent normal vectors are, in general, not the same.

We say that a submanifold M of a space form has *constant principal curvatures* if for any parallel normal vector field $\xi(t)$, along any piecewise differentiable curve, the principal curvatures in direction $\xi(t)$ are constant. We will deal later with such submanifolds, starting from Section 4.4. If in addition the normal bundle of M is flat, one says that the submanifold is *isoparametric* .

Observe that, since the principal curvatures are roots of a polynomial (namely, the characteristic polynomial of A_ξ), they are continuous but do not need to be differentiable. For instance, if M is a surface in $\mathbb{R}^3$, since the principal curvatures can be expressed in terms of the Gaussian curvature K and the (length of the) mean curvature H by

$$\lambda_i = H \pm \sqrt{H^2 - K} , \qquad i = 1, 2 ,$$

it is clear that they are differentiable on the set of nonumbilical points (a point is *umbilical* if there is only one principal curvature at that point). A simple example of a surface where the principal curvature functions are not smooth is the monkey saddle $z = (x^3 - 3xy^2)/3$. Here, the principal curvatures are not smooth in the origin, cf. [46].

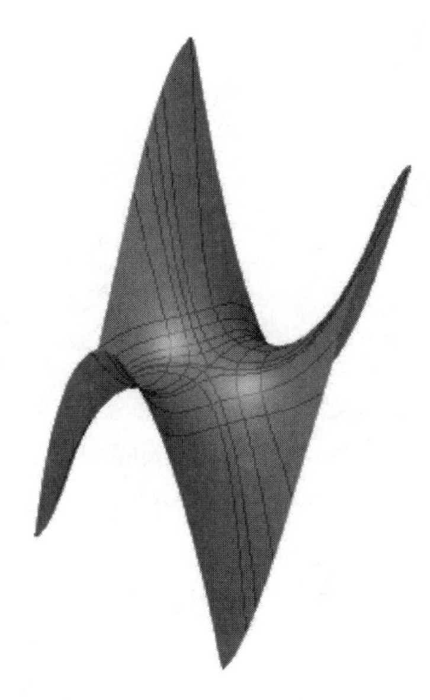

FIGURE 2.1: Principal curvature lines on the monkey saddle $z = (x^3 - 3xy^2)/3$.

However, if the multiplicities of the principal curvatures are constant on the unit normal bundle, then the principal curvatures are smooth functions.

b) Principal curvature distributions and nullity

Let ξ be a local unit normal vector field of M that is defined on a connected open subset U of M. Then A_ξ is smoothly diagonalizable over an open and dense subset of U. On each connected component of this subset we have k smooth eigenvalue functions λ_i with multiplicities m_i, $m = m_1 + \ldots + m_k$. The principal curvature space with respect to λ_i is $E_{\lambda_i} = E_i = \ker\{A_\xi - \lambda_i \mathrm{id}\}$. We also call E_i a *curvature distribution*. Note that, if ξ happens to be a global unit normal vector field on M and the principal curvatures of M are constant with respect to ξ, then each curvature distribution is globally defined on M. A curve in M, all of whose tangent vectors belong to a curvature distribution, is called a *curvature line* of M. Some curvature lines on the monkey saddle are illustrated in Figure 2.1.

More in general, a *curvature surface* is a connected submanifold S of M for which there exists a parallel unit normal vector field ξ such that $T_x S$ is contained in a principal curvature space of the shape operator A_{ξ_x} for all $x \in S$.

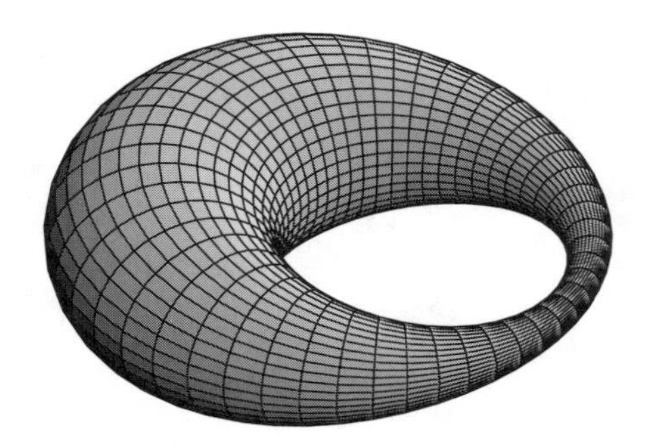

FIGURE 2.2: A cyclides of Dupin. This class of surfaces can be characterized by the fact that their curvature lines are circles or straight lines.

A submanifold M in $\mathbb{R}^n$ or S^n is said to be *Dupin*, if the principal curvatures are constant along all curvature surfaces of M. A Dupin submanifold is called *proper* if the number g of distinct principal curvatures of A_ξ is constant on the unit normal bundle of M. Important examples of Dupin submanifolds are the cyclides of Dupin and isoparametric submanifolds. We will not study Dupin submanifolds in this book; refer to [41,42,45] for more details.

The linear subspace $E_0 = \cap_{\xi \in \nu_p M} \ker A_\xi$ of T_pM is called the *nullity space of M at p*. The collection of all these spaces is called the *nullity distribution* of M. Note that this is actually a distribution only on any connected component of a suitable dense and open subset of M.

2.4 Totally geodesic submanifolds of space forms

a) Definitions

Let M be a submanifold of a Riemannian manifold $\bar{M}$. Suppose γ is a geodesic in M. Then the Gauss formula says that $\alpha(\dot{\gamma}, \dot{\gamma})$ is the second derivative of γ when considered as a curve in the ambient space $\bar{M}$. Since

$$2\alpha(X, Y) = \alpha(X + Y, X + Y) - \alpha(X, X) - \alpha(Y, Y)$$

for all $X, Y \in T_pM$, $p \in M$, we see that the second fundamental form α vanishes precisely if every geodesic in M is also a geodesic in $\bar{M}$. In such a case, M is called a *totally geodesic submanifold* of $\bar{M}$. The basic problems concerning totally geodesic submanifolds deal with existence, classification and congruency. We will deal with these problems more thoroughly in the general context of submanifolds of a Riemannian manifold in Section 8.3.

For submanifolds of space forms, as we will soon see, we have a positive answer regarding the existence problem in the following sense. For each point $p \in \bar{M}$ and every linear subspace $V \subset T_p\bar{M}$ there exists a totally geodesic submanifold M of $\bar{M}$ with $p \in M$ and $T_pM = V$. Moreover, since the exponential map $\exp_p : T_p\bar{M} \to \bar{M}$ maps straight lines through the origin $0 \in T_p\bar{M}$ to geodesics in $\bar{M}$, there is an open neighborhood U of 0 in $T_p\bar{M}$ such that $\exp_p$ maps $U \cap V$ diffeomorphically onto some open neighborhood of p in M. This implies that M is uniquely determined near p, and that any totally geodesic submanifold of $\bar{M}$ containing p and being tangent to V is contained as an open part in a maximal one with this property. This feature is known as rigidity of totally geodesic submanifolds.

b) Classification in space forms

Geodesics are clearly the simplest examples of totally geodesic submanifolds. In the standard models $\bar{M}^n(\kappa)$ of space forms, as discussed in Section 2.2, we can give the following explicit description of geodesics. Let $p \in \bar{M}^n(\kappa)$ and $X \in T_p\bar{M}^n(\kappa)$. The geodesic $\gamma_X : \mathbb{R} \to \bar{M}^n(\kappa)$ with $\gamma_X(0) = p$ and $\dot{\gamma}_X(0) = X$ is given by

$$\gamma_X(t) = \begin{cases} \cos(\sqrt{\kappa}t)p + \frac{1}{\sqrt{\kappa}}\sin(\sqrt{\kappa}t)X & , \kappa > 0 \\ p + tX & , \kappa = 0 \\ \cosh(\sqrt{-\kappa}t)p + \frac{1}{\sqrt{-\kappa}}\sinh(\sqrt{-\kappa}t)X & , \kappa < 0 \end{cases} .$$

This describes the classification of the one-dimensional totally geodesic submanifolds of $\bar{M}^n(\kappa)$.

From this we also easily see that the canonical embeddings $\bar{M}^k(\kappa) \subset \bar{M}^n(\kappa)$, $1 < k < n$, are totally geodesic. The isometry group of $\bar{M}^n(\kappa)$ acts transitively on the pairs (p, V) with $p \in \bar{M}^n(\kappa)$ and V a k-dimensional linear subspace of $T_p\bar{M}^n(\kappa)$. This, together with the uniqueness properties described above, establishes the classification of the totally geodesic submanifolds in the standard space forms.

THEOREM 2.4.1

Let $p \in \bar{M}^n(\kappa)$ and V a k-dimensional linear subspace of $T_p\bar{M}^n(\kappa)$, $0 < k < n$. Then there exists a connected, complete, totally geodesic submanifold M of $\bar{M}^n(\kappa)$ with $p \in M$ and $T_pM = V$. Moreover, M is congruent to the canonical totally geodesic embedding of $\bar{M}^k(\kappa)$ in $\bar{M}^n(\kappa)$. Each connected, totally geodesic submanifold N of $\bar{M}^n(\kappa)$ with $p \in N$ and $T_pN = V$ is an open part of M.

Actually, it is not difficult to show directly that the totally geodesic submanifolds of $\mathbb{R}^n$ are the affine subspaces (see Exercise 2.8.4). Moreover, the connected, complete, totally geodesic submanifolds of $S^n(r) \subset \mathbb{R}^{n+1}$ are precisely the intersections of $S^n(r)$ with the linear subspaces of $\mathbb{R}^{n+1}$. Analogously, the connected, complete, totally geodesic submanifolds of $H^n(r) \subset \mathbb{R}^{n,1}$ are precisely the intersections of $H^n(r)$ with the linear Lorentz subspaces of $\mathbb{R}^{n,1}$. We also propose as an exercise to give a direct proof of this (see Exercises 2.8.6 and 2.8.7).

2.5 Reduction of the codimension

A submanifold M of a Riemannian manifold $\bar{M}$ is said to be *full* if it is not contained in any totally geodesic submanifold of $\bar{M}$ of dimension less than $n = \dim \bar{M}$. If M is not full in $\bar{M}$, one says there is a *reduction of the codimension of* M. For example, M is full in $\mathbb{R}^n$ if and only if it is not contained in any affine hyperplane of $\mathbb{R}^n$. If M is not full in $\mathbb{R}^n$, there exists a smallest affine subspace of $\mathbb{R}^n$ containing M, namely the intersection of all affine subspaces containing M. If k is the dimension of this affine subspace, then one might view M as a full submanifold of $\mathbb{R}^k$. This means that we have reduced the codimension of M by $n - k$ dimensions.

In order to reduce the codimension, it is useful to study a particular subspace of the normal space called the *first normal space*. It is defined as the subspace $\mathcal{N}_p^1$ of $\nu_p M$ spanned by the image of the second fundamental form at p, that is,

$$\mathcal{N}_p^1 = \operatorname{span}\{\alpha(X,Y) \mid X, Y \in T_p M\} \subset \nu_p M .$$

In other words, $\mathcal{N}_p^1$ is the orthogonal complement in $\nu_p M$ of the linear subspace of $\nu_p M$ consisting of all normal vectors ξ at p for which the shape operator A_ξ vanishes. If the dimension of the first normal space does not depend on p, then $\mathcal{N}^1$ is a subbundle of the normal bundle νM.

The following criterion is very useful in this context for submanifold theory in space forms (see [64, Chapter 4], cf. also [78]).

THEOREM 2.5.1 (Reduction of codimension)
Let $f : M \to \bar{M}$ be an isometric immersion from an m-dimensional connected Riemannian manifold M into an n-dimensional standard space form $\bar{M}$. If for some, and hence for any, $p \in M$ the first normal space $\mathcal{N}_p^1$ is invariant under parallel translation with respect to $\nabla^\perp$, and if l denotes the constant dimension of $\mathcal{N}^1$, then there exists an $(m+l)$-dimensional totally geodesic submanifold N of $\bar{M}$ such that f is an isometric immersion of M into N.

PROOF Observe first that the orthogonal complement $\mathcal{N}^\perp$ in νM of the first normal bundle $\mathcal{N} := \mathcal{N}^1$ is invariant under $\nabla^\perp$-parallel transport as well. Note also that $\xi \in \mathcal{N}_p^\perp$ if and only if $A_\xi \equiv 0$. We divide the proof into three separate cases according to the sign of the curvature of $\bar{M}$. It is also clear that we can restrict to the model spaces $\mathbb{R}^n$, S^n and H^n, because a homothetic change of the metric does not affect the assertion.

(1) The case $\bar{M} = \mathbb{R}^n$. Let $p \in M$ and c a curve in M with $c(0) = p$. Let $\xi_0 \in \mathcal{N}_p^\perp$ and ξ the parallel normal vector field along c with $\xi(0) = \xi_0$. Since $\mathcal{N}^\perp$ is invariant under $\nabla^\perp$-parallel translation, the Weingarten formula

$$\bar{\nabla}_{\dot{c}}\xi = -A_\xi \dot{c} + \nabla^\perp_{\dot{c}}\xi = 0$$

implies that ξ is a $\bar{\nabla}$-parallel vector field along c. Since the parallel transport in $\mathbb{R}^n$ along curves is independent of initial and end points of the curve, this implies that the parallel transport of $\xi_0 \in \mathcal{N}^\perp$ is independent of the curve. Hence, for any $\xi_0 \in \mathcal{N}_p^\perp$ the parallel translation of ξ_0 along curves determines a well-defined parallel normal vector field ξ. Thus, there exists a parallel orthonormal frame field $\xi_1, \ldots, \xi_{n-m-l}$ of $\mathcal{N}^\perp$. As we have just seen, each ξ_i is the pull-back to M of a suitable constant vector field on $\mathbb{R}^n$, which we also denote by ξ_i. For any $i \in \{1, \ldots, n-m-l\}$ we define the height function

$$f_i : M \to \mathbb{R} \ , \ p \mapsto \langle f(p), \xi_i \rangle \ .$$

For each $X \in TM$ we then get

$$df_i(X) = \langle f_* X, \xi_i \rangle - \langle f(p), A_{\xi_i} X \rangle = 0 \ .$$

Hence, f_i is constant and it follows that $f(M)$ is contained in the intersection of $n - m - l$ affine hyperplanes of $\mathbb{R}^n$ with pairwise linearly independent normal directions. Such an intersection is isometrically congruent to the totally geodesic $\mathbb{R}^{m+l} \subset \mathbb{R}^n$, by which the assertion is proved.

(2) The case $\bar{M} = S^n$. Consider S^n as the unit sphere in $\mathbb{R}^{n+1}$ with centre at the origin. Let ζ be the unit normal vector field on S^n in $\mathbb{R}^{n+1}$ pointing outwards, that is, $\zeta_p = p$ for all $p \in S^n$. Recall that the Levi Civita connection $\bar{\nabla}$ of S^n is the orthogonal projection onto the tangent spaces of S^n of the directional derivative D of $\mathbb{R}^{n+1}$. Then we get

$$D_X \zeta = X \ , \ D_X \xi = \bar{\nabla}_X \xi$$

for all $X \in TM$ and all normal vector fields ξ of M which are tangent to S^n. Thus, when we consider f as an isometric immersion into $\mathbb{R}^{n+1}$, the corresponding first normal spaces form a subbundle $\bar{\mathcal{N}}^1$ of the normal bundle of M in $\mathbb{R}^{n+1}$, which is the span of $\mathcal{N}$ and $\mathbb{R}\zeta$. We also see that $\bar{\mathcal{N}}^1$ is invariant under $D^\perp$-parallel transport, where $D^\perp$ is the normal connection of M regarded as a submanifold of $\mathbb{R}^{n+1}$. By case *(1)* we now see that f is an isometric immersion into some totally geodesic $\mathbb{R}^{m+l+1} \subset \mathbb{R}^{n+1}$. But, since $\mathbb{R}^{m+l+1}$ contains $\mathbb{R}\zeta$, it also contains the origin of $\mathbb{R}^{n+1}$, and it follows that f is an isometric immersion into some totally geodesic $\mathbb{R}^{m+l+1} \cap S^n = S^{m+l}$.

(3) The case $\bar{M} = H^n$. The proof is similar to case (2) and we therefore sketch it here only. We consider H^n as a hypersurface in Lorentz space $\mathbb{R}^{n,1}$

and denote by ζ the time-like unit normal vector field on H^n given by $\zeta_p = p$ for all $p \in H^n$. As in case (2) we first deduce that f is an isometric immersion into $\mathbb{R}^{n,1}$ whose first normal spaces are invariant under parallel translation with respect to the normal connection in $\mathbb{R}^{n,1}$. We then prove, as in case (1), that f is an isometric immersion into an affine subspace of $\mathbb{R}^{n,1}$ whose linear part is a Lorentz subspace of $\mathbb{R}^{n,1}$. But this affine subspace then contains the origin of $\mathbb{R}^{n,1}$, which eventually implies that f is an isometric immersion into a totally geodesic $H^{m+l} \subset H^n$. ∎

Actually, there exists a more general version of the above theorem (see [64, Proposition 4.1]): it is sufficient to require the existence of a parallel subbundle $\mathcal{L}$ of rank $d < \operatorname{codim} M$ of the normal bundle with the property that $\mathcal{N}^1(x) \subseteq \mathcal{L}(x)$ for any x. Then the codimension reduces to d.

Some necessary and sufficient conditions for the invariance under parallel transport of the first normal bundle were obtained by do Carmo, Colares, Dajczer and Rodriguez and can be found in [64, Section 4.2].

A certain generalization to arbitrary Riemannian manifolds $\bar{M}$ can be found in [192].

REMARK 2.5.2 If a submanifold M of Euclidean or Lorentz space admits a parallel (nonzero) normal vector field ξ such that $A_\xi \equiv 0$ then M is not full (Exercise; what happens in the case of a submanifold of the sphere?). ∎

2.6 Totally umbilical submanifolds of space forms

A submanifold M of a Riemannian manifold $\bar{M}$ is said to be *umbilical in direction* ξ, if the shape operator A_ξ of M in direction of the normal vector ξ is a multiple of the identity.

A normal vector field ξ such that A_ξ is a multiple of the identity is called *umbilical normal vector field (or section)*.

If M is umbilical in any normal direction ξ, then M is called a *totally umbilical submanifold* of $\bar{M}$. M is totally umbilical if and only if

$$\alpha(X, Y) = \langle X, Y \rangle H$$

for all vector fields X, Y on M, where H is the mean curvature vector field. It is obvious that each one-dimensional submanifold and each totally geodesic submanifold is totally umbilical. It is also clear that conformal transformations of $\bar{M}$ preserve totally umbilical submanifolds.

A totally umbilical submanifold with nonzero parallel mean curvature vector field is called an *extrinsic sphere*. In a space form the two concepts of totally umbilical (and non-totally geodesic) submanifolds and extrinsic spheres coincide in dimensions ≥ 2. Indeed, we have the following

LEMMA 2.6.1
Let M be a totally umbilical submanifold of dimension ≥ 2 in a space form. Then the mean curvature vector field H is parallel with respect to the normal connection (i.e., $\nabla^\perp H = 0$) and the normal curvature $R^\perp$ vanishes identically.

PROOF We first observe that the shape operator of M is of the form $A_\xi(X) = \langle \xi, H\rangle X$. The Ricci equation then easily implies $R^\perp = 0$ (see also Exercise 2.8.12). And, since the second fundamental form of M is of the form $\alpha(X,Y) = \langle X, Y\rangle H$, the Codazzi equation implies $\langle Y, Z\rangle \nabla^\perp_X H = \langle X, Z\rangle \nabla^\perp_Y H$.

Since $\dim M \geq 2$ we can choose $Y = Z$ perpendicular to X, and we get $\nabla^\perp H = 0$. ∎

The connected, complete, totally umbilical and non-totally geodesic submanifolds M with $m = \dim M \geq 2$ of $\mathbb{R}^n$, S^n and H^n are as follows (cf. Exercises 2.8.6, 2.8.7):

In $\mathbb{R}^n$: M is a sphere $S^m(r) \subset \mathbb{R}^n$.

In S^n: M is a m-dimensional sphere which is obtained by intersecting S^n with an affine and nonlinear subspace of $\mathbb{R}^{n+1}$.

In H^n: When we consider H^n sitting inside $\mathbb{R}^{n,1}$, M can be obtained by intersecting H^n with an affine and nonlinear subspace of $\mathbb{R}^{n,1}$. In particular, the totally umbilical hypersurfaces are the intersections of H^n with the affine subspaces of $\mathbb{R}^{n,1}$ whose vector part is $(\mathbb{R}a)^\perp$. Moreover:

- If a is a time-like vector in $\mathbb{R}^{n,1}$, in which case $(\mathbb{R}a)^\perp$ is a Euclidean vector space, the totally umbilical hypersurfaces obtained in this way are geodesic hyperspheres. A *geodesic hypersphere* $M_r(p)$ in H^n is the set of all points in H^n with distance $r > 0$ to a point $p \in H^n$.

- If a is a space-like vector in $\mathbb{R}^{n,1}$, in which case $(\mathbb{R}a)^\perp$ is a Lorentzian vector space, the totally umbilical hypersurfaces obtained in this way are the hypersurfaces that are equidistant to a totally geodesic $H^{n-1} \subset H^n$.

- If a is a light-like vector in $\mathbb{R}^{n,1}$, in which case $(\mathbb{R}a)^\perp$ is degenerate, the totally umbilical hypersurfaces obtained in this way are the so-called *horospheres*.

In the Poincaré ball model of H^n, the horospheres are the spheres in the ball that are tangent to the boundary sphere of the ball. In this model, it is clear that horospheres are totally umbilical. Indeed, the identity map from the ball equipped with the Euclidean metric onto the ball equipped with the Poincaré metric is a conformal transformation. Therefore, it sends the spheres tangent to the boundary sphere of the ball, which are totally umbilical, onto totally umbilical submanifolds of H^n.

In the half plane model, the hypersurfaces $x_n = c$, $c > 0$, are horospheres. Actually, $x_n = c$ gives a family of parallel hypersurfaces that are all centred at the same point at infinity. Moreover, in this model, it is easy to see that every horosphere in H^n is isometric to the Euclidean space $\mathbb{R}^{n-1}$, and that they are totally umbilical, because of the description of the geodesics in this model.

We can summarize the above discussion on extrinsic spheres in space forms in the following theorem, which gives us an explicit description.

THEOREM 2.6.2
Let p be a point in a standard space form $\bar{M}^n(\kappa)$, $m \geq 2$, V an m-dimensional linear subspace of the tangent space $T_p\bar{M}^n(\kappa)$ and H a non-zero vector of $T_p\bar{M}^n(\kappa)$ orthogonal to V. Then there exists a unique connected complete extrinsic sphere M of $\bar{M}^n(\kappa)$ with $p \in M$, $T_pM = V$ and $H_p = H$. Moreover, M is a space of constant curvature $\kappa + \langle H, H \rangle$.

A survey about totally umbilical submanifolds in more general ambient spaces, as well as many references, can be found in [50]. We discuss totally umbilical submanifolds and extrinsic spheres in symmetric spaces in Section 9.2.

b) Pseudoumbilical submanifolds

A generalization of totally umbilical submanifolds is that of pseudoumbilical ones. A submanifold M of a Riemannian manifold $\bar{M}$ is called *pseudoumbilical* if it is umbilical in direction of the mean curvature vector field H. This just means

$$\langle \alpha(X,Y), H \rangle = \langle X, Y \rangle ||H||^2$$

for all $X, Y \in T_pM$, $p \in M$. We have the following proposition [54].

PROPOSITION 2.6.3
Let M^m be a pseudoumbilical submanifold of a standard space form $\bar{M}^n(\kappa)$. If the mean curvature vector field H of M is parallel, then either M is a minimal submanifold of $\bar{M}^n(\kappa)$, or M is a minimal submanifold of some extrinsic sphere in $\bar{M}^n(\kappa)$.

PROOF By assumption, $||H||$ is constant. If $H \equiv 0$, then M is minimal in $\bar{M}^n(\kappa)$ and the proposition holds. Let us assume that $H \not\equiv 0$. Then $\xi := H/||H||$ is a $\nabla^\perp$-parallel unit normal vector field of M. If $\bar{M}^n(\kappa) = \mathbb{R}^n$, we consider the vector field Y on M defined by

$$Y_p = p + \frac{1}{||H||}\xi_p$$

for all $p \in M$. For any $X \in TM$, we then get

$$\bar{\nabla}_X Y = X - \frac{1}{||H||} A_\xi X = X - X = 0 \; .$$

Hence, Y is a constant vector field, say $Y = p_o$. Therefore, M is contained in the hypersphere with centre at p_o and radius $1/||H||$. Moreover, since H is orthogonal to this hypersphere, M is minimal in it.

If $\kappa \neq 0$, we regard $\bar{M}^n(\kappa)$ as a hypersurface if $\mathbb{R}^{m+1}$ resp. $\mathbb{R}^{m,1}$, and similar arguments yield the result. □

2.7 Reducibility of submanifolds

a) Submanifold products and extrinsically reducible submanifolds

Let $M_1, \ldots, M_s, \bar{M}_1, \ldots, \bar{M}_s$ be Riemannian manifolds and $f_i : M_i \to \bar{M}_i$, $i = 1, \ldots, s$, be isometric immersions. The product map

$$f = f_1 \times \ldots \times f_n : M_1 \times \ldots \times M_s \to \bar{M}_1 \times \ldots \times \bar{M}_s \; ,$$
$$(p_1, \ldots, p_s) \mapsto (f_1(p_1), \ldots, f_s(p_s))$$

is called the *immersion product* of $f_1, \ldots, f_s$ or the *submanifold product* of M_1, ..., M_s in $\bar{M}_1 \times \ldots \times \bar{M}_s$. There are simple equations relating the second fundamental form and the mean curvature vector field of a submanifold product with those of its factors. Recall that there is a natural isomorphism

$$T_{(p_1,\ldots,p_s)}(M_1 \times \ldots \times M_s) = T_{p_1} M_1 \oplus \ldots \oplus T_{p_s} M_s \; ,$$

which we will use frequently in the following. Denote by α_i and H_i the second fundamental form and the mean curvature vector field of M_i, respectively. Then the second fundamental form α of $M_1 \times \ldots \times M_s$ is given by

$$\alpha\left((X_1, \ldots, X_s), (Y_1, \ldots, Y_s)\right) = ((\alpha_1(X_1, Y_1), \ldots, \alpha_s(X_s, Y_s))$$

for all $X_i, Y_i \in T_{p_i} M_i$. Similarly, the mean curvature vector field H of $M_1 \times \ldots \times M_s$ is given by

$$H = (H_1, \ldots, H_s) \; .$$

More generally, let $M = M_1 \times \ldots \times M_s$ be a submanifold of a Riemannian manifold $\bar{M}$, where $\dim M_i \geq 1$ for all $i = 1, \ldots, s$ and $s \geq 2$. Here, $\bar{M}$ is not necessarily a Riemannian product. We denote by $L_1, \ldots, L_s$ the totally geodesic foliations on M that are canonically induced by the product structure of M. For instance, the leaf $L_1(p)$ of L_1 through $p = (p_1, \ldots, p_s)$ is $M_1 \times \{p_2\} \times \ldots \times \{p_s\}$. Note that $\mathcal{H}_i = TL_i$ is a parallel distribution on M for each $i = 1, \ldots, s$. One says that M is *extrinsically reducible in* $\bar{M}$, or M is an *extrinsic product in* $\bar{M}$, if the second fundamental form α of M satisfies $\alpha(X_i, Y_j) = 0$ for all $X_i \in T_pL_i(p)$, $Y_j \in T_pL_j(p)$, $i \neq j$, $p \in M$. From the above equation for the second fundamental form of submanifold products we immediately see that each submanifold product M in a Riemannian product manifold $\bar{M}$ is extrinsically reducible in $\bar{M}$. We say that a submanifold M of $\bar{M}$ is *locally extrinsically reducible in* $\bar{M}$ *at* $p \in M$ if there exists an open neighborhood of p in M that is extrinsically reducible in $\bar{M}$. Finally, we say that M is *locally extrinsically reducible in* $\bar{M}$, if it is locally extrinsically reducible in $\bar{M}$ at each point in M.

b) Extrinsically reducible submanifolds in Euclidean spaces and spheres

There is a useful criterion for local extrinsic reducibility of submanifolds in Euclidean spaces due to Moore [139].

LEMMA 2.7.1 (Lemma of Moore)
Let M be a submanifold of $\mathbb{R}^n$. If there exists a nontrivial parallel distribution $\mathcal{H}$ on M such that the second fundamental form α of M satisfies $\alpha(\mathcal{H}, \mathcal{H}^\perp) = 0$, then M is locally a submanifold product in $\mathbb{R}^n$ and hence locally extrinsically reducible in $\mathbb{R}^n$.

Moreover, if $f : M \to \mathbb{R}^n$ is a complete simply connected immersed submanifold, then it is a product of immersions.

PROOF Since $\mathcal{H}$ is a parallel distribution on M, $\mathcal{H}^\perp$ is also a parallel distribution on M. Hence, both $\mathcal{H}$ and $\mathcal{H}^\perp$ are integrable with totally geodesic leaves. We choose and fix a point $p \in M$. By the de Rham decomposition theorem there exists an open neighborhood of p in M that is isometric to the Riemannian product $M_1 \times M_2$, where M_1 and M_2 are connected integral manifolds of $\mathcal{H}$ and $\mathcal{H}^\perp$ through p, respectively. We will now prove that $M_1 \times M_2$ is a submanifold product in $\mathbb{R}^n$.

For each point $q = (q_1, q_2) \in M_1 \times M_2$ we put $L_1(q_1) = \{q_1\} \times M_2$ and $L_2(q_2) = M_1 \times \{q_2\}$. We now choose two points $q = (q_1, q_2)$ and $\tilde{q} = (\tilde{q}_1, \tilde{q}_2)$ in $M_1 \times M_2$ and two tangent vectors $X \in T_qL_1(q_1) = \mathcal{H}_q^\perp$ and $Y \in T_{\tilde{q}}L_2(\tilde{q}_2) = \mathcal{H}_{\tilde{q}}$. Let $c : [0, 1] \to L_2(q_2)$ be a smooth curve with $c(0) = q = (q_1, q_2)$ and $c(1) = (\tilde{q}_1, q_2)$. Let E_X be the ∇-parallel vector field along c with $E_X(0) = X$, where ∇ is the Levi Civita connection of M. By construction, $\dot{c}$ is tangent to $\mathcal{H}$ everywhere, and since $X \in \mathcal{H}^\perp$ and $\mathcal{H}^\perp$ is a parallel distribution on M, we see that E_X is tangent to $\mathcal{H}^\perp$ everywhere. Since, by assumption $\alpha(\mathcal{H}, \mathcal{H}^\perp) = 0$, the Gauss formula for $M \subset \mathbb{R}^n$ implies

$$\bar{\nabla}_{\dot{c}} E_X = \nabla_{\dot{c}} E_X + \alpha(\dot{c}, E_X) = 0 \ .$$

Thus, E_X is a $\bar{\nabla}$-parallel vector field along c, and hence $E_X(t) = X$ for all $t \in [0, 1]$, where we identify, as usual, the tangent spaces of $\mathbb{R}^n$ with $\mathbb{R}^n$ in the canonical way. It follows that $E_X(1) = X \in \mathcal{H}^\perp$. Next, let $d : [0, 1] \to L_1(\tilde{q}_1)$

be a smooth curve with $d(0) = \tilde{q} = (\tilde{q}_1, \tilde{q}_2)$ and $d(1) = (\tilde{q}_1, q_2)$, and let E_Y be the ∇-parallel vector field along d with $E_Y(0) = Y$. Similarly, one can show that $E_Y(1) = Y \in \mathcal{H}$. We thus have proved that $T_q L_1(q_1)$ and $T_{\tilde{q}} L_2(\tilde{q}_2)$ are perpendicular to each other for all $q, \tilde{q} \in M_1 \times M_2$.

Since $\mathbb{R}^n$ is homogeneous, we can assume without loss of generality that p is the origin of $\mathbb{R}^n$. Let $\mathbb{R}^{n_1}$ and $\mathbb{R}^{n_2}$ be the linear subspaces of $\mathbb{R}^n$ that are generated by the linear subspaces $T_q L_2(q_2) = \mathcal{H}_q$ and $T_q L_1(q_1) = \mathcal{H}_q^\perp$ for all $q \in M_1 \times M_2$, respectively. We have just proved that $\mathbb{R}^{n_1}$ and $\mathbb{R}^{n_2}$ are perpendicular to each other. By construction, we have $M_1 \times M_2 \subset \mathbb{R}^{n_1} \times \mathbb{R}^{n_2}$, which shows that $M_1 \times M_2$ is a submanifold product in $\mathbb{R}^n$.

The same proof, but using the global de Rham theorem, yields the global version. □

Since S^n is a totally umbilical submanifold of $\mathbb{R}^{n+1}$, the Lemma of Moore implies

COROLLARY 2.7.2
Let M be a submanifold of S^n and consider S^n as a submanifold of $\mathbb{R}^{n+1}$. If there exists a nontrivial parallel distribution $\mathcal{H}$ on M such that the second fundamental form α of $M \subset S^n$ satisfies $\alpha(\mathcal{H}, \mathcal{H}^\perp) = 0$, then M is locally a submanifold product in $\mathbb{R}^{n+1}$ and hence locally extrinsically reducible in S^n.

c) Extrinsically reducible submanifolds in Lorentzian spaces and hyperbolic spaces

The Lorentzian analogue of the Lemma of Moore is not a straightforward generalization. This is due to the fact that, in Lorentzian spaces, there exist degenerate linear subspaces. Recall that a linear subspace V of $\mathbb{R}^{n,1}$ is called *degenerate* if there exists a nonzero vector $v \in V$ such that $\langle v, w \rangle = 0$ for all $w \in V$. Evidently, any such vector v must be light-like.

PROPOSITION 2.7.3
Let M be a Riemannian submanifold of $\mathbb{R}^{n,1}$. If there exists a nontrivial parallel distribution $\mathcal{H}$ on M such that the linear subspaces $\mathcal{H}_q$, $q \in M$, span a nondegenerate linear subspace V of $\mathbb{R}^{n,1}$ and such that the second fundamental form α of M satisfies $\alpha(\mathcal{H}, \mathcal{H}^\perp) = 0$, then M is locally a submanifold product in $\mathbb{R}^{n,1}$ and hence locally extrinsically reducible in $\mathbb{R}^{n,1}$.

PROOF The first part of the proof is analogous to the one in the Euclidean case. (Since we assume M to be a Riemannian submanifold we can apply the de Rham decomposition theorem.) Since V is nondegenerate by assumption, it is either a Lorentzian or a Euclidean subspace of $\mathbb{R}^{n,1}$, and hence $V^\perp$ is either Euclidean or Lorentzian, respectively. It follows that $M_1 \times M_2 \subset V \times V^\perp \subset \mathbb{R}^{n,1}$, which shows that M is locally a submanifold product in $\mathbb{R}^{n,1}$. □

Using the Lorentzian version of Moore's lemma, we now derive a reducibility result for submanifolds in real hyperbolic spaces.

COROLLARY 2.7.4

Let M be a submanifold of H^n and consider H^n as a submanifold of $\mathbb{R}^{n,1}$. If M is not contained in a horosphere of H^n, and if there exists a nontrivial parallel distribution $\mathcal{H}$ on M such that the second fundamental form α of $M \subset H^n$ satisfies $\alpha(\mathcal{H}, \mathcal{H}^\perp) = 0$, then M is locally a submanifold product in $\mathbb{R}^{n,1}$ and hence locally extrinsically reducible in H^n.

PROOF Since H^n is totally umbilical in $\mathbb{R}^{n,1}$, we also have $\tilde{\alpha}(\mathcal{H}, \mathcal{H}^\perp) = 0$, where $\tilde{\alpha}$ is the second fundamental form of $M \subset \mathbb{R}^{n,1}$. Moreover, since M is a submanifold of H^n and H^n is a Riemannian submanifold of $\mathbb{R}^{n,1}$, M is a Riemannian submanifold of $\mathbb{R}^{n,1}$. Let V_1 and V_2 be the span of the linear subspaces $\mathcal{H}_q$ and $\mathcal{H}_q^\perp$, $q \in M$, respectively. It remains to be proven that either V_1 or V_2 is a nondegenerate subspace of $\mathbb{R}^{n,1}$. Assume that both V_1 and V_2 are degenerate. Then, since V_1 is orthogonal to V_2, also $V_1 + V_2$ is a degenerate subspace of $\mathbb{R}^{n,1}$. It follows that M is contained in an affine subspace of $\mathbb{R}^{n,1}$ whose linear part is degenerate. Since the intersection of such an affine subspace with H^n is a horosphere in H^n, we see that M lies in a horosphere of $\mathbb{R}H^n$, which is a contradiction. □

REMARK 2.7.5 Horospheres are isometric to Euclidean space. So, a submanifold of hyperbolic space that is contained in a horosphere can be regarded as a submanifold of Euclidean space. □

2.8 Exercises

Exercise 2.8.1 Let M be a m-dimensional hypersurface of $\mathbb{R}^n$, with $m > 2$. Let $\kappa_1, ..., \kappa_m$ be its principal curvatures. Suppose that at least three principal curvatures are nonzero. Prove that the sectional curvatures determine $\kappa_1, ..., \kappa_m$. Deduce the Beez-Killing Theorem, namely that, for an m-dimensional hypersurface of $\mathbb{R}^n$, with $m > 2$ and with at least three nonzero principal curvatures, the second fundamental form is determined by the first fundamental form (cf. [12], 10.8).

Exercise 2.8.2 Let $f : S^2 \to \mathbb{R}^5$ be given by

$$(x, y, z) \mapsto \left(xy, xz, yz, \frac{1}{2}(x^2 - y^2), \frac{1}{2\sqrt{3}}(x^2 + y^2 - 2z^2) \right) .$$

(a) Verify that f induces an embedding $\tilde{f}$ of the real projective plane into a hypersphere $S^4(1/\sqrt{3})$ of $\mathbb{R}^5$.

(b) Compute the second fundamental form of f (or of $\tilde{f}$), verifying that they are minimal in the sphere.

$\tilde{f}$ is called the Veronese surface.

Exercise 2.8.3 Let $P : S^1(R) \times S^1(R) \to \mathbb{R}^4$ be the Clifford torus, given by

$$(u, v) \mapsto (R \cos u, R \sin u, R \cos v, R \sin v) .$$

Compute the second fundamental form of P.

Exercise 2.8.4 Give a direct proof of the fact that the totally geodesic submanifolds of $\mathbb{R}^n$ are affine subspaces. *Hint:* See the proof of Theorem 2.5.1 (Reduction of codimension).

Exercise 2.8.5 (cf. [217], Corollary 1.5) Let M be a submanifold of a space of constant curvature. Prove that, if there exists a parallel normal field ξ such that the eigenvalues of A_ξ are all distinct, then M has flat normal bundle.

Exercise 2.8.6 Prove that the connected, complete, totally geodesic (resp. totally umbilical) submanifolds M^m, $m \geq 2$, of $S^n(r) \subset \mathbb{R}^{n+1}$ are the intersections of $S^n(r)$ with the linear (resp. affine) subspaces of $\mathbb{R}^{n+1}$.

Exercise 2.8.7 Prove that the connected, complete, totally geodesic (resp. totally umbilical) submanifolds M^m, $m \geq 2$, of $H^n(r) \subset L^n$ are the intersections of $H^n(r)$ with the linear (resp. affine) subspaces of L^n.

Exercise 2.8.8 Prove that a submanifold M of Euclidean space with parallel second fundamental form has parallel first normal space.

Exercise 2.8.9 Prove that two autoparallel distributions that are orthogonally complementary are both parallel. Is this result still true for three autoparallel distributions?

Exercise 2.8.10 Let M be a submanifold of Euclidean space with parallel second fundamental form. Suppose that the shape operator A_H relative to the mean curvature H has two distinct eigenvalues. Prove that M is locally reducible.

Exercise 2.8.11 Suppose M is a totally geodesic submanifold of a space form, and let N be a submanifold of M. Prove that $A^M_\xi X = A^N_\xi X$ for all $\xi \in \nu_p M$, $X \in T_p N$, $p \in N$, where A^M and A^N are the shape operators of M and N, respectively. Prove that the above property (satisfied for any submanifold N of M) characterizes totally umbilical submanifolds.

Exercise 2.8.12 Prove that a totally umbilical submanifold of a space form has flat normal bundle.

Exercise 2.8.13 Prove that a geodesic with initial direction v, $||v|| = 1$, in a horosphere

$$q + (\mathbb{R}\eta)^\perp \cap H^n(-1) ,$$

where $\langle \eta, \eta \rangle = 0$, has the expression

$$\gamma_v(t) = q + tv - \frac{t^2}{2\langle \eta, q \rangle} \eta .$$

Exercise 2.8.14 Prove that a geodesic with initial direction v, $||v|| = 1$, in a totally umbilical lower dimensional hyperbolic space in $H^n(-1)$ given by

$$q + (\mathbb{R}\eta)^\perp \cap H^n(-1),$$

where $\langle \eta, \eta \rangle = -1$, has the expression

$$\gamma_v(t) = q + \frac{1}{2}\langle \eta, q \rangle (\sinh t\theta) v + \frac{1}{2}\langle \eta, q \rangle (\cosh t\theta - 1)\eta ,$$

with $\theta = 2/\langle \eta, q \rangle$.

Exercise 2.8.15 (Ejiri [77], suggested by A. J. Di Scala) Let $f : M \times N \to \mathbb{R}^n$ be an isometric minimal immersion from a product of Riemannian manifolds. Then f is a product of immersions. *Hint:* use Gauss equation for proving that $\alpha(X, Y) = 0$ if X is tangent to M and Y is tangent to N.

Chapter 3

Submanifold geometry of orbits

In this chapter, we investigate submanifolds that arise as orbits of isometric Lie group actions on Riemannian manifolds. These so-called *(extrinsically) homogeneous submanifolds* have the important feature that their geometric invariants, like the second fundamental form, are independent of the point.

In Section 3.1, we start with the general setup and introduce some basic concepts such as orbit types, principal orbits, isotropy and slice representations. The purpose of this section is also to introduce the notation that will be used in the sequel.

In these notes, we are interested in orbits from two different viewpoints: the geometry of a single orbit and the geometry of the entire set of orbits of an action. The orbits of an isometric Lie group action on a Riemannian manifold $\bar{M}$ might be viewed as a singular foliation on $\bar{M}$. For some particular types of representations, like s-representations and polar actions, which we will introduce in Section 3.2, it is of great interest to investigate the entire orbit foliation. Polar actions on $\mathbb{R}^n$ are characterized by the existence of a linear subspace of $\mathbb{R}^n$, a so-called *section*, that intersects each orbit and is perpendicular to orbits at the points of intersection. An s-representation is the isotropy representation of a semisimple symmetric space. An s-representation is polar, since the tangent space to a flat in the symmetric space is a section, and by Dadok's Theorem 3.2.15 [63] it has the same orbits as a polar action. The existence of a section implies that the orbit foliation has remarkable Riemannian geometric properties. The orbits are equidistant with parallel tangent spaces, the normal bundle of a principal orbit is flat with trivial holonomy and the principal curvatures of a principal orbit with respect to any parallel normal field are constant. This motivates the study of isoparametric submanifolds of $\mathbb{R}^n$, which will be discussed later in these notes. We will see that s-representations have a distinguished rôle in submanifold geometry, to many extents comparable to the one of symmetric spaces in intrinsic Riemannian geometry.

In Section 3.3, we reverse our approach. We start with a homogeneous Riemannian manifold and investigate whether it can be viewed as an orbit in another Riemannian manifold or, in other terms, if it admits an equivariant embedding.

In the next sections, we look at the geometry of single orbit. In Sections 3.4 and 3.5, we study homogeneous submanifolds of space forms. In Section 3.6, we describe the extrinsic differential geometry of orbits, and give, among other things, a description of the second fundamental form of the orbit of a representation of a Lie group in terms of the corresponding Lie algebra representation.

Symmetric submanifolds of $\mathbb{R}^n$, and of spaces of constant curvature, form historically one of the first class of homogeneous submanifolds that were studied and classified. Section 3.7 is devoted to this topic.

In Sections 3.8, and 3.9 we study classes of submanifolds that are characterized by important geometric properties of homogeneous submanifolds. The most classical "homogeneous-like" property is the constancy of principal curvatures, which characterizes isoparametric hypersurfaces of space forms. Traditionally, these hypersurfaces are defined as regular level sets of isoparametric functions, so that they determine an orbit-like foliation of the manifold. Isoparametric hypersurfaces are pretty close to being homogeneous, and actually, in many cases, they are. In higher codimension, a natural "homogeneous-like" property is that the algebraic type of the second fundamental form does not depend on the point. This is an extrinsic version of curvature-homogeneous manifolds [225].

3.1 Isometric actions of Lie groups

An important class of submanifolds in a smooth manifold is given by orbits of Lie group actions. In the framework of Riemannian geometry, one is interested in isometric Lie group actions. In this section, we summarize some basic concepts of this topic, like orbit types, principal orbits, isotropy and slice representations. For details and further reading, refer to [28, 74, 111].

a) Basic concepts

Let M be a Riemannian manifold and G a Lie group acting smoothly on M by isometries. Then we have a Lie group homomorphism $\rho : G \to I(M)$ and a smooth map

$$G \times M \to M \ , \ (g, p) \mapsto \rho(g)(p) = gp$$

satisfying $(gg')p = g(g'p)$ for all $g, g' \in G$ and $p \in M$. An isometric action of a Lie group G' on a Riemannian manifold M' is said to be *equivalent* to the action of G on M if there exists a Lie group isomorphism $\psi : G \to G'$ and an isometry $f : M \to M'$ such that $f(gp) = \psi(g)f(p)$ for all $p \in M$ and $g \in G$. For each point $p \in M$ the *orbit* of the action of G through p is

$$G \cdot p := \{gp \mid g \in G\} \ ,$$

and the isotropy group at p is

$$G_p := \{g \in G \mid gp = p\} \ .$$

If $G \cdot p = M$ for some $p \in M$, and hence, for each $p \in M$, the action of G is said to be *transitive* and M is a *homogeneous G-space*. More details about homogeneous G-spaces can be found in Appendix A.3. We assume from now on that the action of

G is not transitive. Each orbit $G \cdot p$ is a submanifold of M, but, in general, not an embedded one.

Example 3.1
Consider the flat torus T^2 obtained from $\mathbb{R}^2$ by factoring out the integer lattice. For each $\omega \in \mathbb{R}_+$ the Lie group $\mathbb{R}$ acts on T^2 isometrically by

$$\mathbb{R} \times T^2 \to T^2 \ , \ (t, [x, y]) \mapsto [x + t, y + \omega t] \ ,$$

where $[x, y]$ denotes the image of $(x, y) \in \mathbb{R}^2$ under the canonical projection $\mathbb{R}^2 \to T^2$. If ω is an irrational number, then each orbit of this action is dense in T^2 and hence cannot be an embedded submanifold. $\square$

Each orbit $G \cdot p$ inherits a Riemannian structure from the ambient Riemannian manifold M. With respect to this structure, $G \cdot p$ is a Riemannian homogeneous space $G \cdot p = G/G_p$ on which G acts transitively by isometries.

DEFINITION 3.1.1 *Let M be a submanifold of a Riemannian manifold $\bar{M}$. We say that M is an (extrinsically) homogeneous submanifold of $\bar{M}$ if, for any two points $p, q \in M$, there exists an isometry g of $\bar{M}$ such that $g(M) = M$ and $g(p) = q$.*

Example 3.2 (Flag manifolds)
A flag in $\mathbb{C}^n$ is a sequence of inclusions $\{0\} \subset V_1 \subset \ldots \subset V_j \subset \mathbb{C}^n$, where V_i is a complex linear subspace of $\mathbb{C}^n$ of fixed dimension. These complex algebraic varieties are classical examples of *complex flag manifolds* or *C-spaces*. If the inclusions are all strict and every dimension between 1 and $n-1$ appears, the variety is called the *full complex flag manifold of* $\mathbb{C}^n$. Any of these flag manifolds can be realized as an orbit of the adjoint representation of $SU(n)$ (see Appendix A.4, page 309), so that it becomes a homogeneous submanifold of some Euclidean space, more precisely, of the Lie algebra $\mathfrak{su}(n)$ of $SU(n)$ equipped with the inner product that is induced from the negative of the Killing form of $\mathfrak{su}(n)$. Indeed, it suffices to take the adjoint orbit of a diagonal element with zero trace in $\mathfrak{su}(n)$. If all entries in the diagonal are different, we get the full complex flag manifold of $\mathbb{C}^n$.

One can generalize this example by taking the orbits of the adjoint representation of a semisimple compact Lie group G. We will refer to such (generalized) complex flag manifolds also as *adjoint orbits*. Note that, via the inner products on the Lie algebra $\mathfrak{g}$ of G and the dual Lie algebra $\mathfrak{g}^*$ that are induced from the Killing form of $\mathfrak{g}$, the adjoint representation can be identified with the coadjoint representation on $\mathfrak{g}^*$. In this setting, adjoint orbits are the same as *coadjoint orbits*.

Complex flag manifolds belong to a larger class of homogeneous submanifolds that will be of great interest for us in the following, namely real flag manifolds. A real flag manifold is an orbit of the isotropy representation of a semisimple Riemannian symmetric space, or briefly, of an s-representation. For details on this, refer to Appendix A.4, page 310.

We will come back to the adjoint orbits of $SU(n)$ later in this chapter as an illustrating example for various concepts that we are going to introduce. □

A homogeneous submanifold M is thus an orbit of some subgroup G of the isometry group of a Riemannian manifold. Although it is often assumed in the literature that G is connected and closed, we will not assume closure unless explicitly stated.

b) The set of orbits

We denote by M/G the set of orbits of the action of G on M and equip M/G with the quotient topology relative to the canonical projection $M \to M/G$, $p \mapsto G \cdot p$. In general, M/G is not a Hausdorff space. For instance, when ω is an irrational number in Example 3.1, then $T^2/\mathbb{R}$ is not a Hausdorff space. This unpleasant behaviour does not occur for so-called proper actions. The action of G on M is *proper* if, for any two distinct points $p, q \in M$, there exist open neighbourhoods U_p and U_q of p and q in M, respectively, such that $\{g \in G \mid gU_p \cap U_q \neq \emptyset\}$ is relatively compact in G. This is equivalent to saying that the map

$$G \times M \to M \times M \ , \ (g, p) \mapsto (p, gp)$$

is a proper map, i.e., the inverse image of each compact set in $M \times M$ is also compact in $G \times M$. Every compact Lie group action is proper, and the action of any closed subgroup of the isometry group of M is proper as well. If G acts properly on M, then M/G is a Hausdorff space, each orbit $G \cdot p$ is closed in M and hence an embedded submanifold, and each isotropy group G_p is compact.

c) Slices

A fundamental feature of proper actions is the existence of slices. A submanifold Σ of M is called a *slice* at $p \in M$ if

(Σ_1) $p \in \Sigma$,

(Σ_2) $G \cdot \Sigma := \{gq \mid g \in G \ , \ q \in \Sigma\}$ is an open subset of M,

(Σ_3) $G_p \cdot \Sigma = \Sigma$,

(Σ_4) the action of G_p on Σ is isomorphic to an orthogonal linear action of G_p on an open ball in some Euclidean space,

(Σ_5) the map

$$(G \times \Sigma)/G_p \to M \ , \ G_p \cdot (g, q) \mapsto gq$$

is a diffeomorphism onto $G \cdot \Sigma$, where $(G \times \Sigma)/G_p$ is the space of orbits of the action of G_p on $G \times \Sigma$ given by $k(g, q) := (gk^{-1}, kq)$ for all $k \in G_p$, $g \in G$ and $q \in \Sigma$. Note that $(G \times \Sigma)/G_p$ is the fibre bundle associated to the principal bundle $G \mapsto G/G_p$ of fibre Σ and hence a smooth manifold.

Montgomery and Yang [142] proved that every proper action admits a slice at each point. It is useful to remark that a slice Σ enables us to reduce the study of the action of G on M in some G-invariant open neighbourhood of p to the action of G_p on the slice Σ.

d) Orbit types

The existence of a slice at each point also enables us to define a partial ordering on the set of orbit types. We say that two orbits $G \cdot p$ and $G \cdot q$ have the same orbit type if G_p and G_q are conjugate in G. This defines an equivalence relation among the orbits of G. We denote the corresponding equivalence class by $[G \cdot p]$, called *orbit type* of $G \cdot p$. By $\mathcal{O}$ we denote the set of all orbit types of the action of G on M. We introduce a partial ordering on $\mathcal{O}$ by saying that $[G \cdot p] \leq [G \cdot q]$ if and only if G_q is conjugate in G to some subgroup of G_p. If Σ is a slice at p, then properties (Σ_4) and (Σ_5) imply that $[G \cdot p] \leq [G \cdot q]$ for all $q \in G \cdot \Sigma$. We assume that M/G is connected. Then there exists a largest orbit type in $\mathcal{O}$. Each representative of this largest orbit type is called a *principal orbit*. In other words, an orbit $G \cdot p$ is principal if and only if for each $q \in M$ the isotropy group G_p at p is conjugate in G to some subgroup of G_q. The union of all principal orbits is a dense and open subset of M. Each principal orbit is an orbit of maximal dimension. A non-principal orbit of maximal dimension is called *exceptional*. An orbit whose dimension is less than the dimension of a *principal orbit* is called *singular*. The *cohomogeneity* of the action is the codimension of a principal orbit.

Exercise 3.10.17 leads to a proof of the existence of principal orbits for isometric actions.

Example 3.2 (continued). The principal orbits of the adjoint action of $SU(3)$ are the full complex flag manifolds given by complete flags $\{0\} \subset V_1 \subset V_2 \subset \mathbb{C}^3$, corresponding, for instance, to the orbit of a diagonal element in $\mathfrak{su}(3)$ (with zero trace and different entries). If two entries are equal, one gets a singular orbit diffeomorphic to the complex projective space. Of course, for $\mathrm{Ad}(SU(n))$-orbits, diagonal elements with different entries determine principal orbits.
As we will see, adjoint representations (and more generally, polar actions) have no exceptional orbit (see Remark 3.2.10). □

e) Isotropy representations and slice representations

We assume from now on that the action of G on M is proper and that M/G is connected. Recall that, for each $g \in G$, the map

$$\varphi_g : M \to M \ , \ p \mapsto gp$$

is an isometry of M. If $p \in M$ and $g \in G_p$, then φ_g fixes p. Therefore, at each point $p \in M$, the isotropy group G_p acts on T_pM by

$$G_p \times T_pM \to T_pM \ , \ (g, X) \mapsto g \cdot X := (\varphi_g)_{*p}X \ .$$

But, since $g \in G_p$ leaves $G \cdot p$ invariant, this action leaves the tangent space $T_p(G \cdot p)$ and the normal space $\nu_p(G \cdot p)$ of $G \cdot p$ at p invariant, too. The restriction

$$\chi_p : G_p \times T_p(G \cdot p) \to T_p(G \cdot p) \ , \ (g, X) \mapsto g \cdot X$$

is called the *isotropy representation* of the action at p, while the restriction

$$\sigma_p : G_p \times \nu_p(G \cdot p) \to \nu_p(G \cdot p) \ , \ (g, \xi) \mapsto g \cdot \xi$$

is called the *slice representation* of the action at p. If $(G_p)_0$ is the connected component of the identity in G_p, the restriction of the slice representation to $(G_p)_0$ will be called *connected slice representation.*

f) Geodesic slices

Let $p \in M$ and $r \in \mathbb{R}_+$ be sufficiently small so that the restriction of the exponential map $\exp_p$ of M at p to $U_r(0) \subset \nu_p(G \cdot p)$ is an embedding of $U_r(0)$ into M. Then $\Sigma = \exp_p(U_r(0))$ is a slice at p, a so-called *geodesic slice*. Geometrically, the geodesic slice Σ is obtained by running along all geodesics emanating orthogonally from $G \cdot p$ at p up to the distance r.

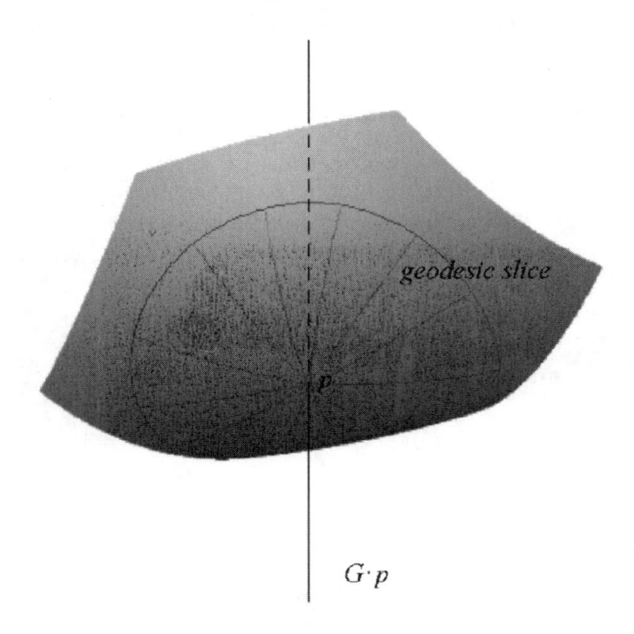

FIGURE 3.1: A geodesic slice.

Since isometries map geodesics to geodesics, it is clear that

$$g\Sigma = \exp_{gp}(g \cdot U_r(0))$$

for all $g \in G$. Thus, $G \cdot \Sigma$ is obtained by sliding Σ along the orbit $G \cdot p$ using the group action. Let $q \in \Sigma$ and $g \in G_q$. Then $gq \in \Sigma$ and hence $g\Sigma = \Sigma$. Since $\Sigma \cap G \cdot p = \{p\}$, it follows that $gp = p$ and hence, $g \in G_p$. Thus we have

LEMMA 3.1.2
If Σ is a geodesic slice at p, then $G_q \subset G_p$ for all $q \in \Sigma$.

Let Σ be a geodesic slice at p. Then $G \cdot \Sigma$ is an open subset of M. As principal orbits form an open and dense subset of M, the previous lemma implies that $G \cdot p$ is a principal orbit if and only if $G_q = G_p$ for all $q \in \Sigma$. On the other hand, each $g \in G_q$ fixes both q and p and therefore, assuming the geodesic slice is sufficiently small, the entire geodesic in Σ connects p and q. Thus, G_q pointwise fixes the one-dimensional linear subspace of $\nu_p(G \cdot p)$ corresponding to this geodesic. This implies the following useful characterization of principal orbits.

THEOREM 3.1.3
An orbit $G \cdot p$ is principal if and only if the slice representation Σ_p is trivial.

g) Killing vector fields and reductive decompositions

Let G be a Lie group acting on $\bar{M}$ isometrically and $p \in \bar{M}$. Then the orbit $M := G \cdot p$ is a Riemannian G-homogeneous space in the induced Riemannian metric. Therefore, we can identify M with the homogeneous space G/K, where $K = G_p$ is the isotropy group at p. As K is compact, the homogeneous space G/K is reductive. Let $\mathfrak{g} = \mathfrak{k} \oplus \mathfrak{m}$ be a reductive decomposition of the Lie algebra $\mathfrak{g}$ of G. Each $X \in \mathfrak{g}$ determines a Killing vector field X^* on $\bar{M}$ by means of

$$X^*_q := \left.\frac{d}{dt}\right|_{t=0} (t \mapsto \operatorname{Exp}(tX)q)$$

for all $q \in \bar{M}$, where Exp denotes the Lie exponential map $\mathfrak{g} \to G$. Note that

$$X^*_p = 0 \Longleftrightarrow X \in \mathfrak{k}\,.$$

Whenever the action of G on M is effective, there is a particularly nice reductive decomposition. The restriction of X^* to M is a Killing vector field on M which we also denote by X^*. Since X^* is a Killing vector field on M, the covariant derivative ∇X^* is a skew-symmetric tensor field on M and hence $(\nabla X^*)_p \in \mathfrak{so}(T_pM)$. Let B be the Killing form of $\mathfrak{so}(T_pM)$, which is a negative definite symmetric bilinear form on $\mathfrak{so}(T_pM)$. We define a symmetric bilinear form on $\mathfrak{g}$ by

$$\langle X, Y\rangle := -B((\nabla X^*)_p, (\nabla Y^*)_p)$$

for all $X, Y \in \mathfrak{g}$. If $X \in \mathfrak{k}$, then $X^*_p = 0$ and hence $X^* = 0$ if and only if $(\nabla X^*)_p = 0$, since a Killing vector field X^* is uniquely determined by the values of X_p^* and $(\nabla X^*)_p$. Therefore, if G acts effectively on M, then $\langle \cdot, \cdot \rangle$ is positive definite on $\mathfrak{k}$. Let $\mathfrak{m}$ be the orthogonal complement of $\mathfrak{k}$ in $\mathfrak{g}$ with respect to $\langle \cdot, \cdot \rangle$. Then $\mathfrak{k} \cap \mathfrak{m} = \{0\}$ and $\mathrm{Ad}(K)\mathfrak{m} \subset \mathfrak{m}$, see for instance [223]. Thus, $\mathfrak{g} = \mathfrak{k} \oplus \mathfrak{m}$ is a reductive decomposition of $\mathfrak{g}$. Since Exp maps open neighbourhoods of $0 \in \mathfrak{g}$ diffeomorphically onto open neighbourhoods of $e \in G$, it follows that

$$T_p M = \{X_p^* \mid X \in \mathfrak{g}\} = \{X_p^* \mid X \in \mathfrak{m}\} .$$

If M_1 is a homogeneous submanifold of $\bar{M}_1$ and M_2 is a homogeneous submanifold of $\bar{M}_2$, then $M_1 \times M_2$ is clearly a homogeneous submanifold of $\bar{M}_1 \times \bar{M}_2$. Conversely, if $M_1 \times M_2$ is a connected homogeneous submanifold of $\bar{M}_1 \times \bar{M}_2$, then M_i is a connected homogeneous submanifold of $\bar{M}_i$, $i = 1, 2$. In fact, assume $M_1 \times M_2 = G \cdot (p_1, p_2)$, where $G \subset I(\bar{M}_1 \times \bar{M}_2)$ is a connected Lie subgroup and $p_i \in \bar{M}_i$. Enlarge G to the connected component $\tilde{G}$ of the group $\{g \in I(\bar{M}_1 \times \bar{M}_2) \mid g(M_1 \times M_2) = M_1 \times M_2\}$ of extrinsic isometries of $M_1 \times M_2$. We will show that $\tilde{G} = \tilde{G}_1 \times \tilde{G}_2 \subset I(\bar{M}_1) \times I(\bar{M}_2)$, which implies homogeneity of M_i. Let $X = (X_1, X_2)$ be a Killing vector field induced by $\tilde{G}$ (X_i a Killing vector field of $\bar{M}_i$). Then $X^1 = (X_1, 0)$ and $X^2 = (0, X_2)$ are Killing vector fields of $\bar{M}_1 \times \bar{M}_2$. Moreover, X^1 and X^2 are both tangent to $M_1 \times M_2$. Thus, X^1, X^2 are both Killing vector fields induced by $\tilde{G}$. It follows that $\tilde{G} = \tilde{G}_1 \times \tilde{G}_2 \subset I(\bar{M}_1) \times I(\bar{M}_2)$. This implies the following

PROPOSITION 3.1.4
A homogeneous submanifold M of a Riemannian product $\bar{M}_1 \times \ldots \times \bar{M}_k$ is always a submanifold product $M_1 \times \ldots \times M_k$ of homogeneous submanifolds $M_i \subset \bar{M}_i$. If, in addition, M is full in $\bar{M}$, then each factor M_i is full in $\bar{M}_i$.

h) Equivariant normal vector fields

If $G \cdot p$ is a principal orbit and $\xi \in \nu_p(G \cdot p)$ then

$$\hat{\xi}_{gp} := g \cdot \xi$$

is a well-defined normal vector field on $G \cdot p$. Indeed, if $gp = g'p$, then $g^{-1}g' \in G_p$ and $g^{-1}g' \cdot \xi = \xi$, that is, $g \cdot \xi = g' \cdot \xi$. The vector field $\hat{\xi}$ will be called *equivariant normal vector field determined by* ξ. Hence, if $G \cdot p$ is a principal orbit and $\xi_1, \ldots, \xi_k$ is an orthonormal basis of $\nu_p(G \cdot p)$, then $\hat{\xi}_1, \ldots, \hat{\xi}_k$ is a global smooth orthonormal frame field of the normal bundle of $G \cdot p$. This just means that the normal bundle of a principal orbit is trivial, that is, it is isomorphic to the trivial bundle $G \cdot p \times \mathbb{R}^k \to G \cdot p$.

Note that, from a given principal orbit $G \cdot p$, one can determine all nearby orbits by using equivariant normal vector fields. Indeed, let $G \cdot p$ be a principal orbit and $\hat{\xi}$ an equivariant normal vector field of $G \cdot p$. Then

$$\exp_{gp}(\hat{\xi}_{gp}) = \exp_{gp}(g \cdot \hat{\xi}_p) = g \exp(\hat{\xi}_p)$$

and hence,

$$M_\xi := \{\exp_q(\hat{\xi}_q) \mid q \in G \cdot p\} = G \cdot \exp_p(\hat{\xi}_p) ,$$

that is, M_ξ is the orbit through $\exp_p(\hat{\xi}_p)$. If M is connected and complete, each orbit of G can be obtained in this manner from a single principal orbit.

3.2 Polar actions and s-representations

a) Polar actions.

On $\mathbb{R}^2 \setminus \{0\}$ consider polar coordinates (ρ, θ). Any point $(\rho, \theta) \in \mathbb{R}^2 \setminus \{0\}$ lies in the orbit of the point $(\rho, 0)$ with respect to the standard action of the special orthogonal group $SO(2)$ on $\mathbb{R}^2$. Therefore, the line $\theta = 0$, and, more generally, any line through the origin meets any $SO(2)$-orbit orthogonally. It is easy to see that the standard action of $SO(n)$ on $\mathbb{R}^n$ by rotations also has this property. Thus, it is natural to consider isometric actions of a Lie group on a Riemannian manifold with this feature.

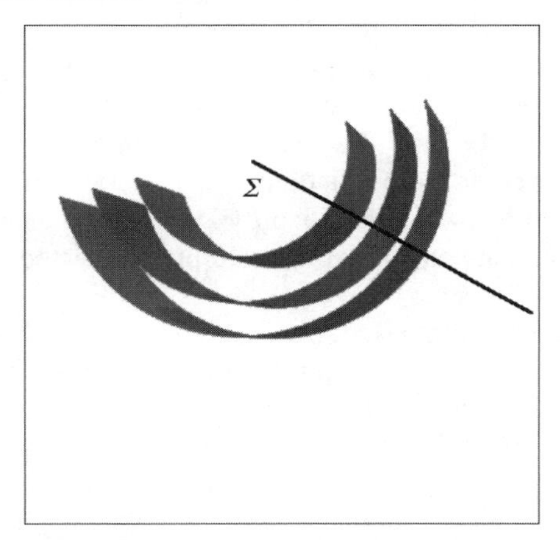

FIGURE 3.2: A section and some orbits.

Let M be a connected complete Riemannian manifold and G a closed (so that the action is proper) subgroup of $I(M)$. A complete, embedded and closed submanifold Σ of M is called a *section* if Σ intersects each orbit of G and is perpendicular to orbits at intersection points. If there exists a section in M, then the action of G is called *polar*. Notice that from a section we can obtain a section that contains any given point by using the group G. If G is disconnected and acts polarly on M, then also the connected component G^o of G containing the identity acts polarly on M.

Let K be a compact Lie group. A representation $\rho : K \to SO(n)$ is called a *polar representation*, if $\rho(K)$ acts polarly on $\mathbb{R}^n$. As seen above, a simple example of a polar representation is given by the standard action of $SO(n)$ on $\mathbb{R}^n$. Note that this action is the isotropy representation of the sphere $S^n = SO(n+1)/SO(n)$.

Actually, as we will show below, a very important class of polar representations is the isotropy representations of symmetric spaces, which are also called s-representations.

A special case is the adjoint representation of a compact Lie group; in this case, using Lie algebra theory, it is easy to see that any Cartan subalgebra provides a section for the action.

Example 3.2 (continued). Consider the adjoint action of $SU(n)$. The Lie algebra $\mathfrak{su}(n)$ of $SU(n)$ is given by $\mathfrak{su}(n) = \{A \in \mathfrak{gl}(n, \mathbb{C}) \mid \bar{A}^t = -A, \mathrm{tr} A = 0\}$. A Cartan subalgebra of $\mathfrak{su}(n)$ is

$$\mathfrak{t} = \left\{ \begin{pmatrix} i\lambda_1 & & 0 \\ & \ddots & \\ 0 & & i\lambda_n \end{pmatrix} \,\middle|\, \sum \lambda_i = 0 \right\}.$$

The fact that $\mathfrak{t}$ meets each $\mathrm{Ad}(SU(n))$-orbit in $\mathfrak{su}(n)$ can be seen from linear algebra: each skew-hermitian matrix can be put in diagonal form with purely imaginary entries by conjugation with a special unitary matrix. Since $\mathfrak{t}$ is a section, each $\mathrm{Ad}(SU(n))$-orbit can be thought of as an orbit of a diagonal element. (We already used this fact while discussing this example.) □

An action is called *hyperpolar* if it admits a flat section. Clearly, every hyperpolar action is polar, and in $\mathbb{R}^n$ these two concepts coincide. The study of hyperpolar actions was initiated by Conlon [56, 57], who called them "representations admitting a K-transversal domain" and showed that they are variationally complete (see [27, p. 974] for this notion). To our knowledge, polar representations were first considered in the early '80s by Szenthe [206–208], who called them isometric actions admitting an orthogonally transversal submanifold. In the late '80s, Palais and Terng [186] discovered an interesting relation between polar representations and isoparametric submanifolds. The polar representations on $\mathbb{R}^n$ were classified by Dadok [63], and Kollross [119] classified the hyperpolar actions on irreducible, simply connected, symmetric Riemannian spaces of compact type up to orbit equivalence.

As we will see, polar representations are important for submanifold geometry in Euclidean space. Indeed, the existence of a section Σ implies that the orbit foliation has remarkable geometric properties. The orbits are equidistant with parallel tangent spaces and, if N is a principal orbit, the normal bundle to N is flat with trivial holonomy, and the principal curvatures of N with respect to any parallel normal field are constant. This leads to the study of isoparametric submanifolds of $\mathbb{R}^n$, which will be carried out later in these notes.

As was first shown by Szenthe [207, 208], a section is necessarily totally geodesic (see also [186]). We now give a proof of this fact using Killing vector fields.

THEOREM 3.2.1
Every section of a polar action is totally geodesic.

PROOF Let Σ be a section. We denote by Σ_r the set of points in Σ which lie on a principal orbit of the action. Let $p \in \Sigma_r$ and $\xi \in \nu_p\Sigma$. Then the group action induces a Killing vector field X on an open neighbourhood of p with $X_p = \xi$. The polarity of the action implies that X is perpendicular to Σ. Let A be the shape operator of Σ. Since X is a Killing vector field, its covariant derivative ∇X is a skewsymmetric tensor field. The Weingarten equation thus implies $\langle A_\xi w, w\rangle = -\langle \nabla_w X, w\rangle = 0$ for all $w \in T_p\Sigma$. Hence, Σ is totally geodesic at points in Σ_r. By Exercise 3.10.9, Σ_r is open and dense in Σ, and hence Σ is totally geodesic. □

The following proposition is due to Dadok [63] for the case of polar representations and to Heintze, Palais, Terng and Thorbergsson [97] for the general case.

PROPOSITION 3.2.2
The connected slice representation of a polar action at any point is polar. Moreover, if Σ is a section of the polar action and $p \in \Sigma$, then $T_p\Sigma$ is a section of the connected slice representation at p.

PROOF We denote by $K = G_p$ the isotropy group at p of the polar action of G on M. By Exercise 3.10.2, the codimension of a principal orbit of the action of K on the normal space $\nu_p(G\cdot p)$ is equal to the dimension of $T_p\Sigma$. If we prove that $T_p\Sigma$ is perpendicular to the orbits of K we are finished, since $T_p\Sigma$ would coincide with the normal space to a principal K-orbit, and so it would intersect all other orbits (see Exercise 3.10.11). The Lie algebra of K may be regarded as the set of skewsymmetric endomorphisms of $\nu_p(G\cdot p)$ of the form $(\nabla X)_p$, where X is a Killing vector field on M induced by G_p. But every Killing vector field X induced by G is always perpendicular to Σ. Therefore, for each $u \in T_p\Sigma$, $\nabla_u X$ is orthogonal to $T_p\Sigma$, since Σ is totally geodesic by 3.2.1 (note that the tangent component of $\nabla_u X$ is equal to $-A_{X(p)}u$, where A is the shape operator of Σ). This shows that the Killing fields induced by K on $\nu_p(G\cdot p)$ are perpendicular to $T_p\Sigma$. □

For any Killing vector field X on M induced by the polar action of G we denote by B_p^X, $p \in M$, the skewsymmetric endomorphism on $\nu_p(G\cdot p)$ defined by

$$\langle B_p^X v, w\rangle = \langle \nabla_v X, w\rangle\ , \quad v, w \in \nu_p(G\cdot p)\ .$$

Let $\mathfrak{h}^p$ be the subalgebra of $\mathfrak{so}(\nu_p(G\cdot p))$, which is generated by all these endomorphisms B_p^X, and denote by H^p the connected Lie subgroup of $SO(\nu_p(G\cdot p))$ associated to $\mathfrak{h}^p$.

LEMMA 3.2.3
The Lie group H^p contains the image $\sigma_p(G_p^o)$ of the connected slice representation at p (regarded as a subgroup of $SO(\nu_p(G\cdot p))$), and the action of H^p on $\nu_p(G\cdot p)$ has the same orbits as the connected slice representation at p.

PROOF Let $\sigma_p : G_p \to SO(\nu_p(G \cdot p))$ be the slice representation at p. If $X \in \mathfrak{g}_p$, then $\sigma_p(\mathrm{Exp}(tX)) = e^{t\sigma_{p*}X}$. But if γ is a curve in $\nu_p(G \cdot p)$ with $\gamma(0) = p$ and $\xi = \gamma'(0)$, then

$$\begin{aligned}\sigma_{p*}X(\xi) &= \frac{D}{dt}\bigg|_{t=0} \sigma_p(\mathrm{Exp}(tX))_*\xi = \frac{D}{dt}\bigg|_{t=0} \frac{d}{ds}\bigg|_{s=0} (\mathrm{Exp}(tX))\gamma(s) = \\ &= \frac{D}{ds}\bigg|_{s=0} \frac{d}{dt}\bigg|_{t=0} (\mathrm{Exp}(tX))\gamma(s) = \frac{D}{ds}\bigg|_{s=0} X^*_{\gamma(s)} = \nabla_\xi X^* = B_p^{X^*}\xi,\end{aligned}$$

since $\nabla_\xi X^* \in \nu_p(G \cdot p)$. This implies that $\sigma_p(G_p^o) \subset H^p$.

For the second part, observe that $T_p\Sigma$ is a section of the connected slice representation at p, where Σ is a section of the G-action at p (see Proposition 3.2.2). Let X be a Killing vector field induced by G, and denote by A the shape operator of Σ. Then we have $\langle B_p^X v, w\rangle = \langle \nabla_v X, w\rangle = \langle A_X v, w\rangle = 0$ for all $v, w \in T_q\Sigma$, since X is perpendicular to Σ and Σ is totally geodesic. So $v \mapsto B_p^X v$ is a linear Killing vector field on $\nu_p M$ that is perpendicular to any section of σ_p. The lemma now follows from Exercise 3.10.15. □

PROPOSITION 3.2.4
Let G act polarly on M and assume that $\xi \in \nu_p(G \cdot p)$ is fixed under the connected slice representation at p. Then ξ extends locally to a G-invariant $\nabla^\perp$-parallel normal vector field to $G \cdot p$.

PROOF Let F be the set of fixed points of G_p^o on $\nu_p M$. Recall that if Σ is a section for the G-action with $p \in \Sigma$, then $T_p\Sigma$ is a section for the connected slice representation. Since H^p has the same orbits as the connected slice representation, we get $H^p\xi = \xi$ for all $\xi \in F$ and, in particular, $B_p^X\xi = 0$ for any Killing vector field X induced by G. We extend ξ to a (local) equivariant normal vector field $\tilde{\xi}$ of $G \cdot p$, and extend $\tilde{\xi}$ to a (local) vector field η on M. Then we get $0 = \langle \nabla_\xi X, \xi\rangle = \langle \nabla_{X_p}\eta - [X,\eta]_p, \xi\rangle = \langle \nabla_{X_p}\eta, \xi\rangle$, since $[X,\eta]_p = \frac{d}{dt}\big|_{t=0} (\Phi^X_{-t})_{*\Phi_t(p)}\eta = 0$ and $\tilde{\xi}$ is equivariant. □

Since the slice representation acts trivially on the normal space of a principal orbit, the proof of the previous proposition readily implies

COROLLARY 3.2.5
Let G act polarly on M. Then every G-equivariant vector field on a principal orbit is $\nabla^\perp$-parallel.

There exists a partial local converse of the previous corollary. Let G be a Lie group acting on a connected Riemannian manifold M, and denote by M_r the open and dense subset of M that is formed by all points in M that lie on principal orbits of the action. We say that G acts locally polar on M if the distribution ν on M_r given by the normal spaces $\nu_p(G \cdot p)$ to the principal orbits is integrable (and hence with totally geodesic leaves, by the same proof as Theorem 3.2.1).

PROPOSITION 3.2.6
Let G be a Lie group acting on a connected Riemannian manifold M. If every equivariant normal vector field on a principal orbit is $\nabla^\perp$-parallel, then G acts locally polar on M.

PROOF Let ξ_1 and ξ_2 be G-equivariant vector fields tangent to ν and X a Killing vector field induced by G. Then $\langle \nabla_{\xi_1}\xi_2, X\rangle = -\langle \xi_2, \nabla_{\xi_1}X\rangle = -\langle \xi_2, \nabla_X\xi_1 - [X,\xi_1]\rangle = 0$, since ξ_1 is parallel and $[X,\xi_1] = 0$ by the same argument as in the proof of Proposition 3.2.4. Thus ν is autoparallel and hence, in particular, integrable. $\square$

The following proposition will be used for the geometric study of the orbits of the isotropy representation of a symmetric space.

PROPOSITION 3.2.7
Let G act polarly on M and let Σ be a section. If S is a connected totally geodesic submanifold of M that intersects all G-orbits orthogonally, then there exists an isometry $g \in G$ such that $g(S) \subset \Sigma$.

PROOF We can assume that S is not a point. Then there exists a point $p \in S$ that lies on a principal orbit. Let $g \in G$ such that $g(p) \in \Sigma$. Then we have $g_*T_pS \subset g_*\nu_p(G\cdot p) = \nu_{g(p)}(G \cdot p) = T_p\Sigma$, which implies $g(S) \subset \Sigma$ since both $g(S)$ and Σ are totally geodesic and connected and Σ is complete. $\square$

REMARK 3.2.8 Let G be a connected Lie subgroup of $SO(n)$ that acts polarly on $\mathbb{R}^n$ and let $G \cdot p$ be an orbit. Then there exists an open neighbourhood V of 0 in the normal space $\nu_p(G\cdot p)$ with the following property: if $v \in V$ and $G\cdot(p+v)$ has the same dimension as $G \cdot p$, then the connected component of the isotropy subgroup of G at p fixes v (and therefore both orbits are parallel). $\square$

PROPOSITION 3.2.9
Let G be a connected Lie subgroup of $SO(n)$ that acts polarly on $\mathbb{R}^n$. If S is a submanifold of $\mathbb{R}^n$ that is locally invariant under the action of G, then G acts locally polar on S.

PROOF Let $p \in S$ such that $\mathfrak{g}\cdot p \subset T_pS$ has maximal dimension. Since the dimension of the G-orbits does not decrease locally, we can assume that all (local) G-orbits in S have the same dimension as $G \cdot p$. Let $V(p) = p + \nu_p(G \cdot p)$ be the affine normal space to the orbit $G \cdot p$ in $\mathbb{R}^n$. We will show that the intersection of $V(p)$ and S (by taking a smaller S if necessary) is a section for the (local) G-action on S (see [187]). As $T_pV(p) + T_pS = \nu_p(G \cdot p) + T_pS = \mathbb{R}^n$, $\Sigma := V(p) \cap S$ is a submanifold of S whose dimension is equal to the codimension of the $G \cdot p$ in S. It remains to prove that $V(p) \cap S$ meets orthogonally local orbits of G in S. Let $q \in V(p) \cap S$ and $v = q - p$. Then the orbit $G \cdot q = G \cdot (p + v)$ has the same dimension as $G \cdot p$. By Remark 3.2.8, both orbits must be parallel (possibly by considering a smaller S). Thus $V(p) = V(q)$ and hence $\Sigma = V(q) \cap S$, which says that $G \cdot q \cap S$ is perpendicular to Σ at q. □

REMARK 3.2.10 An important property of polar actions on $\mathbb{R}^n$ is that they have *no exceptional orbit*. This can be shown directly by arguments presented in this section, but an easier proof will be given in Section 5.4 (Corollary 5.4.3) by using properties of isoparametric submanifolds. □

REMARK 3.2.11 In [91] Gorodski, Olmos and Tojeiro introduced a new invariant for isometric actions of compact Lie groups, which they called *copolarity*. Roughly speaking, it measures how far from being polar the action is. The idea is to generalize the notion of section. They call *minimal k-section* (through a regular point p of the action) the smallest connected, complete, totally geodesic submanifold of the ambient space passing through p that intersects all the orbits and such that, at any intersection point with a principal orbit, its tangent space contains the normal space of that orbit with codimension k. The integer k is called *copolarity* of the isometric action. Polar actions correspond to the case $k = 0$. □

b) s-representations.

We first recall the definition of an s-representation (cf. Appendix A.4). Let S be a simply connected semisimple Riemannian symmetric space. If G is the identity component of the full group of isometries of S, then G acts transitively on S and we can write $S = G/K$, where K is the isotropy subgroup of G at a point $p \in S$. Since S is simply connected and G is connected, K is also connected. The isotropy representation of G/K at p is the Lie group homomorphism $\chi : K \to SO(T_pS)$ given by $\chi(k) = k_{*p}$. Note that χ is injective, since any isometry that fixes p is completely determined by its differential at this point. By an orbit of the isotropy representation of S at p we mean an orbit in T_pS of the group $\chi(K)$. If the base point p is replaced by $q = gp$, $g \in G$, the resulting isotropy representation is equivalent to the one at p. For this reason, we will often omit the base point and will simply speak of the isotropy representation of the symmetric space S. Note that the s-representation of a symmetric space of noncompact type is the same as the one of the corresponding dual simply connected symmetric space of compact type.

Some important properties of s-representations that we will discuss in Chapters 4, 5 and 6 are the following: The group $\chi(K)$ coincides with the holonomy group of S at p. Moreover, the Lie algebra of this holonomy group coincides with the linear span of the set $\{R_p(X,Y) \mid X,Y \in T_pS\}$, where R_p is the curvature tensor of S at p. Therefore, the holonomy representation of a simply connected semisimple symmetric space $S = G/K$ coincides with the isotropy representation of G/K.

REMARK 3.2.12 If $\mathfrak{g} = \mathfrak{k} \oplus \mathfrak{p}$ is the Cartan decomposition associated to the symmetric pair (G, K), then the isotropy representation of S is equivalent to the adjoint representation of K on $\mathfrak{p}$. If the symmetric space is a compact Lie group H, the corresponding symmetric pair is $(H \times H, \triangle H)$, where $\triangle H = \{(h,h) \mid H \in H\}$, and the isotropy representation is equivalent to the adjoint representation of H on its Lie algebra $\mathfrak{h}$. □

The orbits of s-representations are known as *R-spaces*, *real flag manifolds* (see also Appendix A.4, page 310), or *generalized real flag manifolds*, or more precisely, as *standard embeddings of R-spaces* (cf. [115, 117, 118, 146]). They play an important role in geometry, topology and representation theory and have been extensively studied.

We will now prove (see also [27] and [56]) that the isotropy representation of $S = G/K$ is polar, that is, there exists a linear subspace Σ of $\mathfrak{p}$ which meets all $\mathrm{Ad}(K)$-orbits, and is perpendicular to the orbits at the points of intersection.

THEOREM 3.2.13

Let (G, K) be a Riemannian symmetric pair of a simply connected semisimple Riemannian symmetric space S, and let $\mathfrak{g} = \mathfrak{k} \oplus \mathfrak{p}$ be the corresponding Cartan decomposition of the Lie algebra $\mathfrak{g}$ of G. Then the adjoint representation of K on $\mathfrak{p}$ is polar and every maximal Abelian subspace of $\mathfrak{p}$ is a section.

PROOF We can assume that S is irreducible and of compact type, and that the Riemannian metric on S is normalized in such a way that the inner product $\langle \cdot, \cdot \rangle$ on $\mathfrak{p} \cong T_pS$ that is induced from the Riemannian metric on S is equal to the negative of the Killing form B of $\mathfrak{g}$. We choose a point $v \in \mathfrak{p}$ such that the orbit $\mathrm{Ad}(K) \cdot v$ is principal, and denote by Σ the affine normal space to this orbit at v, that is, $\Sigma = \{v + \xi \mid \xi \in \nu_v(\mathrm{Ad}(K) \cdot v)\}$. Note that $0 \in \Sigma$, since $-v \in \nu_v(\mathrm{Ad}(K) \cdot v)$. Thus $\Sigma = \nu_v(\mathrm{Ad}(K) \cdot v)$, where we regard the normal space $\nu_v(\mathrm{Ad}(K) \cdot v)$ as a linear subspace of $\mathfrak{p}$.

i) Σ meets every orbit: Let $\mathrm{Ad}(K) \cdot w$ be another orbit. By compactness of K and homogeneity of $\mathrm{Ad}(K) \cdot v$ we can assume that the distance (induced from $\langle \cdot, \cdot \rangle$) between v and w is equal to the distance between the two compact orbits. It is easy to see that $w - v$ belongs to the normal space $\nu_v(\mathrm{Ad}(K) \cdot v)$, and hence $w \in v + \nu_v(\mathrm{Ad}(K) \cdot v) = \Sigma$. This proves that Σ meets every orbit.

ii) Σ is an Abelian subspace of $\mathfrak{p}$: If $u \in \Sigma$, then $[u, v] \in \mathfrak{k}$, since $[\mathfrak{p}, \mathfrak{p}] \subset \mathfrak{k}$.

If $x \in \mathfrak{k}$ is arbitrary, then $B([u,v],x) = -B(u,[x,v]) = 0$, since $[x,v] \in T_v(\mathrm{Ad}(K)\cdot v$ and $u \in \Sigma = \nu_v(\mathrm{Ad}(K)\cdot v)$. This implies $[u,v] = 0$ since B is non-degenerate. For $u, w \in \Sigma$ we compute $[[u,w],v] = [[u,v],w] + [u,[w,v]] = 0$, since $[u,v] = 0 = [w,v]$. This shows that $[u,w]$ lies in the isotropy subalgebra $\mathfrak{k}_v$ of $\mathfrak{k}$ at v. But $[u,w]$ must also be in the isotropy subalgebra of $\mathfrak{k}$ at any point in Σ, since v is a point in a principal orbit (note that Σ is a geodesic slice, cf. Lemma 3.1.2). In particular, $[[u,w],w] = 0$. Moreover, $[u,w] = 0$, since $B([u,w],[u,w]) = -B([u,w],w],u) = 0$.

iii) Σ meets orbits orthogonally: If $u \in \Sigma$ and $z \in T_u\Sigma$, then $B(T_u(\mathrm{Ad}(K)\cdot u, z) = B([\mathfrak{k},u],z) = B(\mathfrak{k},[u,z]) = 0$, since Σ is Abelian. This finishes the proof that the adjoint action of K on $\mathfrak{p}$ is polar.

iv) The section Σ is a maximal Abelian subspace: Suppose that

$$\Sigma = \nu_v(\mathrm{Ad}(K)v) \subset \Sigma'$$

and Σ' is a maximal Abelian subspace of $\mathfrak{p}$. Then one can show, as in the proof of (iii), that Σ' is perpendicular to the orbit $\mathrm{Ad}(K)\cdot v$ at v. Hence, $\Sigma' \subset \nu_v(\mathrm{Ad}(K)\cdot v)$, which implies $\Sigma = \Sigma'$. □

Example 3.3 Real flag manifolds

We consider the symmetric space $S = SL(n,\mathbb{R})/SO(n)$. The Lie algebra $\mathfrak{sl}(n,\mathbb{R})$ of $SL(n,\mathbb{R})$ has the Cartan decomposition $\mathfrak{sl}(n,\mathbb{R}) = \mathfrak{so}(n) \oplus \mathcal{S}_n$, where $\mathcal{S}_n$ denotes the real vector space of all traceless symmetric $n \times n$-matrices with real coefficients. The isotropy representation of $SO(n)$ is just conjugation on $\mathcal{S}_n$ by matrices in $SO(n)$. A section is given by the diagonal matrices with trace zero. The orbits are the standard embeddings of the real flag manifolds, that is, the varieties of all possible arrangements $\{0\} \subset V_1 \subset \ldots \subset V_j \subset \mathbb{R}^n$, where V_i is a linear subspace of $\mathbb{R}^n$ of fixed dimension. A special case is the orbit through the diagonal matrix with entries $(1,0,\ldots,0)$, which is the Veronese embedding of the real projective space $\mathbb{R}P^{n-1}$ in $\mathcal{S}_n$. □

A *flat* in a symmetric space S is a connected, complete, totally geodesic, flat submanifold. Every complete geodesic is a one-dimensional flat. The tangent spaces to flats at p are in one-to-one correspondence with Abelian subspaces of $\mathfrak{p}$ (via the exponential map and the usual identification of T_pS with $\mathfrak{p}$). A maximal flat in S is a flat of maximal dimension. A well-known result by E. Cartan asserts that any two maximal flats in S are conjugate by some element of G. The infinitesimal, and equivalent, version of this is the following classical result that follows from the above theorem and Proposition 3.2.7.

THEOREM 3.2.14

If $\mathfrak{a}$ and $\mathfrak{a}'$ are maximal Abelian subspaces of $\mathfrak{p}$, then there exists an element $k \in K$ such that $\mathrm{Ad}(k)\mathfrak{a} = \mathfrak{a}'$.

We recall that the dimension of a maximal flat in the symmetric space S, or equivalently, the dimension of a maximal Abelian subspace of $\mathfrak{p}$, is the rank of the symmetric space. The above discussion shows that the rank is equal to the cohomogeneity of the isotropy representation of $S = G/K$.

We now study more thoroughly the geometry of an orbit of an s-representation and relate it to the root system of the symmetric space $S = G/K$.

Example 3.2 (continued). We begin by investigating the adjoint orbits of a compact semisimple Lie group G. We equip its Lie algebra $\mathfrak{g}$ with inner product $\langle \cdot, \cdot \rangle$ given by the negative of the Killing form B of $\mathfrak{g}$. In this situation, any Cartan subalgebra $\mathfrak{t}$ of $\mathfrak{g}$ provides a section. Let $G \cdot X = \mathrm{Ad}(G) \cdot X$, $X \neq 0$, be an orbit of the adjoint action of G on $\mathfrak{g}$. By polarity, we can choose $X \in \mathfrak{t}$.

Note first that $T_X(G \cdot X) = \mathrm{im}\,\mathrm{ad}(X)$, the image of the endomorphism $\mathrm{ad} : \mathfrak{g} \to \mathfrak{g}$, and hence, we get a reductive decomposition

$$\mathfrak{g} = \mathrm{im}\,\mathrm{ad}(X) \oplus \ker \mathrm{ad}(X) = T_X(G \cdot X) \oplus \mathfrak{g}_X ,$$

where $\mathfrak{g}_X = \{Y \in \mathfrak{g} \mid [X, Y] = 0\}$ is the Lie algebra of the isotropy subgroup G_X. It is easy to see that $\mathfrak{g}_X$ is orthogonal to $\mathrm{im}\,\mathrm{ad}(X) = T_X(G \cdot X)$, and thus $\mathfrak{g}_X = \nu_X(G \cdot X)$. Clearly,

$$\mathfrak{t} \subseteq \nu_X(G \cdot X) = \mathfrak{g}_X = \{Y \in \mathfrak{g} \mid [X, Y] = 0\} ,$$

with equality if and only if $G \cdot X$ is a principal orbit. Note that $G \cdot X$ is a principal orbit if and only if X is a *regular element* of $\mathfrak{g}$, that is, if $\mathfrak{g}_X$ has minimal dimension. Recall the following well-known fact from Lie algebra theory: An element $X \in \mathfrak{g}$ is regular if and only if $\mathfrak{g}_X$ is Abelian (a Cartan subalgebra coinciding with $\mathfrak{t}$, for $X \in \mathfrak{t}$).

One can get more geometric information by considering the root space decomposition

$$\mathfrak{g}^{\mathbb{C}} = \mathfrak{g}_0^{\mathbb{C}} \oplus \bigoplus_{\alpha \in \Delta} \mathfrak{g}_\alpha^{\mathbb{C}}$$

of $\mathfrak{g}^{\mathbb{C}}$ with respect to $\mathfrak{t}^{\mathbb{C}}$. We equip $\mathfrak{t}^{\mathbb{C}}$ with a notion of positivity and denote by Δ_+ the resulting subset of positive roots. A crucial fact for us is that, for each $\alpha \in \Delta$, there exist vectors $X_\alpha, Y_\alpha \in \mathfrak{g}$ such that

$$[T, X_\alpha] = \alpha(T) Y_\alpha \quad \text{and} \quad [T, Y_\alpha] = -\alpha(T) X_\alpha \tag{3.1}$$

for all $T \in \mathfrak{t}$. In other words, X_α and Y_α are eigenvectors of $\mathrm{ad}(T)^2$ for all $T \in \mathfrak{t}$. Thus, $X_\alpha + iY_\alpha \in \mathfrak{g}_\alpha^{\mathbb{C}}$ and $X_\alpha - iY_\alpha \in \mathfrak{g}_{-\alpha}^{\mathbb{C}}$. This induces the decomposition

$$\mathfrak{g} = \mathfrak{t} \oplus \sum_{\alpha \in \Delta} (\mathbb{R} X_\alpha \oplus \mathbb{R} Y_\alpha) .$$

We define $\Delta_0 = \{\lambda \in \Delta \mid \lambda(X) = 0\}$ and $\Delta_+ = \{\lambda \in \Delta \mid \lambda(X) > 0\}$, which gives

$$\mathfrak{g} = \mathfrak{t} \oplus \sum_{\alpha \in \Delta_0} \{\mathbb{R} X_\alpha \oplus \mathbb{R} Y_\alpha\} \oplus \sum_{\beta \in \Delta_+} \{\mathbb{R} X_\beta \oplus \mathbb{R} Y_\beta\}$$

Geometrically, $\sum_{\beta\in\Delta_+}\{\mathbb{R}X_\beta \oplus \mathbb{R}Y_\beta\}$ coincides with $T_X(G\cdot X)$, and this decomposition of the tangent space is actually related to the shape operator, as we will see in Section 3.6. Moreover, $\mathfrak{t}\oplus\sum_{\alpha\in\Delta_0}\{\mathbb{R}X_\alpha\oplus\mathbb{R}Y_\alpha\}$ coincides with the Lie algebra $\mathfrak{g}_X$ of the isotropy subgroup at X.

The set of orbits of the adjoint action of G can be parametrized by the closure of a fundamental Weyl chamber C in the Cartan subalgebra $\mathfrak{t}$. The points in C represent the principal orbits and are full flag manifolds G/T, where T is a maximal torus in G. The points in $\bar{C}\setminus C$ represent the singular orbits. □

We now turn to the general case of an orbit of an s-representation. We assume that the semisimple symmetric space G/K is noncompact. Let $A_0\in\mathfrak{p}$ and consider the orbit $M=\mathrm{Ad}(K)\cdot A_0 = K\cdot A_0$. Let $\mathfrak{a}$ be a maximal Abelian subspace of $\mathfrak{p}$ containing A_0, and let $\Delta\subseteq\mathfrak{a}^*\setminus\{0\}$ be the set of restricted roots of the symmetric space with respect to $\mathfrak{a}$ (cf. Section A.2). Notice that, unlike in the case of root systems of compact Lie groups, the root system of a symmetric space can be nonreduced. Recall that $\lambda\in\Delta$ if and only if $0\neq\lambda\in\mathfrak{a}^*$ and $\mathfrak{g}_\lambda=\{X\in\mathfrak{g}\mid [A,X]=\lambda(A)X$ for all $A\in\mathfrak{a}\}\neq 0$. Then we have

$$\mathfrak{g}=\mathfrak{m}\oplus\mathfrak{a}\oplus\sum_{\lambda\in\Delta}\mathfrak{g}_\lambda\ ,$$

where $\mathfrak{m}$ is the centralizer of $\mathfrak{a}$ in $\mathfrak{k}$. Since $\mathrm{ad}(A)$, $A\in\mathfrak{a}$, interchanges $\mathfrak{k}$ and $\mathfrak{p}$, one gets $\mathfrak{g}_\lambda\oplus\mathfrak{g}_{-\lambda}=\mathfrak{k}_\lambda\oplus\mathfrak{p}_\lambda$, where

$$\begin{aligned}\mathfrak{k}_\lambda &= \{X\in\mathfrak{k}\mid \mathrm{ad}(A)^2X=\lambda^2(A)X \text{ for all } A\in\mathfrak{a}\}\ ,\\ \mathfrak{p}_\lambda &= \{X\in\mathfrak{p}\mid \mathrm{ad}(A)^2X=\lambda^2(A)X \text{ for all } A\in\mathfrak{a}\}\ .\end{aligned}$$

Note that $\mathfrak{k}_\lambda=\mathfrak{k}_{-\lambda}$ and $\mathfrak{p}_\lambda=\mathfrak{p}_{-\lambda}$.

Let $\Delta_0=\{\lambda\in\Delta\mid\lambda(A_0)=0\}$ and $\Delta_+=\{\lambda\in\Delta\mid\lambda(A_0)>0\}$. Then we have

$$\mathfrak{g}=\mathfrak{m}\oplus\sum_{\lambda\in\Delta}\mathfrak{k}_\lambda\oplus\mathfrak{a}\oplus\sum_{\lambda\in\Delta}\mathfrak{p}_\lambda=\mathfrak{k}_{A_0}\oplus\mathfrak{k}_+\oplus\mathfrak{p}_{A_0}\oplus\mathfrak{p}_+\ ,$$

where

$$\mathfrak{k}_{A_0}=\mathfrak{m}\oplus\sum_{\lambda\in\Delta_0}\mathfrak{k}_\lambda=\{Z\in\mathfrak{k}\mid[A_0,Z]=0\}$$

is the centralizer of A_0 in $\mathfrak{k}$,

$$\mathfrak{p}_{A_0}=\mathfrak{a}\oplus\sum_{\lambda\in\Delta_0}\mathfrak{p}_\lambda=\{Z\in\mathfrak{p}\mid[A_0,Z]=0\}\ ,$$

and $\mathfrak{k}_+=\sum_{\lambda\in\Delta_+}\mathfrak{k}_\lambda$ and $\mathfrak{p}_+=\sum_{\lambda\in\Delta_+}\mathfrak{p}_\lambda$.

The geometric interpretation of these splittings is explained by

$$T_{A_0}M=\mathfrak{p}_+\ ,\ \nu_{A_0}M=\mathfrak{p}_{A_0}\ . \tag{3.2}$$

Indeed, if we endow $\mathfrak{g}$ with the standard inner product equal to $-B$ on $\mathfrak{k}$ and B on $\mathfrak{p}$, where B is the Killing form of G, we obtain an $\mathrm{Ad}(K)$-invariant inner product $(\cdot,\cdot)$ on $\mathfrak{g}$. Thus, for any $Y \in \mathfrak{p}$ and $\xi \in \nu_{A_0}M$, we have

$$0 = (\xi, [A_0, Y]) = ([\xi, A_0], Y) ,$$

and so $\nu_{A_0}M = \{\xi \in \mathfrak{p} \mid [\xi, A_0] = 0\} = \mathfrak{p}_{A_0}$. Note that $\mathrm{ad}(A_0) : \mathfrak{k}_+ \to \mathfrak{p}_+$ and $\mathrm{ad}(A_0) : \mathfrak{p}_+ \to \mathfrak{k}_+$ are isomorphisms since $\mathrm{ad}(A_0)^2 = \lambda^2(A_0)I$.

An element A is regular if and only if $\dim \mathfrak{p}_A < \dim \mathfrak{p}_B$ for all $B \in \mathfrak{p}$, or equivalently, if $\dim \mathfrak{k}_A < \dim \mathfrak{k}_B$ for all $B \in \mathfrak{p}$. Note that an element A is regular if and only if $\mathfrak{p}_A$ is a (maximal) Abelian subspace of $\mathfrak{p}$. Recall that the dimension of $\mathfrak{a}$ is the rank of the symmetric space.

We have just proved that an s-representation is polar. The converse is also true to some extent, as we will now explain. Two representations $\rho_1 : G_1 \to SO(n)$ and $\rho_2 : G_2 \to SO(n)$ are said to be *orbit-equivalent* if there is an isometry $A : \mathbb{R}^n \to \mathbb{R}^n$ that maps the orbits of G_1 to the orbits of G_2, that is, $A(G_1(x)) = G_2(Ax)$ for all $x \in \mathbb{R}^n$. Dadok [63] classified all polar representations. He used Proposition 3.2.2 for an inductive argument. Then he saw from the list that polar representations are all orbit-equivalent to isotropy representations of semisimple symmetric spaces. That is, there is an s-representation with the same orbits. Hence, he proved the following result.

THEOREM 3.2.15 (Dadok)

Every polar representation on R^n is orbit-equivalent to an s-representation.

It is clear that every orthogonal representation of cohomogeneity one is polar. In fact, it must be transitive on spheres. Every cohomogeneity two action on $\mathbb{R}^n$ is polar as well, since it is polar on the spheres in $\mathbb{R}^n$, and the cohomogeneity of the action on any sphere is one. These actions were first classified by Harvey and Lawson in [102]. A geometric proof of Dadok's result was obtained by Heintze and Eschenburg [81] for cohomogeneity at least three.

A simple example of a polar representation that is not the isotropy representation of a symmetric space is the standard inclusion of $SU(n)$ into $SO(2n)$. The orbits of $SU(n)$ in $\mathbb{R}^{2n} \cong \mathbb{C}^n$ are hyperspheres centred about the origin. Therefore the orbits of $SU(n)$ coincide with the orbits of the action of $SO(2n)$ on $\mathbb{R}^{2n}$, which is the isotropy representation of the sphere $S^{2n} = SO(2n+1)/SO(2n)$. In Dadok's list, examples appear that are not transitive on the sphere, and it also contains an example of a (reducible) polar representation of a simple Lie group whose orbits are products of spheres.

3.3 Equivariant maps

a) Definitions

Let M, M' be smooth manifolds and G, G' Lie groups acting smoothly on M and M', respectively. A smooth map $f : M \to M'$ is called *equivariant* (with respect to these two actions) if there exists a Lie group homomorphism $\psi : G \to G'$ such that

$$f(gp) = \psi(g)f(p)$$

for all $p \in M$ and $g \in G$. The basic feature of equivariant maps is that they map orbits of the G-action on M into orbits of the G'-action on M'. In the framework of Riemannian geometry of particular interest for us will be equivariant immersions and embeddings. If M and M' are Riemannian manifolds, $G' = I(M')$, and f is an isometric immersion or embedding, then we speak of a G-equivariant isometric immersion or embedding. If, in particular, $M' = \mathbb{R}^n$, $G' = I(\mathbb{R}^n) = O(n) \rtimes \mathbb{R}^n$, and the G'-action on $\mathbb{R}^n$ is the standard one

$$\varphi : (O(n) \rtimes \mathbb{R}^n) \times \mathbb{R}^n \to \mathbb{R}^n \ , \ ((A, a), p) \mapsto Ap + a \ ,$$

then an equivariant isometric immersion $f : M \to \mathbb{R}^n$ is called a *linearization of the G-action on M*.

REMARK 3.3.1 If G is compact, then the closed Lie subgroup $\psi(G)$ of $I(\mathbb{R}^n)$ has a fixed point. Thus, ψ might be regarded as a homomorphism of G into $O(n)$, or $SO(n)$ if G is connected. Indeed, let $x \in \mathbb{R}^n$ and consider the orbit $\psi(G) \cdot x$. Let y be the barycenter of $\psi(G) \cdot x$ (defined by means of integration over $\psi(G) \cdot x$, which is compact). Since $\psi(G) \cdot x$ is invariant under the action of $\psi(G)$, so is the barycenter y. Hence $y = \psi(G) \cdot y$, that is, y is the common fixed point. □

A classical problem is the existence of linearizations of isometric Lie group actions on Riemannian manifolds or, more generally, the existence of equivariant immersions or embeddings into a given Riemannian manifold.

b) Existence of equivariant embeddings of compact Riemannian homogeneous spaces into Euclidean spaces

A well-known result by Nash asserts that any Riemannian manifold can be isometrically embedded in some Euclidean space [164]. Moreover, a result by Mostow and Palais (see e.g. [28], page 315) states that if a compact Lie group G acts on a compact manifold M, then M admits an equivariant embedding into some Euclidean space. Moore in [140] gave an answer to the question whether a compact Riemannian homogeneous space admits an embedding that is both equivariant and isometric.

THEOREM 3.3.2 (Moore)
Every compact Riemannian homogeneous space admits an isometric embedding into some Euclidean space that is equivariant with respect to the full isometry group.

There are important examples of equivariant immersions of compact Riemannian homogeneous spaces that can be constructed from the eigenvalues of their Laplace operator.

c) The standard minimal isometric immersions of compact Riemannian homogeneous spaces into spheres

Let $M = G/K$ be a compact Riemannian homogeneous space. Since M is compact, its Laplace operator has a discrete spectrum $0 = \lambda_0 < \lambda_1 < \ldots \to +\infty$. We denote by V_i the eigenspace of λ_i in the Hilbert space $L^2(M)$ of all L^2-functions on M and by m_i the dimension of V_i, which is always a positive integer. We assume $i \geq 1$ from now on. The Laplacian on any Riemannian manifold is invariant under the action of the isometry group. Thus, we get a representation ψ of G on V_i by means of

$$\psi(g)f := f \circ g^{-1}$$

for all $g \in G$ and $f \in V_i$. We equip V_i with a G-invariant inner product

$$\langle f_1, f_2 \rangle := \frac{m_i}{\mathrm{vol}_g(M)} \int_M f_1 f_2 \, d\mathrm{vol}_g \ ,$$

where $\mathrm{vol}_g(M)$ denotes the volume of M with respect to the Riemannian metric g on M and $d\mathrm{vol}_g$ is the volume element on M determined by g. Then ψ becomes an orthogonal representation of G on the m_i-dimensional Euclidean vector space $(V_i, \langle \cdot, \cdot \rangle)$. Let $F_1, \ldots, F_{m_i}$ be an orthonormal basis of V_i consisting of smooth functions $M \to \mathbb{R}$. We identify V_i with $\mathbb{R}^{m_i}$ by means of this basis. The smooth map

$$F^i := (F_1, \ldots, F_{m_i}) : M \to \mathbb{R}^{m_i}$$

is an equivariant immersion with respect to $\psi : G \to O(\mathbb{R}^{m_i})$. Thus, F^i is a linearization of the G-action on M.

Moreover, the image $F^i(M)$ is contained in the unit sphere $S^{m_i - 1} \subset \mathbb{R}^{m_i}$. We give a short proof of this fact. We first remark that, for each $g \in G$,

$$(F_1 \circ g^{-1}, \ldots, F_{m_i} \circ g^{-1}) = (\psi(g)F_1, \ldots, \psi(g)F_{m_i}) = \psi(g)F^i$$

is also an orthonormal basis of V_i. Thus there exists a matrix $(A_{ij}(g)) \in O(m_i)$ such that

$$\psi(g)F^i = \sum_{i=1}^{m_i} A_{ji}(g)F_j \ .$$

Note that $(A_{ij}(g))$ is the matrix associated to $\psi(g)$ with respect to the basis $F_1, \ldots,$ F_{m_i}. Since $(A_{ij}(g))$ is orthogonal, we get

$$\sum_{i=1}^{m_i} (\psi(g)F_i)^2 = \sum_{i,j,k=1}^{m_i} A_{ji}(g)A_{ki}(g)F_jF_k = \sum_{i=1}^{m_i}(F_i)^2 ,$$

and hence, $\sum_{i=1}^{m_i}(F_i)^2$ is constant on M. By integrating, and using the orthonormality of $F_1, \ldots, F_{m_i}$, we get

$$\sum_{i=1}^{m_i}(F_i)^2 \ \mathrm{vol}_g(M) = \sum_{i=1}^{m_i} \int_M (F_i)^2 d\mathrm{vol}_g = \frac{\mathrm{vol}_g(M)}{m_i} m_i = \mathrm{vol}_g(M) ,$$

that is, $\sum_{i=1}^{m_i}(F_i)^2 = 1$.

If $M = G/K$ is isotropy irreducible, then M has a unique G-invariant Riemannian metric up to a constant factor. Therefore, since the induced metric by F^i on M is G-invariant, F^i is homothetic. Thus, by possibly rescaling the Riemannian metric by a constant factor, F^i can be assumed to be isometric. In this situation, the map

$$F^i : M \to S^{m_i - 1} \subset \mathbb{R}^{m_i}$$

is called *i-th standard isometric immersion* of M into S^{m_i-1} resp. $\mathbb{R}^{m_i}$. This is an equivariant isometric immersion into the sphere S^{m_i-1} together with its standard action of $O(m_i)$. Moreover, F^i is a minimal isometric immersion into S^{m_i-1}, (see, e.g., [210]). Indeed, since $\Delta F^i = \lambda_i F^i$ and ΔF^i is proportional to the mean curvature vector field of the isometric immersion $F^i : M \to \mathbb{R}^{m_i}$, the mean curvature vector field of the isometric immersion $F^i : M \to S^{m_i-1}$ vanishes. These particular immersions have been widely studied, see, for instance, [72, 134, 135, 221, 235].

Thus, we have explicitly proved that every compact isotropy irreducible Riemannian homogeneous space G/K admits a G-equivariant isometric immersion into some Euclidean space (or into some sphere). Similarly, we might say that every transitive isometric action of a compact Lie group on a Riemannian homogeneous space admits a linearization.

d) The Veronese surface

In general, the standard isometric immersions are not embeddings. Consider, for instance, $S^2 = SO(3)/SO(2)$, which is a space form with constant curvature one, and consider the second standard isometric immersion. Explicitly, an orthogonal basis of the eigenspace V_2 corresponding to the second eigenvalue λ_2 of the Laplace operator on S^2 is given by

$$\left(xy, xz, yz, \frac{1}{2}(x^2 - y^2), \frac{1}{2\sqrt{3}}(x^2 + y^2 - 2z^2) \right) .$$

Any such eigenfunction is a homogeneous polynomial P of degree two in three variables that is harmonic, that is,

$$\partial^2 P/\partial x^2 + \partial^2 P/\partial y^2 + \partial^2 P/\partial z^2 = 0 .$$

An easy way to see directly that this is an orthogonal basis is to observe that V_2 is an irreducible $SO(2)$-module. By Schur's Lemma, all inner products on V_2 are homothetic. An $SO(2)$-invariant inner product on the space V_2 of harmonic polynomials of degree two is given by summing the products of all monomial coefficients.

The immersion $F^2 : S^2 = SO(3)/SO(2) \to \mathbb{R}^5$ is given by

$$F^2 : S^2 \to S^4\left(\frac{1}{\sqrt{3}}\right) \subseteq \mathbb{R}^5 ,$$
$$(x, y, z) \mapsto \left(xy, xz, yz, \frac{1}{2}(x^2 - y^2), \frac{1}{2\sqrt{3}}(x^2 + y^2 - 2z^2)\right) .$$

Since $S^2 = SO(3)/SO(2)$ is isotropy irreducible, we see that F^2 is minimal in the hypersphere $S^4(\frac{1}{\sqrt{3}})$. Moreover, as one can easily see, F^2 maps a point and its antipodal point on S^2 to the same point. Thus F^2 induces an isometric immersion of the real projective plane $\mathbb{R}P^2$ into S^4. This immersed surface is also known as the *Veronese surface* in S^4.

The action ψ of $SO(3)$ on V_2 is given by

$$\psi(g)P(v) = P(g^{-1}v) , \; P \in V_2 , \; g \in SO(3) , \; v \in \mathbb{R}^3 . \tag{3.3}$$

We observe that ψ induces a representation $\psi_* : \mathfrak{so}(3) \to \mathfrak{so}(V_2)$ of the corresponding Lie algebras, which maps a matrix

$$\begin{pmatrix} 0 & -\alpha & -\gamma \\ \alpha & 0 & -\beta \\ \gamma & \beta & 0 \end{pmatrix} \in \mathfrak{so}(3)$$

to the matrix

$$\begin{pmatrix} 0 & \alpha & \gamma & -\beta & \beta \\ -\alpha & 0 & \beta & \gamma & \gamma \\ -\gamma & -\beta & 0 & 2\alpha & 0 \\ \beta & -\gamma & -2\alpha & 0 & 0 \\ -\beta & -\gamma & 0 & 0 & 0 \end{pmatrix} \in \mathfrak{so}(V_2) .$$

e) Nonexistence of equivariant immersions of symmetric space of noncompact type

For noncompact Riemannian homogeneous spaces there is no general result for the existence of equivariant immersions. In fact, for Riemannian symmetric spaces of noncompact type one has the following negative result [232].

THEOREM 3.3.3 (Vargas)
A Riemannian symmetric space of noncompact type has no equivariant isometric immersion into a Euclidean space.

This result generalizes a classical result of Bieberbach ([23], Section IV) asserting that the real hyperbolic plane has no equivariant isometric embedding into a Euclidean space.

3.4 Homogeneous submanifolds of Euclidean space

Recall that if $\rho : G \to I(\mathbb{R}^n)$ is a representation of a compact Lie group, then the elements of G have a common fixed point q (Remark 3.3.1), and hence, we can assume that $G \subseteq O(n+1)$ ($G \subseteq SO(n+1)$ if G is connected) and that each orbit lies in a sphere. The case of orbits of a representation $\rho : G \to O(n+1)$ (*orthogonal representation*) has been extensively studied, and we will be mainly concerned with this situation. More generally, for a homogeneous submanifold of $\mathbb{R}^n$, we have the following result by Di Scala [68].

THEOREM 3.4.1 (Di Scala)
Let $M = G \cdot v$ be a homogeneous irreducible submanifold of $\mathbb{R}^n$, where G is a connected Lie subgroup of the isometry group $I(\mathbb{R}^n)$ of $\mathbb{R}^n$. Then the universal covering group $\tilde{G}$ of G is isomorphic to the direct product $K \times \mathbb{R}^k$, where K is a simply connected compact Lie group. Moreover, the induced representation ρ of $K \times \mathbb{R}^k$ is equivalent to $\rho_1 \oplus \rho_2$, where ρ_1 is a representation of $K \times \mathbb{R}^k$ into $SO(d)$ and ρ_2 is a linear map of $\mathbb{R}^k$ into $\mathbb{R}^e$ ($n = d + e$), regarding $\mathbb{R}^e$ as a group of translations.

Roughly speaking, this result says that the homogeneous submanifolds of Euclidean spaces that are not contained in a sphere are generalized helicoids.

PROOF We divide the proof into three steps. Steps (1) and (2) were given in [176, Appendix].

(1) *The universal covering group $\tilde{G}$ of G is isomorphic to the direct product $K \times \mathbb{R}^k$, where K is a simply connected compact Lie group.*

The subgroup H of G consisting of all translations is a normal subgroup. The tangent spaces to the orbits of H in M define a parallel distribution $\mathcal{D}$ on M (being the restriction of a parallel distribution on $\mathbb{R}^n$). Moreover, it is easy to see that $\mathcal{D}$ is invariant under the shape operator of M, and therefore by the Lemma of Moore (Lemma 2.7.1) M splits, unless $\mathcal{D} = 0$ (or M is a straight line). Therefore, the canonical projection from G into $SO(n)$ is a Lie group homomorphism with discrete kernel and consequently an immersion. The Lie algebra of G is then isomorphic to a Lie subalgebra of $\mathfrak{so}(n)$ and admits a biinvariant metric, and hence, $\tilde{G} = K \times \mathbb{R}^k$, with K compact.

(2) *The induced representation of $K \times \mathbb{R}^k$ into $I(\mathbb{R}^n)$ is equivalent to $\tilde{\rho}_1 \oplus \tilde{\rho}_2$, where $\tilde{\rho}_1$ is a representation of $K \times \mathbb{R}^k$ into $SO(d)$, $\tilde{\rho}_2$ is a representation (not necessarily linear) of $\mathbb{R}^k$ into $I(\mathbb{R}^e)$, and $d + e = n$.*

The orbits of K in M are compact submanifolds of $\mathbb{R}^n$, and each of them has a well-defined barycenter in $\mathbb{R}^n$. Let B be the affine subspace of $\mathbb{R}^n$ generated by all these barycenters. Since K is a normal subgroup of G, the group $\mathbb{R}^k$ acts on B, which is pointwise fixed by K. Let V be the orthogonal complement to B at some point. It is easy to see that the representation of ρ in $I(\mathbb{R}^n)$ can be written in the form $\tilde{\rho}(k,w)(v,b) = (\tilde{\rho}_1(k,w)(v), \tilde{\rho}_2(w)(b))$, where $(k,w) \in K \times \mathbb{R}^k = \tilde{G}$, $(v,b) \in V \times B = \mathbb{R}^n$. Moreover, $\tilde{\rho}_1$ is a representation of $\tilde{G}$ into $SO(V)$. (In the above characterization we can replace B by the bigger subset consisting of all points of $\mathbb{R}^n$ fixed by K.)

(3) *The induced representation ρ of $K \times \mathbb{R}^k$ is equivalent to $\rho_1 \oplus \rho_2$, where ρ_1 is a representation of $K \times \mathbb{R}^k$ into $SO(d)$ and ρ_2 is a linear map of $\mathbb{R}^k$ into $\mathbb{R}^e$ ($n = d + e$), regarding $\mathbb{R}^e$ as a group of translations.*

By step (2) we only need to consider the case $\tilde{G} = \mathbb{R}^k$. Let $\rho : \mathbb{R}^k \to I(\mathbb{R}^n)$ be the canonical representation. We will show that $\rho = \rho_1 \oplus \rho_2$, where ρ_1 is (up to equivalence) a representation of $\mathbb{R}^k$ into $SO(\mathbb{R}^d)$ and ρ_2 is a linear map of $\mathbb{R}^k$ into $\mathbb{R}^e$ ($n = d + e$). The Lie algebra $\mathcal{L}(I(\mathbb{R}^n))$ is the semidirect product of $\mathfrak{so}(n)$ and $\mathbb{R}^n$ with $[(A,v),(B,u)] = (AB - BA, A(u) - B(v))$, and the Lie exponential map is $\operatorname{Exp}(t(A,v))(p) = e^{tA}(p - c) + c + td$, where $d \in \ker(A)$ and $v = d - A(c)$. We will show that there exists a common c for the rotational part of the Lie algebra $\mathcal{L}(\rho(\mathbb{R}^k))$. Let $\mathcal{R}$ be the image of the projection of $\mathcal{L}(\rho(\mathbb{R}^k))$ to $\mathfrak{so}(n)$. Then $\mathcal{R}$ is a commuting family of skewsymmetric endomorphisms that can be simultaneously diagonalized over $\mathbb{C}$. Let $\lambda_1, \ldots, \lambda_r$ be the different nonzero complex eigenfunctionals on $\mathcal{R}$. The subset $\mathcal{O} = \{R \in \mathcal{R} \mid \lambda_i(R) \neq 0, i = 1, \ldots, r\}$ is open and dense in $\mathcal{R}$. It is not hard to see that the density of $\mathcal{O}$ yields the existence of a basis $w_1 = (R_1, d_1 - R_1(c_1)), \ldots, (R_k, d_k - R_k(c_k))$ of $\mathcal{L}(\rho(\mathbb{R}^n))$, where $d_i \in \ker(R_i)$ and $c_i \in (\ker(R_i))^\perp$, such that $R_i \in \mathcal{O}$ for $i = 1, \ldots, k$. From the definition of $\mathcal{O}$, the R_i's all have the same kernel, say V. Since $[R_i, R_j] = 0$, the bracket formula yields $R_i(R_j(c_i - c_j)) = 0$, which implies $c_i = c_j$ for all i, j. Fixing the origin at c, we have that ρ is orthogonally equivalent to $\rho_1 \oplus \rho_2$, where ρ_1 is a representation of $\mathbb{R}^k$ into $SO(V^\perp)$ and ρ_2 is a linear map of $\mathbb{R}^k$ into V. □

THEOREM 3.4.2 (Di Scala)
Every minimal homogeneous submanifold of $\mathbb{R}^n$ is totally geodesic.

PROOF Without loss of generality we can assume that the homogeneous submanifold, say $G \cdot p$, is irreducible. By Theorem 3.4.1 and its proof there exists a basis of $\mathcal{L}(\rho(G))$ of the form $(A_1, d_1), \ldots, (A_n, d_n)$, where all vectors d_i belong to the intersection V of the kernels of $A_1, \ldots, A_n$. We assume that the first $l < n$ vectors are in the isotropy subalgebra at p and that $A_{l+1}p + d_{l+1}, \ldots, A_n p + d_n$ form an orthonormal basis of $T_p(G \cdot p)$. Now decompose $p = p_1 + p_2$, where $p_1 \in V^\perp$ and $p_2 \in V$ ($A_i p_2 = 0$ and $d_i = 0$ for $i = 1, \ldots, l$). For $i = l+1, \ldots, n$, put $\gamma_i(t) = \operatorname{Exp}(t(A_i, d_i))p = e^{tA_i}p_1 + td_i + p_2$. Observe that p_1 belongs to the normal space of $G \cdot p$ at p, since $d_i \in V$ and A_i is skewsymmetric. We claim that p_1 must be 0. Using minimality, we get $0 = \sum_{i=l+1}^n \langle \gamma_i''(0), p_1 \rangle = \sum_{i=l+1}^n \langle A_i^2 p_1, p_1 \rangle = \sum_{i=1}^n \langle A_i^2 p_1, p_1 \rangle =$

$\sum_{i=1}^{n}\langle A_i p_1, A_i p_1\rangle$. Hence, $A_i p_1 = 0$ for all $i = 1, \ldots, n$ and $p_1 \in V$. This implies $p_1 = 0$ since p_1 also belongs to $V^{\perp}$. It is now clear that $G \cdot p = G \cdot (0, p_2)$ coincides with the linear span of $d_{l+1}, \ldots, d_n$ and is totally geodesic. □

This result is sharp in the sense that there exist minimal submanifolds of $\mathbb{R}^n$ with codimension one (for instance, minimal surfaces of revolution). By Calabi's Rigidity Theorem, any holomorphic isometry of a complex submanifold of $\mathbb{C}^n$ extends to $\mathbb{C}^n$. On the other hand, any complex submanifold is minimal. So we have the following:

COROLLARY 3.4.3
Every complex homogeneous submanifold of $\mathbb{C}^n$ is totally geodesic.

3.5 Homogeneous submanifolds of hyperbolic spaces

In this section, we outline the results of [69] about homogeneous submanifolds of the real hyperbolic space $H^n = SO^o(n,1)/SO(n)$. We first introduce some notation. Let $(V, \langle\cdot,\cdot\rangle)$ be a real vector space endowed with a nondegenerate symmetric bilinear form of signature $(n,1)$. We can identify V with the Lorentzian space $\mathbb{R}^{n,1}$, whence $\mathrm{Aut}(V, \langle\cdot,\cdot\rangle) \cong O(n,1)$. It is well known that H^n can be identified with a connected component of the set of points $p \in \mathbb{R}^{n,1}$ with $\langle p, p\rangle = -1$. Observe that the identity component of $O(n,1)$ acts transitively on H^n by isometries. An affine subspace W of V is called Riemannian, Lorentzian or degenerated if the restriction of $\langle\cdot,\cdot\rangle$ to the vector part of W is positive definite, has signature $(\dim W - 1, 1)$ or is degenerate, respectively. A *horosphere* in H^n is the submanifold that is obtained by intersecting H^n with an affine degenerate hyperplane. Recall that the ideal boundary $H^n(\infty)$ is the set of equivalence classes of asymptotic geodesics. Thus $H^n(\infty)$ can be regarded as the set of *light* lines through the origin, that is, $H^n(\infty) \equiv \{z \in \mathbb{R}^{n,1} \mid \langle z, z\rangle = 0\}$. Observe that a point z at infinity defines a foliation of H^n by parallel horospheres. We say that the horosphere Q is centred at $z \in H^n(\infty)$ if Q is a leaf of that foliation. An action of a subgroup G of $O(n,1)$ is called *weakly irreducible* if it leaves invariant degenerate subspaces only.

The classification of homogeneous submanifolds of H^n is basically given by the following result.

THEOREM 3.5.1 [69]
Let G be a connected (not necessarily closed) Lie subgroup of $I^o(H^n) = SO^o(n,1)$. Then one of the following statements holds:

(i) G has a fixed point;

(ii) G has a unique nontrivial totally geodesic orbit (possibly H^n);

(iii) All orbits are contained in horospheres centered at the same point at infinity.

This result is also a tool for the proof of the next result, which shows how the theory of homogeneous submanifolds of H^n can be used to obtain general results about the action of a connected Lie subgroup of $O(n,1)$ on the Lorentzian space $\mathbb{R}^{n,1}$.

THEOREM 3.5.2 [69]
Let G be a connected (not necessarily closed) Lie subgroup of $SO^o(n,1)$ and assume that the action of G on the Lorentzian space $\mathbb{R}^{n,1}$ is weakly irreducible. Then G acts transitively either on H^n or on a horosphere of H^n. Moreover, if G acts irreducibly, then $G = SO^o(n,1)$.

Theorem 3.5.2 has an immediate consequence, which provides a purely geometric answer to a question posed in [9].

COROLLARY 3.5.3 **(*M. Berger*** [10,11]***)***
Let M^n be a Lorentzian manifold. If the restricted holonomy group of M^n acts irreducibly, then it coincides with $SO^o(n,1)$. In particular, if M^n is locally symmetric, then M^n has constant sectional curvature.

We will now present some ideas for the proof of Theorem 3.5.2. The fundamental tools for the proof of Theorem 3.5.2 are Theorem 3.5.1 and the following uniqueness result.

LEMMA 3.5.4 [69]
Let G be a connected Lie subgroup of $SO^o(n,1)$. If the action of G on H^n has a totally geodesic orbit, then no other orbit of positive dimension is minimal.

PROOF of Lemma 3.5.4 Suppose $G \cdot p$ is a totally geodesic orbit, and consider another orbit $G \cdot q$ with $G \cdot q \neq \{q\}$. Let γ be a geodesic in H^n minimizing the distance between q and $G \cdot p$. Without loss of generality we may assume that γ meets $G \cdot p$ at p and that γ is parametrized so that $\gamma(0) = p$ and $\gamma(1) = q$. Since γ minimizes the distance between q and $G \cdot p$, $\dot\gamma(0)$ is perpendicular to $G \cdot p$ at p, that is, $\langle \dot\gamma(0), X_p^* \rangle = 0$ for all $X \in \mathfrak{g}$, where X^* denotes as usual the Killing vector field generated by X. Since X^* is a Killing vector field, we have $\langle \nabla_{\dot\gamma(t)} X^*, \dot\gamma(t) \rangle = 0$ for all t. Hence, $\frac{d}{dt}\langle X^*_{\gamma(t)}, \dot\gamma(t) \rangle = 0$ and therefore $\langle X^*_{\gamma(t)}, \dot\gamma(t) \rangle = 0$ for all t. This means that $\dot\gamma(t)$ is orthogonal to $G \cdot \gamma(t)$ at $\gamma(t)$ for all t.

Now assume that $X \in \mathfrak{g}$ satisfies $X_q^* \neq 0$, and let $\Phi_s^{X^*}$ be the one-parameter group of isometries generated by X^*. Let $h : [0,1] \times \mathbb{R} \to H^n$ be defined by $h_s(t) = \Phi_s^{X^*}(\gamma(t))$. Note that $X^*_{h_s(t)} = \frac{\partial}{\partial s} h_s(t)$, and that $h_s(t)$ is a geodesic for each s. Let $A_{\dot\gamma(t)}$ be the shape operator of the submanifold $G \cdot \gamma(t)$ at $\gamma(t)$ with respect to $\dot\gamma(t)$, and define

$$f(t) = -\langle A_{\dot\gamma(t)} X^*_{\gamma(t)}, \dot\gamma(t)\rangle = \langle \frac{D}{\partial s}\frac{\partial}{\partial t} h_s(t), X^*_{h_s(t)}\rangle|_{s=0} \ .$$

Then the derivative $f'(t)$ of $f(t)$ is given by

$$\begin{aligned} f'(t) &= \langle \tfrac{D}{\partial t}\tfrac{D}{\partial s}\tfrac{\partial}{\partial t} h_s(t), X^*_{h_s(t)}\rangle|_{s=0} + \langle \tfrac{D}{\partial s}\tfrac{\partial}{\partial t} h_s(t), \tfrac{D}{\partial t} X^*_{h_s(t)}\rangle|_{s=0} \\ &= \langle R(\tfrac{\partial}{\partial t} h_s(t), \tfrac{\partial}{\partial s} h_s(t))\tfrac{\partial}{\partial t} h_s(t), X^*_{h_s(t)}\rangle|_{s=0} + \langle \tfrac{D}{\partial t}\tfrac{\partial}{\partial s} h_s(t), \tfrac{D}{\partial t} X^*_{h_s(t)}\rangle|_{s=0} \\ &= \langle R(\dot\gamma(t), X^*_{\gamma(t)})\dot\gamma(t), X^*_{\gamma(t)}\rangle + \|\nabla_{\dot\gamma(t)} X^*\|^2 \\ &= \|\dot\gamma(t)\|^2 \|X^*_{\gamma(t)}\|^2 + \|\nabla_{\dot\gamma(t)} X^*\|^2 \ . \end{aligned}$$

Therefore, $f'(t) \geq 0$ and $f'(1) > 0$ since $X_q^* \neq 0$. As $G \cdot p$ is totally geodesic, we have $f(0) = 0$, and hence, $\langle A_{\dot\gamma(1)} X_q^*, X_q^*\rangle = -f(1) < 0$. Thus, $A_{\dot\gamma(1)}$ is negative definite, which shows that $G \cdot q$ cannot be minimal. □

A simple consequence of Theorem 3.5.1 and Lemma 3.5.4 is that every minimal homogeneous submanifold of H^n is totally geodesic. As we saw in Theorem 3.4.2, the analogous statement for $\mathbb{R}^n$ is also true.

A key fact in the proof of Theorem 3.5.1 is the following observation: If a normal subgroup H of G has a totally geodesic orbit $H \cdot p$ of positive dimension, then $G \cdot p = H \cdot p$. This is because G permutes the H-orbits, and hence $H \cdot p = G \cdot p$ by Lemma 3.5.4.

The next step for proving Theorem 3.5.1 is to study the two following cases separately: G is semisimple (and of noncompact type) and G is not semisimple. In the latter case one proves the statement first for Abelian groups. Applying the previous observation to a normal Abelian subgroup of G, three possibilities can occur: G translates a geodesic, G fixes a point at infinity, or G has a proper totally geodesic orbit. As a consequence, a connected Lie subgroup G of $O(n,1)$ that acts irreducibly on $\mathbb{R}^{n,1}$ must be semisimple. Finally, one shows that if G has a fixed point z at infinity, then G has a totally geodesic orbit (possibly the entire H^n), or G has a fixed point in H^n, or all of its orbits are contained in the horospheres centered at z. The idea is that if G has neither a fixed point nor orbits in horospheres, then there exists a codimension one subgroup H of G such that every H-orbit is contained in the horosphere foliation determined by z. Then H acts isometrically on horospheres, and one can use the fact that H must have a totally geodesic orbit in each horosphere, because each horosphere is a Euclidean space. At last, it is not hard to show that the union of all these totally geodesic orbits over all horospheres is a totally geodesic G-invariant submanifold of H^n.

Where G is a semisimple Lie group we choose an Iwasawa decomposition $G = KAN$. Then it is possible to prove that the solvable subgroup AN of G has a minimal orbit that is also a G-orbit. One first chooses a fixed point p of the compact group K, which always exists by Cartan's Fixed Point Theorem. It is possible to prove that the isotropy subgroup G_p of G at p coincides with K. Then the mean curvature vector field H of $G \cdot p = AN \cdot p$ is invariant under G_p. If H does not vanish, then the G-orbits through points on normal K-invariant geodesics are homothetic to the orbit $G \cdot p$. Observe that these orbits are also AN-orbits. The volume element of these orbits can be controlled by Jacobi vector fields, eventually proving that there exists a minimal G-orbit that is also an AN-orbit.

An induction argument involving n and the dimension of G completes the proof of Theorem 3.5.1.

The idea for the proof of Theorem 3.5.2 is as follows: If G acts in a weakly irreducible way, then the G-orbits must be contained in horospheres and, if an orbit is a proper submanifold of some horosphere, one can construct a proper totally geodesic G-invariant submanifold as the union of orbits parallel to totally geodesics orbits of the action of G restricted to the horosphere. This is a contradiction because totally geodesic submanifolds are obtained by intersecting the hyperbolic space H^n with Lorentzian subspaces.

If G acts irreducibly, then it must act transitively on H^n. By a previous observation, we already know that G is semisimple and of noncompact type. The second part of the theorem follows from the theory of Riemannian symmetric spaces of noncompact type, once we show that the isotropy group at some point is a maximal compact subgroup of G.

3.6 Second fundamental form of orbits

Let G be a Lie group acting isometrically on a Riemannian manifold $\bar{M}$. Let $p \in \bar{M}$ and A be the shape operator of the orbit $M = G \cdot p$. Then we have

$$A_{g_* \xi} g_* X = g_* A_\xi X$$

for all $g \in G$, $X \in T_p M$ and $\xi \in \nu_p M$. In particular, if M is a principal orbit and $\hat{\xi}$ is the equivariant normal vector field on M determined by ξ, then

$$A_{\hat{\xi}_{gp}} g_* X = A_{g_* \xi} g_* X = g_* A_\xi X$$

for all $g \in G$, $X \in T_p M$ and $\xi \in \nu_p M$. Therefore we have

$$A_{\hat{\xi}_{gp}} = g_* A_\xi g_*^{-1} , \tag{3.4}$$

and hence,

PROPOSITION 3.6.1
The principal curvatures of a principal orbit with respect to an equivariant normal vector field are constant.

Let now X^* be a Killing vector field on $\bar{M}$ that is induced by the action of G. Recall that

$$T_pM = \{X_p^* \mid X \in \mathfrak{g}\} = \{X_p^* \mid X \in \mathfrak{m}\} ,$$

where $\mathfrak{g} = \mathfrak{k} \oplus \mathfrak{m}$ is a reductive decomposition. Since X^* is a Killing vector field on $\bar{M}$, for any normal vector field ξ we have $(\bar{\nabla}_{X^*}\xi)^T = (\bar{\nabla}_\xi X^*)^T$, and the Weingarten formula tells us

$$A_\xi X_p^* = -((\bar{\nabla} X^*)_p \xi)^T ,$$

where $(\cdot)^T$ denotes the orthogonal projection from $T_p\bar{M}$ onto T_pM. Note that, X^* being a Killing vector field on $\bar{M}$, the covariant derivative $(\bar{\nabla} X^*)_p$ of X^* at p is a skewsymmetric endomorphism of $T_p\bar{M}$. We summarize this in

PROPOSITION 3.6.2
Let G be a Lie group acting isometrically on a Riemannian manifold $\bar{M}$. Then, for each $p \in \bar{M}$ the tangent space of $G \cdot p$ at p is given by

$$T_p(G \cdot p) = \{X_p^* \mid X \in \mathfrak{g}\} = \{X_p^* \mid X \in \mathfrak{m}\} ,$$

where $\mathfrak{g} = \mathfrak{k} \oplus \mathfrak{m}$ is a reductive decomposition of $\mathfrak{g}$ and $\mathfrak{k}$ is the Lie algebra of the isotropy subgroup $K = G_p$ at p. If $\xi \in \nu_p(G \cdot p)$ and $X \in \mathfrak{g}$, the shape operator A_ξ of $G \cdot p$ at p with respect to ξ is given by

$$A_\xi X_p^* = -((\bar{\nabla} X^*)_p \xi)^T ,$$

where $(\cdot)^T$ denotes the orthogonal projection from $T_p\bar{M}$ onto $T_p(G \cdot p)$.

If $\bar{M} = \mathbb{R}^n$ and $G \subset SO(n)$, then the elements in $\mathfrak{g} \subset \mathfrak{so}(n)$ can be identified with skewsymmetric endomorphisms of $\mathbb{R}^n$. Using this identification the previous proposition can be rephrased as

COROLLARY 3.6.3
Let $G \subset SO(n)$ be a Lie group acting isometrically on $\mathbb{R}^n$. Then, for each $p \in \mathbb{R}^n$ the tangent space of $G \cdot p$ at p is given by

$$T_p(G \cdot p) = \{Xp \mid X \in \mathfrak{m}\} ,$$

where $\mathfrak{g} = \mathfrak{k} \oplus \mathfrak{m}$ is a reductive decomposition of $\mathfrak{g}$ and $\mathfrak{k}$ is the Lie algebra of the isotropy subgroup $K = G_p$ at p. If $\xi \in \nu_p(G \cdot p)$ and $X \in \mathfrak{m}$, the shape operator A_ξ of $G \cdot p$ at p with respect to ξ is given by

$$A_\xi Xp = -(X\xi)^T ,$$

where $(\cdot)^T$ denotes the orthogonal projection from $\mathbb{R}^n$ onto $T_p(G \cdot p)$.

Example 3.2 (continued). Let G be a compact semisimple Lie group, $\mathfrak{g}$ its Lie algebra, and $\langle \cdot, \cdot \rangle$ the inner product on $\mathfrak{g}$ given by the negative of the

Killing form of $\mathfrak{g}$. For $X \in \mathfrak{g}$ we consider the orbit $M = \mathrm{Ad}(G) \cdot X = G \cdot X$. Suppose that the element X is regular, that is, the isotropy subalgebra $\mathfrak{g}_X$ has minimal dimension. Recall that, in this case, $\mathfrak{t} = \mathfrak{g}_X$ is a Cartan subalgebra of $\mathfrak{g}$ and the orbit $G \cdot X$ is principal. We shall now compute the shape operator of a principal orbit $M = G \cdot X$, $X \in \mathfrak{t}$, in terms of the roots of $\mathfrak{g}$ with respect to $\mathfrak{t}$ (see also the discussion of this example on page 49).

First of all, since $M = G \cdot X$ is principal and the adjoint action is polar, Corollary 3.2.5 implies that any equivariant normal vector field of M is $\nabla^\perp$-parallel, and hence, the normal curvature tensor of M vanishes. It follows from the Ricci equation that the shape operators of M commute with each other and therefore are simultaneously diagonalizable. We have $\nu_X M = \mathfrak{t}$ and $T_X M = \{[Z, X] \mid Z \in \mathfrak{g}\}$. For each $\xi \in \mathfrak{t}$, Corollary 3.6.3 gives

$$A_\xi [Z, X] = -[Z, \xi]^T = -[Z, \xi] \ . \tag{3.5}$$

We now compute (3.5) for the elements $X_\alpha, Y_\alpha \in \mathfrak{g}$ defined on page 49. By (3.1) we get

$$A_\xi X_\alpha = -\frac{\alpha(\xi)}{\alpha(X)} X_\alpha \quad , \quad A_\xi Y_\alpha = -\frac{\alpha(\xi)}{\alpha(X)} Y_\alpha \ . \tag{3.6}$$

This implies that the curvature distributions are given by $E_\alpha = \{\mathbb{R}X_\alpha \oplus \mathbb{R}Y_\alpha\}$, $\alpha \in \Delta^+$. Note that all eigendistributions are even-dimensional. Since the equivariant normal vector fields $\hat{\xi}$ determined by $\xi \in \mathfrak{t} = \nu_X M$ are $\nabla^\perp$-parallel, the equivariance now implies that *the principal curvatures with respect to the parallel normal vector fields $\hat{\xi}$ are constant.* Submanifolds with this property and flat normal bundle are called *isoparametric* (see, for instance, page 18). □

Example 3.4 Orbits of s-representations

Just as for principal orbits of the adjoint representation of a compact semisimple Lie group, one can compute shape operator of the principal orbits of s-representations by using restricted roots λ. We use the same notation as in Section 3.2 b). Let $M = K \cdot A$, $A \in \mathfrak{a}$, be such a principal orbit. Then, for the shape operator A_ξ of M with respect to $\xi \in \mathfrak{a} = \nu_A M$, we get the expression

$$A_\xi X = -\frac{\lambda(\xi)}{\lambda(A)} X$$

with $X \in \mathfrak{p}_\lambda \subseteq \mathfrak{p}_+ = T_A M$. Thus, the common eigenspaces of the shape operator of M are

$$E_\lambda = \mathfrak{p}_\lambda + \mathfrak{p}_{2\lambda} \ ,$$

where $\mathfrak{p}_{2\lambda} = \{0\}$ if 2λ is not a restricted root. Also in this case, the action is polar, and so the normal bundle of M is flat and the principal curvatures with respect to parallel normal vector fields are constant.

If $M = K \cdot A$, $A \in \mathfrak{a}$, is a singular orbit, then $\mathfrak{a} \subset \nu_A M$ and, using the same methods, one gets that the shape operator A_ξ of M, with respect to $\xi \in \mathfrak{a} \subset \nu_A M$, is given by

$$A_\xi X = -\frac{\lambda(\xi)}{\lambda(A)} X$$

for all $X \in \mathfrak{p}_\lambda$. □

3.7 Symmetric submanifolds

a) Motivation and definition

Symmetric submanifolds are, in a certain sense, analogous to symmetric spaces for submanifold theory. Indeed, they always come equipped with a symmetry at each point, namely the geodesic reflection in the corresponding normal submanifold. Some real flag manifolds, the so-called symmetric R-spaces, can be embedded as symmetric submanifolds in Euclidean spaces. We will see that essentially all symmetric submanifolds in Euclidean spaces arise from symmetric R-spaces.

The study of symmetric submanifolds in Euclidean spaces, and of the closely related submanifolds with parallel second fundamental form, started around 1970. To our knowledge, the origin of these studies goes back to the paper by Chern, do Carmo and Kobayashi [55] on minimal submanifolds of spheres with second fundamental form of constant length. In this paper the condition $\nabla^\perp \alpha = 0$ is explicitly stated. Further studies were undertaken by Vilms [233] and Walden [234]. Then Ferus [83–85], systematically studied submanifolds of $\mathbb{R}^n$ with parallel second fundamental form. He achieved a complete classification of these submanifolds, and, as a consequence of his result, it turns out that such submanifolds are locally extrinsic symmetric. A direct proof of this latter fact was presented by Strübing [204], whose result we will discuss in part b) below.

The precise general definition of a symmetric submanifold is as follows. A submanifold M of a Riemannian manifold $\bar{M}$ is called a *symmetric submanifold* if, for each $p \in M$, there exists an isometry σ_p of $\bar{M}$ with

$$\sigma_p(p) = p \ , \ \sigma_p(M) = M \text{ and } (\sigma_p)_*(v) = \begin{cases} -v & , \ v \in T_pM \\ v & , \ v \in \nu_pM \end{cases} .$$

In this section, we deal with symmetric submanifolds of space forms. The general case will be discussed in Section 8.5.

b) Symmetric submanifolds and parallel second fundamental form

We begin by investigating the relation between symmetric submanifolds and parallel second fundamental form. We will see that for submanifolds of space forms these two concepts lead to the same theory, whereas, in more general Riemannian manifolds, symmetric submanifolds have parallel second fundamental form, but not vice versa.

Let M be a symmetric submanifold of a space form $\bar{M}$. Any isometry of $\bar{M}$ is an affine map with respect to the Levi Civita connection. Using the Gauss formula we obtain

$$(\nabla^{\perp}_X \alpha)(Y,Z) = \sigma_{p*}(\nabla^{\perp}_X \alpha)(Y,Z) = (\nabla^{\perp}_{\sigma_{p*}X}\alpha)(\sigma_{p*}Y, \sigma_{p*}Z) = -(\nabla^{\perp}_X \alpha)(Y,Z)$$

for all $p \in M$ and $X, Y, Z \in T_pM$. Thus, the second fundamental form of a symmetric submanifold is parallel.

PROPOSITION 3.7.1
The second fundamental form of a symmetric submanifold of a space form is parallel.

A natural question arising from this proposition is whether parallelity of the second fundamental form implies symmetry of the submanifold. Since the first condition is local, whereas the second is global, this question makes sense only for some kind of local symmetry. To make this precise, we introduce the notion of a *locally symmetric submanifold* M of a Riemannian manifold $\bar{M}$ by requiring that for each $p \in M$ there exists a local isometry σ_p of $\bar{M}$ with

$$\sigma_p(p) = p\ ,\ \sigma_p(U) = U \text{ and } (\sigma_p)_*(v) = \begin{cases} -v & ,\ v \in T_pM \\ v & ,\ v \in \nu_pM \end{cases}$$

for some open neighbourhood U of p in M. Our aim now is to show that, for a submanifold of a space of constant curvature, local symmetry is equivalent to parallelity of the second fundamental form (see also [204]).

THEOREM 3.7.2
A submanifold of a space of constant curvature is locally symmetric if and only if its second fundamental form is parallel.

We mention straight away that this result does not generalize to more general Riemannian manifolds. As we will see later, a totally geodesic real projective space $\mathbb{R}P^k$ in complex projective space $\mathbb{C}P^n$ is not a locally symmetric submanifold for $k < n$, but obviously has parallel second fundamental form.

PROOF of Theorem 3.7.2 Let $\bar{M}$ be a space of constant curvature and M a submanifold of $\bar{M}$. If M is locally symmetric, the parallelity of the second fundamental form follows by the same argument as in the proof of Proposition 3.7.1. Conversely, suppose that M has parallel second fundamental form in $\bar{M}$ and let $p \in M$. Since $\bar{M}$ has constant curvature, there exist an open neighbourhood W of p in $\bar{M}$ and an isometry $f : W \to W$ with $f(p) = p$, $f_{*p}X = -X$ for all $X \in T_pM$, and $f_{*p}\xi = \xi$ for all $\xi \in \nu_pM$. Note that, for the existence of such a local isometry f, we use the hypothesis that $\bar{M}$ has constant curvature. We have to show that $f(M) \subseteq M$.

Let γ be a geodesic in M with $\gamma(0) = p$ and parametrized by arc length. We will show that

$$f(\gamma(t)) = \gamma(-t) \tag{3.7}$$

for all t with $|t|$ sufficiently small, which then implies that f is a local symmetry at p, leaving M invariant.

We will use the technique that has been used by Strübing [204], but also introduce some simplifications as discussed in [178]. Let $e_1 = \dot{\gamma}(0), e_2, \ldots, e_n$ be a Darboux frame at p, that is, $e_1, \ldots, e_m$ is an orthonormal basis of T_pM and $e_{m+1}, \ldots, e_n$ is an orthonormal basis of ν_pM. Let $E_1, \ldots, E_m$ be the ∇-parallel vector fields along γ induced from $e_1, \ldots, e_m$ and $E_{m+1}, \ldots, E_n$ the $\nabla^\perp$-parallel vector fields along γ induced from $e_{m+1}, \ldots, e_n$. We denote by E_i' the covariant derivative of E_i with respect to the Levi Civita connection of $\bar{M}$. By construction we have $\langle E_i', E_j\rangle = 0$ if $i, j \in \{1, \ldots, m\}$ or $i, j \in \{m+1, \ldots, n\}$. Since the second fundamental form α of M is parallel, we have

$$\langle E_i', E_j\rangle = \langle \alpha(E_1, E_i), E_j\rangle = \langle \alpha(e_1, e_i), e_j\rangle = a_{ij} \in \mathbb{R}$$

for all $i \in \{1, \ldots, m\}$ and $j \in \{m+1, \ldots, n\}$. This implies $\langle E_i', E_j\rangle = -\langle E_i, E_j'\rangle = -a_{ji}$ for all $i \in \{m+1, \ldots, n\}$ and $j \in \{1, \ldots, m\}$. Thus $\gamma, E_1, \ldots, E_n$ satisfy the system of linear differential equations

$$\dot{\gamma} = E_1 \ , \ E_i' = \sum_{j=1}^{n} a_{ij}E_j \quad (i = 1, \ldots, n) \tag{3.8}$$

with initial conditions $\gamma(0) = p$ and $E_i(0) = e_i$ $(i = 1, \ldots, n)$.

Now consider the Darboux frame $f_*e_1, \ldots, f_*e_n$ at p, and extend this frame as above to orthonormal frame fields $F_1, \ldots, F_n$ along $f \circ \gamma$ and $G_1, \ldots, G_n$ along $\tilde{\gamma} : t \mapsto \gamma(-t)$. Then $f \circ \gamma, F_1, \ldots, F_n$ and $\tilde{\gamma}, G_1, \ldots, G_n$ satisfy the system (3.8) of linear differential equations, and $f \circ \gamma(0) = \tilde{\gamma}(0)$ and $F_i(0) = G_i(0)$, $i = 1, \ldots, n$, holds. This implies (3.7). □

If $\bar{M}$ is a standard space of constant curvature there is a global version of Theorem 3.7.2, since the local symmetry defined in the proof can be extended to a global isometry.

THEOREM 3.7.3

Let M be a connected complete submanifold of a standard space of constant curvature. Then M is a symmetric submanifold if and only if M has parallel second fundamental form.

c) Construction methods in standard space forms

In this part we describe two methods for constructing new symmetric submanifolds from given symmetric submanifolds in standard space forms. These methods are quite elementary, but useful for classification purposes.

Method 1: Extrinsic products of symmetric submanifolds.

Let $M = M_1 \times \ldots \times M_s$ be an extrinsic product in a standard space form $\bar{M}$, and suppose that each leaf $L_i(p)$ of the induced totally geodesic foliations L_i, $i = 1, \ldots, s$, on M has parallel second fundamental form $\alpha_{i,p}$ in $\bar{M}$. By definition, the second fundamental form α of M satisfies $\alpha(X_i, Y_j) = 0$ for all $X_i \in T_pL_i(p)$, $Y_j \in T_pL_j(p)$, $i \neq j$, $p \in M$. Since M is a Riemannian product, it follows from this definition that $(\nabla^\perp_{X_i}\alpha)(Y_j, Z_k) = 0$ for all $X_i \in T_pL_i(p)$, $Y_j \in T_pL_j(p)$ and $Z_k \in T_pL_k(p)$ whenever $j \neq k$. Using the Codazzi equation, this implies $(\nabla^\perp_{X_i}\alpha)(Y_j, Z_k) = 0$ for all $X_i \in T_pL_i(p)$, $Y_j \in T_pL_j(p)$ and $Z_k \in T_pL_k(p)$ whenever two of the three indices i, j, k are distinct. On the other hand, if all three indices i, j, k coincide,

$$(\nabla^\perp_{X_i}\alpha)(Y_j, Z_k) = (\nabla^\perp_{X_i}\alpha_{i,p})(Y_j, Z_k) = 0 ,$$

since, by assumption, $\alpha_{i,p}$ is parallel. Altogether, it follows that M has parallel second fundamental form in $\bar{M}$.

Now assume that each leaf $L_i(p)$ is a symmetric submanifold of $\bar{M}$. Since each symmetric submanifold has parallel second fundamental form, the previous discussion shows that M has parallel second fundamental form in $\bar{M}$. From Theorem 3.7.3 we deduce that M is a symmetric submanifold of $\bar{M}$. We summarize this in

LEMMA 3.7.4
The extrinsic product of symmetric submanifolds of a standard space form is also a symmetric submanifold.

Method 2: Prolongation of symmetric submanifolds via totally umbilical submanifolds.

Another method for constructing a new symmetric submanifold from a given symmetric submanifold of a standard space form is the so-called prolongation via a totally umbilical submanifold.

LEMMA 3.7.5
Let M be a connected submanifold of a standard space form M', and suppose that M' is embedded as a totally umbilical submanifold in another standard space form $\bar{M}$. Then M is a symmetric submanifold of $\bar{M}$ if and only if M is a symmetric submanifold of M'.

PROOF We denote by α the second fundamental form of M in $\bar{M}$, by α' the second fundamental form of M in M', and by H the mean curvature vector field of M' in $\bar{M}$. The Gauss formula implies

$$\alpha(X,Y) = \alpha'(X,Y) + \langle X,Y \rangle H$$

for all vector fields X, Y tangent to M. From Lemma 2.6.1 we know that H is parallel in the normal bundle of M'. This implies that α is parallel if and only if α' is parallel. The assertion then follows from Theorem 3.7.3. □

The following example illustrates how these two methods can be used to construct new symmetric submanifolds of spheres from some given symmetric submanifolds of spheres. Let M_1 be a symmetric submanifold of $S^{n_1-1}(r_1)$ and M_2 a symmetric submanifold of $S^{n_2-1}(r_2)$. Since $S^{n_1-1}(r_1)$ sits totally umbilically inside $\mathbb{R}^{n_1}$, it follows from Lemma 3.7.5 that M_1 is a symmetric submanifold of $\mathbb{R}^{n_1}$. Analogously, M_2 is a symmetric submanifold of $\mathbb{R}^{n_2}$. We now apply Lemma 3.7.4 to see that the extrinsic product $M_1 \times M_2$ is a symmetric submanifold of $\mathbb{R}^n$, where $n = n_1 + n_2$. But, by construction, $M_1 \times M_2$ sits inside $S^{n_1-1}(r_1) \times S^{n_2-1}(r_2)$, which is a submanifold of $S^{n-1}(r)$, where $r = \sqrt{r_1^2 + r_2^2}$. Since $S^{n-1}(r)$ is a totally umbilical submanifold of $\mathbb{R}^n$, it follows from Lemma 3.7.5 that $M_1 \times M_2$ is a symmetric submanifold of $S^{n-1}(r)$.

d) Examples of symmetric submanifolds in standard space forms

We will now use Theorem 3.7.3 to present some examples of symmetric submanifolds in standard space forms. In fact, according to this result, it is sufficient to find connected complete submanifolds with parallel second fundamental form.

Example 3.5 Totally geodesic submanifolds
Every totally geodesic submanifold has vanishing second fundamental form and hence also parallel second fundamental form. The connected complete totally geodesic submanifolds of standard space forms have been classified in Theorem 2.4.1. □

Example 3.6 Extrinsic spheres
The second fundamental form α of an extrinsic sphere is of the form

$$\alpha(X,Y) = \langle X,Y \rangle H \ ,$$

where the mean curvature vector field H is parallel in the normal bundle. Therefore, the second fundamental form of an extrinsic sphere is parallel.

We now discuss briefly the existence problem for extrinsic products of extrinsic spheres or totally geodesic submanifolds in standard space forms. Let $M = M_1 \times \ldots \times M_s$ be an extrinsic product in a standard space form $\bar{M}^n(\kappa)$, and suppose that each leaf $L_i(p)$ of the induced totally geodesic foliations L_i,

$i = 1, \ldots, s$, on M is an extrinsic sphere or is totally geodesic in $\bar{M}^n(\kappa)$. We fix a point $p \in M$ and denote by H_i the mean curvature vector at p of the extrinsic sphere $L_i(p)$ in $\bar{M}^n(\kappa)$. Hence, the second fundamental form α_i of $L_i(p)$ at p is given by

$$\alpha_i(X,Y) = \langle X,Y \rangle H_i$$

for all $X, Y \in T_pL_i(p)$. Let $X_i \in T_pL_i(p)$ and $Y_j \in T_pL_j(p)$ be unit vectors. Since M is an extrinsic product in $\bar{M}^n(\kappa)$, the Gauss equation yields

$$0 = \langle R(X_i, Y_j)Y_j, X_i \rangle = \kappa + \langle H_i, H_j \rangle \quad (i \neq j) .$$

Therefore, if $\kappa \neq 0$, none of the leaves $L_i(p)$ can be totally geodesic. We define

$$\kappa_i = \kappa + \langle H_i, H_i \rangle .$$

If $\dim L_i(p) \geq 2$ then $L_i(p)$ is a space of constant curvature κ_i, which follows easily from the Gauss equation. If $H_1, \ldots, H_s$ are mutually distinct, we get

$$0 < \langle H_i - H_j, H_i - H_j \rangle = \langle H_i, H_i \rangle + \langle H_j, H_j \rangle - 2\langle H_i, H_j \rangle = \kappa_i + \kappa_j .$$

Consequently, at most one of the numbers κ_i is nonpositive. This implies, for instance, that such an extrinsic product in real hyperbolic space can never contain two real hyperbolic spaces, or some real hyperbolic space together with a Euclidean space.

Conversely, let $p \in \bar{M}^n(\kappa)$ and V a linear subspace of $T_p\bar{M}^n(\kappa)$. Let $V = V_1 \oplus \ldots \oplus V_s$ be an orthogonal decomposition of V with $s \geq 2$ and $H_1, \ldots, H_s \in T_p\bar{M}^n(\kappa)$ be mutually distinct and perpendicular to V. Does there exist an extrinsic product $M = M_1 \times \ldots \times M_s$ of extrinsic spheres and totally geodesic submanifolds in $\bar{M}^n(\kappa)$ with $p \in M$, $T_pL_i(p) = V_i$, and such that H_i is the mean curvature vector of $L_i(p)$ at p? Note that at most one factor can be totally geodesic since we assume the vectors H_i to be mutually distinct. We have seen above that we necessarily need $\langle H_i, H_j \rangle = -\kappa$ for all $i \neq j$. In fact, it can be shown that this condition is also sufficient for the existence of such an extrinsic product. This can be proved easily when $\kappa = 0$. In this case, the vectors $H_1, \ldots, H_s$ are pairwise orthogonal. Let M_i be the extrinsic sphere or totally geodesic submanifold in $\mathbb{R}^n$ with $p \in M_i$, $T_pM_i = V_i$ and for which the mean curvature vector at p is equal to H_i. These submanifolds are contained in mutually perpendicular Euclidean subspaces of $\mathbb{R}^n$ and hence their Riemannian product yields the extrinsic product we are looking for. The construction in S^n can be done by viewing S^n as a totally umbilical submanifold in $\mathbb{R}^{n+1}$ and using the previous construction method. The case of real hyperbolic space is a little more involved. For further details we refer to [6]. We summarize this in

THEOREM 3.7.6

Let $p \in \bar{M}^n(\kappa)$ and V a linear subspace of $T_p\bar{M}^n(\kappa)$. Let $V = V_1 \oplus \ldots \oplus V_s$ be an orthogonal decomposition of V with $s \geq 2$ and $H_1, \ldots, H_s \in T_p\bar{M}^n(\kappa)$

be mutually distinct and perpendicular to V. Then there exists a submanifold product $M = M_1 \times \ldots \times M_s$ of extrinsic spheres and totally geodesic submanifolds in $\bar{M}^n(\kappa)$ with $p \in M$, $T_pL_i(p) = V_i$, and such that H_i is the mean curvature vector of $L_i(p)$ at p if and only if $\langle H_i, H_j \rangle = -\kappa$ for all $i \neq j$.

This finishes the example of extrinsic spheres. □

Example 3.7 Standard embeddings of symmetric R-spaces
For details on real flag manifolds we refer to Section A.4. Let $\bar{M}$ be a connected, simply connected, semisimple Riemannian symmetric space, $G = I^o(\bar{M})$, $o \in \bar{M}$ and K the isotropy subgroup of G at o. Note that K is connected since we assume $\bar{M}$ to be simply connected. Each orbit of the isotropy representation of K on $T_o\bar{M}$ is a real flag manifold and its realization as a submanifold of $T_o\bar{M}$ is called the standard embedding of the real flag manifold. A real flag manifold that is also a symmetric space is also called a symmetric R-space or, if G is simple, an irreducible symmetric R-space. Let $M = K \cdot X$ be the orbit of the action of K through $X \in T_o\bar{M}$, $X \neq 0$. We will prove below that each standard embedding of a symmetric R-space is a symmetric submanifold of the Euclidean space $T_o\bar{M}$.

Recall that there is a convenient way to describe the isotropy representation. Let $\mathfrak{g} = \mathfrak{k} \oplus \mathfrak{p}$ be the Cartan decomposition of the Lie algebra $\mathfrak{g}$ of G. Then $\mathfrak{p}$ is canonically isomorphic to $T_o\bar{M}$, and, via this identification, the isotropy representation is isomorphic to the adjoint representation $\mathrm{Ad} : K \to SO(\mathfrak{p})$. The orbit $M = K \cdot X$ is a symmetric space if and only if the eigenvalues of the transformation $\mathrm{ad}(X) : \mathfrak{g} \to \mathfrak{g}$ are $\pm c, 0$ for some $c > 0$. Without loss of generality we can assume that X is normalized so that $c = 1$. We decompose the semisimple Lie algebra $\mathfrak{g}$ into the direct sum $\mathfrak{g} = \mathfrak{g}_1 \oplus \ldots \oplus \mathfrak{g}_k$ of simple Lie algebras $\mathfrak{g}_i$ and put $\mathfrak{k}_i = \mathfrak{k} \cap \mathfrak{g}_i$ and $\mathfrak{p}_i = \mathfrak{p} \cap \mathfrak{g}_i$. Then $\mathfrak{p} = \mathfrak{p}_1 \oplus \ldots \oplus \mathfrak{p}_k$, and, by means of this decomposition, we can write $X = (X_1, \ldots, X_k)$. We denote by K_i the connected Lie subgroup of G_i with Lie algebra $\mathfrak{k}_i$. Then $M = K \cdot X$ is isometric to the Riemannian product

$$M = K \cdot X = K_1 \cdot X_1 \times \ldots \times K_k \cdot X_k \ .$$

Viewing M as a submanifold of $\mathfrak{p}$ it is clear that M is the extrinsic product of the submanifolds $K_i \cdot X_i$ of $\mathfrak{p}_i$. In particular, the standard embedding of any symmetric R-space decomposes as the extrinsic product of the standard embeddings of some irreducible symmetric R-spaces.

Let $M = K \cdot X$ be a symmetric R-space regarded as an embedded submanifold of $\mathfrak{p}$. We will now show explicitly that M is a symmetric submanifold of $\mathfrak{p}$. Since $K \subset I^o(M)$, the Cartan decomposition of the Lie algebra of $I^o(M)$ induces a reductive decomposition $\mathfrak{k} = \mathfrak{k}_X \oplus \mathfrak{m}$, where $\mathfrak{k}_X$ is the Lie algebra of the isotropy subgroup of K at X. General theory about symmetric spaces says that for each $U \in \mathfrak{m}$ the curve

$$\gamma : \mathbb{R} \to M \ , \ t \mapsto \mathrm{Ad}(\mathrm{Exp}(tU))X$$

is the geodesic in M with $\gamma(0) = X$ and $\dot\gamma(0) = U$, where we identify $T_X M$ and $\mathfrak{m}$ in the usual way. On the other hand, viewing γ as a curve in $\mathfrak{p}$, we have

$$\begin{aligned}\dot\gamma(0) &= \frac{d}{dt}\Big|_{t=0} (t \mapsto \mathrm{Ad}(\mathrm{Exp}(tU))X) \\ &= \frac{d}{dt}\Big|_{t=0} (t \mapsto e^{\mathrm{ad}(tU)}X) = \mathrm{ad}(U)X = [U, X] \ .\end{aligned}$$

This implies

$$T_X M = \{[X, U] \mid U \in \mathfrak{m}\} = \mathrm{ad}(X)(\mathfrak{m}) \ .$$

Since the inner product on $\mathfrak{p}$ comes from the Killing form of $\mathfrak{g}$, and since $\mathrm{ad}(X)$ is skewsymmetric with respect to the Killing form, this implies

$$\nu_X M = \{\xi \in \mathfrak{p} \mid [X, \xi] = 0\} \ .$$

We denote by $\mathfrak{g}_\nu$ the eigenspace of $\mathrm{ad}(X)$ with respect to $\nu \in \{-1, 0, +1\}$. Thus, we have a vector space direct sum decomposition

$$\mathfrak{g} = \mathfrak{g}_{-1} \oplus \mathfrak{g}_0 \oplus \mathfrak{g}_1 \ .$$

The Jacobi identity implies that

$$[\mathfrak{g}_\nu, \mathfrak{g}_\mu] \subset \mathfrak{g}_{\nu+\mu} \ ,$$

where we put $\mathfrak{g}_{-2} = \mathfrak{g}_2 = \{0\}$. This just says that the eigenspaces of $\mathrm{ad}(X)$ turn $\mathfrak{g}$ into a graded Lie algebra. In order to define an extrinsic symmetry of M at X in $\mathfrak{p}$ we consider the transformation $\mathrm{ad}(X)^2 : \mathfrak{g} \to \mathfrak{g}$. This transformation has two eigenvalues 0 and $+1$ with corresponding eigenspaces $\mathfrak{g}^0 := \mathfrak{g}_0$ and $\mathfrak{g}^1 := \mathfrak{g}_{-1} \oplus \mathfrak{g}_1$, leading to the vector space direct sum decomposition

$$\mathfrak{g} = \mathfrak{g}^0 \oplus \mathfrak{g}^1 \ .$$

Since $[\mathfrak{p}, \mathfrak{k}] \subset \mathfrak{p}$, $[\mathfrak{p}, \mathfrak{p}] \subset \mathfrak{k}$ and $X \in \mathfrak{p}$, it follows that $\mathrm{ad}(X)^2\mathfrak{k} \subset \mathfrak{k}$ and $\mathrm{ad}(X)^2\mathfrak{p} \subset \mathfrak{p}$. In particular, this implies

$$\mathfrak{k} = (\mathfrak{k} \cap \mathfrak{g}^0) \oplus (\mathfrak{k} \cap \mathfrak{g}^1) \ , \ \mathfrak{p} = (\mathfrak{p} \cap \mathfrak{g}^0) \oplus (\mathfrak{p} \cap \mathfrak{g}^1) \ .$$

The subspaces in these decompositions are precisely

$$\mathfrak{k}_X = \mathfrak{k} \cap \mathfrak{g}^0 \ , \ \mathfrak{m} = \mathfrak{k} \cap \mathfrak{g}^1 \ , \ \nu_X M = \mathfrak{p} \cap \mathfrak{g}^0 \ , \ T_X M = \mathfrak{p} \cap \mathfrak{g}^1 \ .$$

Next, we define an involution

$$\rho : \mathfrak{g} = \mathfrak{g}^0 \oplus \mathfrak{g}^1 \to \mathfrak{g} = \mathfrak{g}^0 \oplus \mathfrak{g}^1 \ , \ Z = Z^0 + Z^1 \mapsto Z^0 - Z^1 \ .$$

It leaves $\mathfrak{k}$ and $\mathfrak{p}$ invariant, and the above gradation of $\mathfrak{g}$ shows that $[\mathfrak{g}^0, \mathfrak{g}^0] \subset \mathfrak{g}^0$, $[\mathfrak{g}^0, \mathfrak{g}^1] \subset \mathfrak{g}^1$ and $[\mathfrak{g}^1, \mathfrak{g}^1] \subset \mathfrak{g}^0$, which implies that ρ is an automorphism of

$\mathfrak{g}$. Thus ρ is an involutive automorphism of $\mathfrak{g}$ that commutes with the Cartan involution of $\mathfrak{g}$ corresponding to the decomposition $\mathfrak{g} = \mathfrak{k} \oplus \mathfrak{p}$. We claim that

$$\sigma_X : \mathfrak{p} \to \mathfrak{p} \ , \ W \mapsto \rho W$$

is an extrinsic symmetry of M at X. By construction, σ_X is an involutive isometry of $\mathfrak{p}$ with $\sigma_X X = X$, $\sigma_{X*} W = W$ for all $W \in \nu_X M = \mathfrak{p} \cap \mathfrak{g}^0$, and $\sigma_{X*} W = -W$ for all $W \in T_X M = \mathfrak{p} \cap \mathfrak{g}^1$. Thus, it remains to show that $\sigma_X M = M$. For this we consider once again the geodesics $\gamma(t) = \mathrm{Ad}(\mathrm{Exp}(tU))X$ in M. Since $U \in \mathfrak{m} = \mathfrak{k} \cap \mathfrak{g}^1$ and $\rho^{-1} X = \rho X = X$, we obtain

$$\begin{aligned} \sigma_X(\gamma(t)) &= \rho(\mathrm{Ad}(\mathrm{Exp}(tU))X) = \rho(\mathrm{Ad}(\mathrm{Exp}(tU))\rho^{-1}X) \\ &= \mathrm{Ad}(\mathrm{Exp}(t\rho(U)))X = \mathrm{Ad}(\mathrm{Exp}(-tU))X = \gamma(-t) \end{aligned}$$

for all $t \in \mathbb{R}$. Since γ is a curve in M this shows that σ_X leaves M invariant. At any other point, $\mathrm{Ad}(k)X$ of M the isometry $k\sigma_X k^{-1}$ of $\mathfrak{p}$ defines an extrinsic symmetry of M at $\mathrm{Ad}(k)X$. So we can now conclude that M is a symmetric submanifold of $\mathfrak{p}$.

It is clear from the construction that M lies in the sphere $S^{n-1}(r) \subset \mathfrak{p}$ with radius $r = \|X\|$. Since this sphere is totally umbilical in $\mathfrak{p}$, it follows from Lemma 3.7.5 that M is also a symmetric submanifold of $S^{n-1}(r)$. Note that, by a suitable homothety, we can realize M as a symmetric submanifold of the unit sphere S^{n-1}, as well. We summarize the previous discussion in

PROPOSITION 3.7.7
The standard embedding of any symmetric R-space $K \cdot X$ is a symmetric submanifold both of the Euclidean space $\mathfrak{p}$ and of the sphere $S^{n-1}(\|X\|)$.

This finishes the discussion about the standard embeddings of symmetric R-spaces. □

e) Classification of symmetric submanifolds of standard space forms

We now classify the symmetric submanifolds of the standard space forms. Roughly speaking, we will show that the examples given above exhaust all possibilities. Although it is possible to formulate just one classification theorem for all standard space forms (see [6]), we will investigate the cases of zero, positive and negative curvature separately for the sake of simplicity.

Classification in Euclidean spaces.

As a first step, we classify the locally symmetric submanifolds of $\mathbb{R}^n$. This classification is due to Ferus and can be found in the papers [83], [86], [84] and [85]. The most concise proof was given in [86] using the algebraic framework of Jordan triple systems. Here, we prefer to adopt the more geometric approach of [83] and [84], following also [212] and [80].

THEOREM 3.7.8
Let M be a locally symmetric submanifold of $\mathbb{R}^n$. Then

(a) M is locally a submanifold product

$$\mathbb{R}^{m_0} \times M_1 \times \cdots \times M_s \to \mathbb{R}^{m_0} \times \mathbb{R}^{m_1} \times \cdots \times \mathbb{R}^{m_s} \subseteq \mathbb{R}^n,$$

where M_i is a full immersion into $\mathbb{R}^{m_i}$ that is minimal in a hypersphere of $\mathbb{R}^{m_i}$.

(b) Each M_i as in (a) is locally a standard embedding of an irreducible symmetric R-space (which is, in particular, an orbit of an s-representation).

REMARK 3.7.9 Of course, it may happen that $m_0 = 0$, that is, M has no Euclidean factor. Another possible case is that $\sum m_i = n$, that is, M is full in $\mathbb{R}^n$. Observe further that the statement in the theorem is a local result. If we assume M to be complete we get a global result, namely: *a complete symmetric submanifold of $\mathbb{R}^n$ is covered by a submanifold product $\mathbb{R}^{m_0} \times M_1 \times \cdots \times M_s \to \mathbb{R}^{m_0} \times \mathbb{R}^{m_1} \times \cdots \times \mathbb{R}^{m_s} \subseteq \mathbb{R}^n$, where M_i is a standard embedding of an irreducible symmetric R-space.* □

In the rest of this subsection, we will be concerned with the proof of Theorem 3.7.8. Since M is a locally symmetric submanifold, its second fundamental form α is parallel. It follows that the nullity distribution E_0 (see Section 2.3) on M has constant rank. Moreover, we have

$$\alpha(\nabla_X Y, Z) = \nabla^\perp_X \alpha(Y, Z) - \alpha(Y, \nabla_X Z) = 0$$

for all sections Y in E_0 and all vector fields X, Z on M. This shows that E_0 is a parallel subbundle of TM. Since $\alpha(E_0, E_0^\perp) = 0$ by definition of E_0, the Lemma of Moore now implies that M is locally a submanifold product of $\mathbb{R}^n$ with a Euclidean factor, a leaf of E_0, if the rank of E_0 is nonzero. Without loss of generality we can assume from now on that the nullity distribution on M is trivial.

Since α is parallel, the mean curvature vector field H of M is a parallel normal vector field and the shape operator A_H is a parallel selfadjoint tensor field on M. Therefore, the principal curvatures $\lambda_1, \ldots, \lambda_s$ of M with respect to H are constant and hence, since A_H is parallel, the corresponding principal curvature spaces form parallel distributions $E_1, \ldots, E_s$ on M.

Since $R^\perp(X, Y)H = 0$ for all $X, T \in T_pM$, $p \in M$, the Ricci equation implies $[A_H, A_\xi] = 0$ for all normal vector fields ξ of M. Therefore, each eigendistribution E_i is invariant under all shape operators, that is, $A_\xi E_i \subset E_i$ for all normal vector fields ξ of M. This implies in particular that

$$\alpha(E_i, E_j) = 0 \text{ , for all } i \neq j \text{ .} \tag{3.9}$$

Since the bundles E_i are parallel, (3.9) and Moore's Lemma 2.7.1 implies that M is locally a submanifold product $M = M_1 \times \ldots \times M_s$, where M_i is an integral manifold of E_i and a submanifold of a suitable $\mathbb{R}^{\bar{m}_i} \subset \mathbb{R}^n$. We denote by α_i the second fundamental form of $M_i \subset \mathbb{R}^{\bar{m}_i}$ and by $\pi_i : M \to M_i$ the canonical projection. Then we have

$$\alpha(X,Y) = (\alpha_1(\pi_{1*}X, \pi_{1*}Y), \ldots, \alpha_s(\pi_{s*}X, \pi_{s*}Y))$$

for all $X, Y \in T_pM$, $p \in M$. This implies that α_i is parallel as well. By the theorem on the reduction of codimension (Theorem 2.5.1) we can reduce the codimension of each M_i, since the distribution of the first normal spaces is parallel (see, for instance, Exercise 2.8.8). Thus, for each M_i we get a full immersion $M_i \to \mathbb{R}^{m_i} \subseteq \mathbb{R}^{\bar{m}_i}$. Let H_i be the mean curvature vector field of M_i. Then $H = (H_1, \ldots, H_s)$ and

$$\langle \alpha_i(X,Y), H_i \rangle = \langle \alpha(X,Y), H \rangle = \lambda_i \langle X, Y \rangle$$

for all $X, Y \in T_pM_i$, $p \in M_i$. If $\lambda_i \neq 0$, this shows that M_i is a pseudoumbilical submanifold of $\mathbb{R}^{m_i}$ with parallel mean curvature vector field and $\lambda_i = ||H_i||$. By Proposition 2.6.3 we get that M_i is minimal in a hypersphere of $\mathbb{R}^{m_i}$. Part (a) of Theorem 3.7.8 then follows from the following lemma and the assumption that the nullity distribution of M is trivial.

LEMMA 3.7.10

Every connected minimal submanifold M with parallel second fundamental form in a standard space form $\bar{M}^n(\kappa)$ with $\kappa \leq 0$ is totally geodesic.

PROOF If the submanifold is complete, then it is a symmetric submanifold by Theorem 3.7.3 and therefore homogeneous, which implies that it is totally geodesic by the results in sections 3.4 and 3.5. We will see in Lemma 3.7.13 that every locally symmetric submanifold is an open subset of a symmetric submanifold, which implies the result.

There are also more direct proofs in [212] and [83]. We briefly sketch the one in [212]. We may assume that the dimension of the submanifold M is greater than one. In this situation, some calculations lead to (see also [55])

$$\langle \Delta\alpha, \alpha \rangle = n\kappa||\alpha||^2 - ||\alpha \circ \alpha^t||^2 - ||R^\perp||^2 , \tag{3.10}$$

where Δ is the Laplace operator on tensor fields defined by $\mathrm{tr}(\nabla^2)$. Since $\nabla^\perp \alpha = 0$ we have $\Delta\alpha = 0$ and therefore

$$n\kappa||\alpha||^2 = ||\alpha \circ \alpha^t||^2 + ||R^\perp||^2 ,$$

and if $\kappa \leq 0$ this gives $\alpha \equiv 0$. ∎

We continue with the proof of part (b) of Theorem 3.7.8. Without loss of generality, we can assume that M is a full irreducible symmetric submanifold of $\mathbb{R}^{n+1}$.

Recall from part (a) that M is minimal in a hypersphere S^n. We will assume that this sphere is centered at the origin and has radius $\sqrt{2m}$ with $m = \dim M$.

The idea of the proof is the following: First, we associate to M a Lie algebra $\mathfrak{k}$, which can be seen as the Lie algebra of all isometries of $\mathbb{R}^{n+1}$ that leave M invariant. Then we define a Lie bracket on the vector space $\mathfrak{g} = \mathfrak{k} \oplus \mathbb{R}^{n+1}$ in such a way that $(\mathfrak{g}, \mathfrak{k})$ is a symmetric pair. This proof follows original ideas by Ferus [84] and is completely algebraic. It will turn out that the definition of $\mathfrak{g}$ depends only on the value of the second fundamental form of M at a single point. Hence, as a by-product, we also get that a symmetric submanifold is uniquely determined by the value of its second fundamental form at one point.

We fix a point $p \in M$ and put $V = T_pM$ and $W = \nu_pM$. Since the immersion is full, we have $W = \operatorname{im} \alpha_p = \operatorname{span}\{\alpha_p(v,v) \mid v \in V\}$, where the latter equality follows by polarization of α.

We now introduce an operator which, although encoding the same piece of information as the second fundamental form or the shape operator, turns out to be useful, especially when dealing with homogeneous submanifolds. For each $x \in V$ the *infinitesimal transvection* φ_x is the endomorphism

$$\varphi_x : V \oplus W \to V \oplus W \ , \ X \mapsto \alpha(x, X^T) - A_{X^\perp} x \ .$$

We denote by $\mathfrak{m}$ the real vector space that is spanned by the infinitesimal transvections $\{\varphi_x \mid x \in V\}$.

For $x, y \in V$ we denote by $\bar{R}_{x,y}$ the endomorphism on $V \oplus W$ given by $\bar{R}_{xy}v = R(x,y)v$ for all $v \in V$ and $\bar{R}_{xy}\xi = R^\perp(x,y)\xi$ for all $\xi \in W$, where R is the Riemannian curvature tensor of M and $R^\perp$ is the normal curvature tensor of M. From the equations by Gauss and Ricci it follows that $[\varphi_x, \varphi_y] = \bar{R}_{xy}$. We define a subalgebra $\mathfrak{h}$ of $\mathfrak{so}(V) \oplus \mathfrak{so}(W) \subset \mathfrak{so}(V \oplus W) = \mathfrak{so}(n+1)$ by

$$\begin{aligned} \mathfrak{h} &= \{B \in \mathfrak{so}(V) \oplus \mathfrak{so}(W) \mid B \cdot \alpha = 0\} \\ &= \{B \in \mathfrak{so}(V) \oplus \mathfrak{so}(W) \mid [B, A_\xi] = A_{B\xi} \text{ for all } \xi \in W\} \\ &= \{B \in \mathfrak{so}(V) \oplus \mathfrak{so}(W) \mid [B, \varphi_x] = \varphi_{Bx} \text{ for all } x \in V\} \ . \end{aligned} \tag{3.11}$$

Here $B \cdot \alpha$ means that B acts on α as a derivation. One should think of the elements in $\mathfrak{h}$ as the infinitesimal isometries that are generated by one-parameter groups of isometries of $\mathbb{R}^{n+1}$ leaving M invariant and fixing p. These infinitesimal isometries are just the differentials of $F_t \in O(n+1)$ such that

$$F_t\alpha(x,y) = \alpha(F_tx, F_ty). \tag{3.12}$$

In other words, if $B \in \mathfrak{h}$, then $F_t = \operatorname{Exp}(tB)$ satisfies (3.12). Notice also that $\bar{R}_{xy} \in \mathfrak{h}$.

The direct sum $\mathfrak{k} = \mathfrak{h} \oplus \mathfrak{m}$ is a subalgebra of $\mathfrak{so}(n+1)$ with bracket relations

$$[\varphi_x, \varphi_y] = \bar{R}_{xy} \ , \ [A, \varphi_x] = \varphi_{Ax} \ (x, y \in V, A \in \mathfrak{h}).$$

Let K be the connected Lie subgroup of $SO(n+1)$ with Lie algebra $\mathfrak{k}$. We now define a Lie algebra structure on the vector space $\mathfrak{g} = \mathfrak{k} \oplus \mathbb{R}^{n+1} = \mathfrak{k} \oplus V \oplus W$. For this purpose, we consider the adjoint operator $R^{\perp *} : \Lambda^2 W \to \Lambda^2 V$ of $R^\perp$, which is characterized by

$$\langle R^{\perp *}(\xi \wedge \zeta), x \wedge y\rangle = \langle R^\perp(x,y)\xi, \zeta\rangle = \langle [A_\xi, A_\zeta]x, y\rangle$$

for $x, y \in V$ and $\xi, \zeta \in W$, and hence we can define $R^{\perp *}(\xi \wedge \zeta)x = [A_\xi, A_\zeta]x$. For $\xi, \zeta \in W$ we define an endomorphism $\mathfrak{R}(\xi \wedge \zeta)$ on $V \oplus W$ by $\mathfrak{R}(\xi \wedge \zeta)x = R^{\perp *}(\xi \wedge \zeta)x$ and $\mathfrak{R}(\xi \wedge \zeta)\eta$ equal to the unique element $\rho \in W$ such that

$$A_\rho = [R^{\perp *}(\xi \wedge \zeta), A_\eta] = [[A_\xi, A_\zeta], A_\eta]\,. \tag{3.13}$$

Note that ρ is uniquely determined by (3.13), since M is full in $\mathbb{R}^{n+1}$ and hence $\rho \mapsto A_\rho$ has trivial kernel. By definition we have $[\mathfrak{R}(\xi \wedge \zeta), A_\eta] = A_{\mathfrak{R}(\xi\wedge\zeta)\eta}$, and thus $\mathfrak{R}(\xi \wedge \zeta) \in \mathfrak{h}$.

We are now able to define the Lie algebra structure on $\mathfrak{g}$. For two elements in $\mathfrak{k}$ we use the Lie algebra structure on $\mathfrak{k}$. If $B \in \mathfrak{k}$ and $X \in V \oplus W$, then we define $[B, X] = BX$. Finally, for $v, w \in V$ and $\xi, \zeta \in W$ we define

$$[v, w] = -[\varphi_v, \varphi_w]\ ,\ [v, \xi] = -\varphi_{A_\xi v}\ ,\ [\xi, \zeta] = \mathfrak{R}(\xi \wedge \zeta)\ .$$

We have to verify that the Jacobi identity holds. This is obvious if all elements lie in $\mathfrak{k}$. Since the bracket on $\mathfrak{g}$ is equivariant with respect to the action of K on $\mathfrak{g}$, it follows by differentiation that the Jacobi identity holds if at least one element lies in $\mathfrak{k}$ only. We are thus left to verify the Jacobi identity for elements in $V \oplus W$, that is, we have to show that $\mathfrak{S}[X,[Y,Z]] = 0$ for all $X, Y, Z \in V \oplus W$, where $\mathfrak{S}$ denotes the cyclic sum. This is clear if all the three elements lie in V, since then the Jacobi identity is just the first Bianchi identity for R. If $x \in V$ and $\xi, \zeta \in W$, then

$$\begin{aligned}\mathfrak{S}[x,[\xi,\zeta]] &= [x,[A_\xi, A_\zeta]] + [\xi, \varphi_{A_\zeta x}] - [\zeta, \varphi_{A_\xi x}]\\ &= -[A_\xi, A_\zeta]x + A_\xi A_\zeta x - A_\zeta A_\xi x = 0\,.\end{aligned}$$

If $x, y \in V$ and $\xi \in W$, then

$$\begin{aligned}\mathfrak{S}[x,[y,\xi]] &= -[x, \varphi_{A_\xi y}] + [y, \varphi_{A_\xi x}] - [\xi, [\varphi_x, \varphi_y]]\\ &= \alpha(A_\xi y, x) - \alpha(A_\xi x, y) + R^\perp(x,y)\xi = 0\end{aligned}$$

by the Ricci equation. Finally, we prove the Jacobi identity for $\xi, \zeta, \eta \in W$. Since M is full in $\mathbb{R}^{n+1}$, it suffices to prove that $A_{\mathfrak{S}[\xi,[\zeta,\eta]]} = 0$. But this follows from $A_{[\xi,[\zeta,\eta]]} = -A_{[\zeta,\eta]\xi} = -[[A_\zeta, A_\eta], A_\xi]$ by definition of $\mathfrak{R}$. Note, in particular, that the Lie algebra structure on $\mathfrak{g}$ is completely described by the shape operator of M.

We now define $\eta = -2mH \in W$, where H is the mean curvature vector of M at p. Since M is minimal in the sphere with radius $\sqrt{2m}$, we have that $A_\eta = -\mathrm{id}_V$.

LEMMA 3.7.11

The endomorphism $\mathrm{ad}(\eta) : \mathfrak{g} \to \mathfrak{g}$ *is semisimple and its eigenvalues are* $0, \pm 1$.

PROOF For $B \in \mathfrak{h}$ we have $A_{B\eta} = [B, A_\eta] = [B, -\mathrm{id}_V] = 0$, and since M is full in $\mathbb{R}^{n+1}$ this implies $\mathrm{ad}(\eta)B = [\eta, B] = -B\eta = 0$. This shows $\mathfrak{h} \subseteq \ker \mathrm{ad}(\eta)$. For $\xi \in W$ we have

$$\mathfrak{R}(\eta \wedge \xi)x = [A_\eta, A_\xi]x = 0 \ , \ \mathfrak{R}(\eta \wedge \xi)\zeta = [[A_\eta, A_\xi], A_\zeta] = 0$$

for all $x \in V$ and $\zeta \in W$ since $A_\eta = -\mathrm{id}_V$. As $\mathrm{ad}(\eta)\xi = \mathfrak{R}(\eta \wedge \xi)$ this implies $W \subset \ker \mathrm{ad}(\eta)$. A simple computation shows that

$$\mathrm{ad}(\eta)|\{x \pm \varphi_x \mid x \in V\} = \mp \mathrm{id} \ .$$

Altogether, it now follows that $\mathrm{ad}(\eta)$ is semisimple and its eigenvalues are 0 and ± 1. ∎

LEMMA 3.7.12

The Lie algebra $\mathfrak{g}$ is semisimple and $\mathfrak{g} = \mathfrak{k} \oplus \mathbb{R}^{n+1}$ is a Cartan decomposition. Moreover, the Killing form of $\mathfrak{g}$ restricted to $V \oplus W = \mathbb{R}^{n+1}$ coincides with the inner product on $\mathbb{R}^{n+1}$.

PROOF Since M is full, the action of K cannot fix a nonzero vector in $\mathbb{R}^{n+1}$. Let $\mathfrak{z}$ be the centralizer of $\mathbb{R}^{n+1}$ in $\mathfrak{g}$. Note that $\mathfrak{z}$ is a commutative ideal and contained in $\mathbb{R}^{n+1}$ (exercise, cf. [243, page 235]). Since $[\mathfrak{z}, \eta] = 0$ we have $\mathfrak{z} \subseteq W$. Let $\xi \in \mathfrak{z}$. Then we have $0 = \langle [x, \xi]y, \eta \rangle = -\langle \varphi_{A_\xi x} y, \eta \rangle = \langle \alpha(x, y), \xi \rangle$ for all $x, y \in V$. Thus $\xi = 0$ and $\mathfrak{z} = \{0\}$, which implies that $\mathfrak{g}$ is semisimple (exercise, cf. [243, page 235]).

The restriction of the Killing form B of $\mathfrak{g}$ to $\mathfrak{k}$ is negative definite, and $\mathfrak{g}$ decomposes as a direct sum of simple ideals $\mathfrak{g} = \mathfrak{g}_0 \oplus \cdots \oplus \mathfrak{g}_t$, where $\mathfrak{g}_i = [\mathfrak{q}_i, \mathfrak{q}_i] + \mathfrak{q}_i$ with $\mathfrak{q}_i = \mathfrak{g}_i \cap \mathbb{R}^{n+1}$ and $B = \lambda_i \langle \cdot, \cdot \rangle$ on $\mathfrak{q}_i$ for some $\lambda_i \neq 0$. We decompose $\eta = \eta_1 + \ldots + \eta_t$ according to the decomposition of $\mathfrak{g}$. Then $(\mathrm{ad}_{\mathfrak{g}_i}(\eta_i))^2 = (\mathrm{ad}(\eta))^2|\mathfrak{g}_i$ is semisimple with eigenvalues 0 and 1. Hence $\lambda_i \langle \eta_i, \eta_i \rangle = B(\eta_i, \eta_i) = \mathrm{tr}(\mathrm{ad}_{\mathfrak{g}_i} \eta_i)^2 \geq 0$. But $\eta_i = 0$ would imply $\mathfrak{q}_i \subset W$. Then, for any $\xi \in \mathfrak{q}_i$ and $x \in V$, we would have $-A_\xi x = \varphi_x \xi = [\varphi_x, \xi] \in \mathfrak{g}_i \cap V = \{0\}$, whence $\xi = 0$ and $\mathfrak{q}_i = \mathfrak{g}_i = \{0\}$. Thus, $\lambda_i > 0$ for all i and the restriction of B to $\mathbb{R}^{n+1}$ is positive definite. This proves that $\mathfrak{k} \oplus \mathbb{R}^{n+1}$ is a Cartan decomposition.

Next, we show that the Killing form of $\mathfrak{g}$ restricted to $V \oplus W = \mathbb{R}^{n+1}$ coincides with the inner product on $\mathbb{R}^{n+1}$. For $x, y \in V$ we have

$$B(x, y) = 2\mathrm{tr}\,(\mathrm{ad}(x)\mathrm{ad}(y)) \ .$$

(see [132, page 140]). We choose orthonormal bases (x_i) of V and (ξ_j) of W

with respect to the inner product on $\mathbb{R}^{n+1}$. Then

$$\begin{aligned}
\tfrac{1}{2}B(x,y) &= \textstyle\sum_i \langle \mathrm{ad}(x)\mathrm{ad}(y)x_i, x_i\rangle + \sum_j \langle \mathrm{ad}(x)\mathrm{ad}(y)\xi_j, \xi_j\rangle \\
&= -\textstyle\sum_i \langle [\varphi_{x_i}, \varphi_y]x, x_i\rangle + \sum_j \langle \varphi_{A_{\xi_j}y}x, \xi_j\rangle \\
&= -\langle R(x_i, y)x, x_i\rangle + \textstyle\sum_j \langle A_{\xi_j}x, A_{\xi_j}y\rangle \\
&= -\textstyle\sum_i \langle \alpha(x, x_i), \alpha(y, x_i)\rangle + \sum_i \langle \alpha(x_i, x_i), \alpha(x,y)\rangle \\
&\qquad + \textstyle\sum_j \langle A_{\xi_j}x, A_{\xi_j}y\rangle \\
&= -\langle \alpha(x,y), \tfrac{1}{2}\eta\rangle = \tfrac{1}{2}\langle x, y\rangle \ .
\end{aligned}$$

On the other hand, if $x, \bar{x}, y, \bar{y} \in V$, we get

$$\begin{aligned}
B(\alpha(\bar{x}, x), \alpha(\bar{y}, y)) &= B(\varphi_{\bar{x}}x, \varphi_{\bar{y}}y) = B([\varphi_{\bar{x}}, x], [\varphi_{\bar{y}}, y]) \\
&= -B([\varphi_{\bar{y}}, [\varphi_{\bar{x}}, x]], y) = -\langle [\varphi_{\bar{y}}, [\varphi_{\bar{x}}, x]], y\rangle = \langle \varphi_{\bar{x}}x, \varphi_{\bar{y}}y\rangle \\
&= \langle \alpha(\bar{x}, x), \alpha(\bar{y}, y)\rangle \ .
\end{aligned}$$

Since W is spanned by $\alpha(V,V)$, we have $B = \langle \cdot, \cdot \rangle$ on W. Finally, we have

$$B(V,W) = B(\varphi_V \eta, W) = B([\varphi_V, \eta], W) = B(\varphi_V, [\eta, W]) = 0 \ .$$

This completes the proof. □

Note that the above lemma also shows that the isotropy representation can be identified with the adjoint action of K on $\mathbb{R}^{n+1}$.

According to the general construction (as explained in Section 3.7 d), the orbit $K\cdot\eta$ is a symmetric submanifold or, more precisely, a standard embedded symmetric R-space. Since M is, by assumption, full and irreducible, the remarks of Section 3.7, page 70, imply that $K \cdot \eta$ is actually a standard embedded irreducible symmetric R-space. Moreover, the tangent space at η to $K \cdot \eta$ is easily seen to coincide with V and, using Corollary 3.6.3, one can see that the second fundamental form of $K \cdot \eta$ at η coincides with the one of M at p. Thus, the normal space to $K \cdot \eta$ at η is equal to W.

LEMMA 3.7.13
The locally symmetric submanifold M is an open part of $K \cdot \eta$.

PROOF Let $p \in M$. Recall that an isometry F of $\mathbb{R}^{n+1}$ is uniquely determined by $F(p)$ and the differential F_{*p} at p. We can thus define an isometry F of $\mathbb{R}^{n+1}$ by $F(p) = \eta$ and requiring that $F_{*p}|T_pM$ coincides with the above identification of V and $T_\eta(K \cdot \eta)$, and $F_{*p}|\nu_pM$ coincides with the above identification of W and $\nu_\eta(K \cdot \eta)$. We will show that F maps the connected component of M containing p to $K \cdot \eta$. The method we use is similar to the one we used for the proof of Theorem 3.7.2. The set

$$\begin{aligned}
C = \{q \in M \mid F(q) \in K \cdot \eta \ , \ F_{*q}(T_qM) = T_{F(q)}K \cdot \eta \ , \\
F_{*q}\varphi_v = \varphi_{F_{*q}v} \text{ for all } v \in T_qM\} \ .
\end{aligned}$$

is clearly a closed subset of M and, since $p \in C$, C is nonempty. To see that C is the connected component of M containing p, it is enough to show that C is an open subset of M. With this in mind we will prove that, for each $q \in C$, there is an open neighbourhood of q in M that is contained in C. As a consequence of the Gauss equation, the linear isometry

$$F_{*q}|T_qM : T_qM \to T_{F(q)}K \cdot \eta$$

preserves the curvature tensors of M and $K \cdot \eta$ at q and $F(q)$, respectively. Hence, it can be extended to a local isometry f of $U \subset M$ to $K \cdot \eta$. Let γ be a geodesic in M with $\gamma(0) = q$ and $v_1 = \dot\gamma(0), v_2, \ldots, v_n$ a Darboux frame at q. Parallel translation of this frame along γ (with respect to the Levi Civita and normal connections) yields $\gamma, V_1, \ldots, V_n$, which satisfies the system (3.8) of linear differential equations.

Let $\bar\gamma = f \circ \gamma$ and $\bar v_i = F_{*q}v_i$, and parallel translate this frame along $\bar\gamma$. Moreover, set $\tilde\gamma = F \circ \gamma$ and in an analoguous way consider $\tilde\gamma, \tilde V_1, \ldots, \tilde V_n$. Just as in the proof of Theorem 3.7.2, bearing in mind $F_{*q}\varphi_v = \varphi_{F_{*q}v}$, we have that $\bar\gamma, \bar V_1, \ldots, \bar V_n$ and $\tilde\gamma, \tilde V_1, \ldots, \tilde V_n$ both satisfy the system (3.8) of linear differential equations with the same initial conditions. Thus $\bar\gamma = \tilde\gamma$, that is, $F(\gamma(t)) = f(\gamma(t))$. If W is a normal neighbourhood of q, then $F|(U \cap W) = f|(U \cap W)$ and $U \cap W \subset C$. This shows that C is open in M, and completes the proof. □

Classification in spheres.

The classification of symmetric submanifolds of $S^n(r)$ is a simple consequence of the one of symmetric submanifolds of $\mathbb{R}^n$. An important observation is that a symmetric submanifold of S^n cannot have have a Euclidean factor, because otherwise it could not be contained in a sphere.

THEOREM 3.7.14

Let M be a locally symmetric submanifold of $S^n(r)$. Then:

(a) If M is not contained in any extrinsic sphere of $S^n(r)$, then M is locally a submanifold product

$$M_1 \times \ldots \times M_s \to S^{m_1-1}(r_1) \times \ldots \times S^{m_s-1}(r_s) \to S^n(r)\ ,$$

where $M_i \to S^{m_i-1}(r_i)$ is the standard embedding of an irreducible symmetric R-space M_i, the second map is the submanifold product of extrinsic spheres as in Theorem 3.7.6, $n = \sum m_i + s - 1$, and $\sum r_i^2 = r^2$.

(b) If M is contained in an extrinsic sphere $S^m(r')$ of $S^n(r)$, then $M \to S^n(r)$ factors as $M \to S^m(r') \to S^n(r)$ with $r'^{-2} = r^{-2} + ||H||^2$ and $M \to S^m(r')$ as in (a).

PROOF It is easy to see that M has parallel second fundamental form in $\mathbb{R}^n$ if and only if M has parallel second fundamental form in $S^n(r)$. Moreover, M is not contained in any extrinsic sphere of $S^n(r)$ if and only if M is full in $\mathbb{R}^n$. If M is contained in an extrinsic sphere of $S^n(r)$, it suffices to apply Theorem 2.6.2. The result is then a direct consequence of Theorem 3.7.8. □

Classification in hyperbolic spaces.

The classification of symmetric submanifolds of $H^n(r)$ was carried out independently by Takeuchi [212] and Backes-Reckziegel [6]. In the latter paper, Jordan triple systems were used for the proof (as in [86]). Here, we adopt the more geometric approach of [212]. What comes out from the classification in hyperbolic spaces is that there is no "interesting" new example. This is a consequence of Lemma 3.7.10 and actually follows from the more general results in [69] (see Lemma 3.5.4).

THEOREM 3.7.15

Let M be a locally symmetric submanifold of $H^n(r) \subset \mathbb{R}^{n,1}$.

(a) If M is full in $\mathbb{R}^{n,1}$, then M is locally a submanifold product

$$M^{m_0}(r_0) \times M_1 \times \ldots \times M_s \to M^{m_0}(r_0) \times S^{n-m-1}(r') \to H^n(r) \ ,$$

where $r_0 < 0$, $r' > 0$, $r_0^{-2} - r'^{-2} = r^{-2}$, and $M_1 \times \ldots \times M_s$ is a symmetric submanifold of $S^{n-m-1}(r')$ as in Theorem 3.7.14 (a).

(b) If M is contained in an extrinsic sphere $\bar{M}^m(\kappa)$ of $H^n(r)$, then $M \to H^n(r)$ factors as $M \to \bar{M}^m(\kappa) \to H^n(r)$ with $\kappa = r^{-2} + ||H||^2$, and $M \to \bar{M}^m(\kappa)$ is as in (a) above, as described by Theorem 3.7.14 (a), or as in Theorem 3.7.8 (according to the sign of κ).

PROOF We regard M as a submanifold of $\mathbb{R}^{n,1}$. Recall that M is not contained in any extrinsic sphere of $H^n(r)$ if and only if M is full in $\mathbb{R}^{n,1}$.

(a) In this case, we can proceed as in the proof of Theorem 3.7.8 (a) by applying the version of Moore's Lemma for submanifolds of Lorentzian space (Lemma 2.7.4). Then M is locally a submanifold product $M_0 \times M_1 \times \ldots \times M_s \to \mathbb{R}^{m_0,1} \times \mathbb{R}^{m_1+1} \times \ldots \times \mathbb{R}^{m_s+1}$, where M_0 is a minimal submanifold of a necessarily negatively curved hypersphere of $\mathbb{R}^{m_0,1}$. According to Lemma 3.7.10, M_0 is totally geodesic in this hypersphere, so it must coincide with it, for M is full in $\mathbb{R}^{n,1}$. The other factors M_i are minimal in a hypersphere of $\mathbb{R}^{m_i+1}$, so they can be treated as above.

(b) It suffices to apply Theorem 2.6.2 to reduce this case to case (a). □

3.8 Isoparametric hypersurfaces in space forms

Isoparametric submanifolds are one of the main topics of the present book. In this section, we concentrate on isoparametric hypersurfaces. These hypersurfaces were introduced at the beginning of the 20th century, motivated by questions in geometrical optics, and studied by Segre, Levi Civita and É. Cartan, among others. The generalization to higher codimension came much later, starting from the 1980s. For more details about the historical development see [220].

Isoparametric hypersurfaces of space forms can be characterized by the property of their principal curvatures being constant. They are defined as regular level sets of isoparametric functions, so that they determine an orbit-like foliation of the space form. Isoparametric hypersurfaces share many properties with homogeneous hypersurfaces.

a) Transnormal functions

Let $\bar{M}$ be a connected Riemannian manifold. A *transnormal function* on $\bar{M}$ is a nonconstant smooth function $f : \bar{M} \to \mathbb{R}$ such that $||\text{grad}\, f||^2 = a \circ f$ for some smooth function $a : I \to \mathbb{R}$, where $I := f(\bar{M})$ is an interval in $\mathbb{R}$. Basic properties of transnormal functions were derived by Wang [239]. First of all, f has no critical values in the interior I^o of I. Therefore, if $c \in I^o$, the level set $M_c := \{p \in \bar{M} \mid f(p) = c\}$ is a smooth hypersurface of $\bar{M}$. If $c_1, c_2 \in I^o$, then M_{c_1} and M_{c_2} are *equidistant* to each other, that is, $\bar{d}(p_1, M_{c_1}) = \bar{d}(M_{c_2}, p_2)$ for every $p_1 \in M_{c_1}$ and $p_2 \in M_{c_2}$. So regular level sets of a transnormal function $f : \bar{M} \to \mathbb{R}$ form a foliation by equidistant hypersurfaces on $\bar{M}$, except possibly for one or two singular level sets. One often calls this a *transnormal system*.

b) Isoparametric functions and isoparametric hypersurfaces

An *isoparametric function* is a transnormal function $f : \bar{M} \to \mathbb{R}$ such that $\Delta f = b \circ f$ for some continuous function $b : I \to \mathbb{R}$, where $\Delta f = \text{div}(\text{grad}\, f)$ is the Laplacian of f. Suppose f is a transnormal function on $\bar{M}$ and M_c is a regular level set. Then $\xi := \text{grad}\, f/\sqrt{a(c)}$ is a unit normal vector field on M_c. Let $E_1, \ldots, E_{n-1}$ be a local orthonormal frame field of M_c. Using the Weingarten formula, the mean curvature h_c of M_c is expressed by

$$\begin{aligned} h_c &= \frac{1}{n-1}\sum_{i=1}^{n-1}\langle A_\xi E_i, E_i\rangle = -\frac{1}{(n-1)\sqrt{a(c)}}\sum_{i=1}^{n-1}\langle \bar{\nabla}_{E_i}\text{grad}\, f, E_i\rangle \\ &= -\frac{1}{(n-1)\sqrt{a(c)}}\Delta f\,. \end{aligned}$$

Taking into account that a transnormal function has, at most, two nonregular values, we can now conclude

PROPOSITION 3.8.1
Let $f : \bar{M} \to \mathbb{R}$ be a transnormal function. Then f is isoparametric if and only if the regular level hypersurfaces of f have constant mean curvature.

Each connected component of a regular level hypersurface of an isoparametric function is called an *isoparametric hypersurface*. The transnormal system induced by an isoparametric function is called an *isoparametric system*. By the previous proposition, any isoparametric hypersurface belongs to a family of equidistant hypersurfaces with constant mean curvature.

c) Homogeneous hypersurfaces

Let $\bar{M}$ be a connected complete Riemannian manifold and $I(\bar{M})$ its isometry group. Suppose G is a connected closed subgroup of $I(\bar{M})$ acting on $\bar{M}$ with cohomogeneity one. We equip the orbit space $\bar{M}/G$ with the quotient topology relative to the canonical projection $\bar{M} \to \bar{M}/G$. Then $\bar{M}/G$ is a one-dimensional Hausdorff space homeomorphic to the real line $\mathbb{R}$, the circle S^1, the half-open interval $[0, \infty)$, or the closed interval $[0, 1]$. This was proved by Mostert [143] for the compact case and by Bérard Bergery [8] in the general case. The following basic examples illustrate the four cases. Consider a one-parameter group of translations in $\mathbb{R}^2$. The orbits are parallel lines in $\mathbb{R}^2$, and the space of orbits is homeomorphic to $\mathbb{R}$. Rotating a torus around its vertical axis through the center leads to an orbit space homeomorphic to S^1, whilst rotating a sphere around some axis through its center yields an orbit space homeomorphic to $[0, 1]$. Eventually, rotating a plane around some fixed point leads to an orbit space homeomorphic to $[0, \infty)$.

If $\bar{M}/G$ is homeomorphic to $\mathbb{R}$ or S^1, each orbit of the action of G is principal and the orbits form a codimension one Riemannian foliation on $\bar{M}$. In the case $\bar{M}/G$ is homeomorphic to $[0, \infty)$ or $[0, 1]$, there exist one or two singular orbits, respectively. If a singular orbit has codimension greater than one, then each regular orbit is geometrically a tube around this singular one. And if the codimension of a singular orbit is one, then each regular orbit is an equidistant hypersurface to it. Suppose that, in addition, $\bar{M}$ is simply connected. If $\bar{M}$ is compact, then, for topological reasons, $\bar{M}/G$ is homeomorphic to $[0, 1]$ and each singular orbit has codimension greater than one. Thus, each principal orbit is a tube around any of the two singular orbits, and each singular orbit is a focal set of any principal orbit. If $\bar{M}$ is noncompact, then $\bar{M}/G$ must be homeomorphic to $\mathbb{R}$ or $[0, \infty)$. In the latter case, the singular orbit must have codimension greater than one, and each principal orbit is a tube around the singular one.

It is not difficult to deduce from the previous discussion that the orbits of G form a transnormal system on $\bar{M}$. According to Proposition 3.6.1, each principal orbit of the action of G has constant principal curvatures, hence in particular constant mean curvature. Thus, using Proposition 3.8.1, we get

PROPOSITION 3.8.2
Let $\bar{M}$ be a connected complete Riemannian manifold and G a connected closed

subgroup of the isometry group of $\bar{M}$. Then the orbits of the action of G on $\bar{M}$ form an isoparametric system on $\bar{M}$ whose principal orbits have constant principal curvatures.

d) Isoparametric hypersurfaces

Clearly, the condition of constant principal curvatures is stronger than just having constant mean curvature. A natural question is whether any isoparametric hypersurface has constant principal curvatures. Élie Cartan [32] gave an affirmative answer for the case that $\bar{M}$ is a space form.

THEOREM 3.8.3
Each isoparametric hypersurface in a space of constant curvature has constant principal curvatures.

The proof can be deduced by the more general result in the next chapter. In particular, it is a special instance of Exercise 4.6.5.

The previous result does not extend to more general Riemannian manifolds. In fact, Wang [237] gave an example of an isoparametric hypersurface in complex projective space $\mathbb{C}P^n$ with nonconstant principal curvatures. Further examples are provided by distance spheres in Damek-Ricci spaces. The story briefly goes as follows (for details see [21]): Using the Iwasawa decomposition of semisimple real Lie groups, the complex hyperbolic space $\mathbb{C}H^{n+1}$ can be realized as a solvable Lie group S equipped with a left-invariant Riemannian metric. As a group, S is the semidirect product of $\mathbb{R}$ and the $(2n+1)$-dimensional Heisenberg group. In this construction one can replace the Heisenberg group by a so-called generalized Heisenberg group. For certain generalized Heisenberg groups this yields the quaternionic hyperbolic spaces and the Cayley hyperbolic plane, but, in all other cases, one gets a nonsymmetric homogeneous Hadamard manifold, a so-called Damek-Ricci space. These manifolds are named after Damek and Ricci, who proved that these spaces provide counterexamples to the Lichnerowicz Conjecture, stating that any harmonic manifold is locally isometric to a two-point homogeneous space. There are various ways to define or characterize harmonic manifolds. One characterization is that a Riemannian manifold is harmonic if and only if its geodesic hyperspheres have constant mean curvature. Hadamard's Theorem implies that in a Hadamard manifold the square of the distance function to a point is a well-defined transnormal function. The result of Damek and Ricci says that in a Damek-Ricci space this function is even isoparametric. It was then proved by Tricerri and Vanhecke that the corresponding isoparametric hypersurfaces, which are geodesic hyperspheres, have non-constant principal curvatures. This lead Tricerri and Vanhecke to pose the

Conjecture 3.8.4 *A Riemannian manifold is locally isometric to a two-point homogeneous space if and only if its (sufficiently small) geodesic hyperspheres have constant principal curvatures.*

Two-point homogeneous spaces are precisely the Euclidean spaces, spheres and projective and hyperbolic spaces over the normed real division algebras $\mathbb{R}$, $\mathbb{C}$, $\mathbb{H}$ and $\mathbb{O}$. Equivalently, two-point homogeneous spaces are precisely the Riemannian symmetric spaces of rank one and Euclidean spaces. The above conjecture is known to be true in all dimensions different from 8 and 16. This is because it is closely related to the Osserman Conjecture whose answer is not known in these two dimensions only (see Nikolayevsky [167]). The Osserman Conjecture states that a Riemannian manifold is locally isometric to a two-point homogeneous space if and only if the spectrum of its Riemannian Jacobi operator is independent of the point and of the direction. Again, the relevant information about this and many references can be found in [21].

e) Historical remarks: Cartan's fundamental formula

We now turn to the classification problem of isoparametric hypersurfaces in standard space forms. Recall that any homogeneous hypersurface is isoparametric, so the classification of isoparametric hypersurfaces includes the classification of homogeneous hypersurfaces, or equivalently, of cohomogeneity one actions up to orbit equivalence. For this reason, we also discuss the classification problem of homogeneous hypersurfaces. A crucial step in this context is the so-called *Cartan's fundamental formula*. To state it, denote by g the number of distinct principal curvatures of a given hypersurface M with constant principal curvatures, and by $\lambda_1, \ldots, \lambda_g$ the principal curvatures of M with corresponding multiplicities $m_1, \ldots, m_g$.

THEOREM 3.8.5
Let M be a hypersurface with constant principal curvatures in a Riemannian manifold $\bar{M}$ of constant curvature κ. Then

$$\sum_{\substack{j=1 \\ \lambda_j \neq \lambda_i}}^{g} m_j \frac{\kappa + \lambda_i \lambda_j}{\lambda_i - \lambda_j} = 0$$

for all $i = 1, \ldots, g$.

We give a direct proof along the same lines as Cartan's original proof, although Cartan used differential forms rather than vector fields. It is worthwhile to mention that Nomizu [169] and, independently, Münzner [144] observed that the sum on the left-hand side of the above equation corresponds to the mean curvature of the focal set of M, which is determined by the principal curvature λ_i. This yields the geometrical interpretation of Cartan's fundamental formula in terms of minimality of the focal sets in an isoparametric system. In Section 5.2 we will actually give another proof for isoparametric hypersurfaces in spheres using the Coxeter groups that are associated to isoparametric submanifolds. An even simpler purely geometric proof of the minimality of the focal manifolds can be derived from the theory that we are going to develop in the next chapter (see Section 4.4 and also page 153).

PROOF Let $E_1, \ldots, E_{n-1}$ be a local orthonormal frame field of M such that $AE_i = \lambda_i E_i$, $i = 1, \ldots, n-1$. Note that we change notation here and write down principal curvatures as many times as their multiplicities. Then the Codazzi equation implies

$$\begin{aligned}\langle \nabla_{E_k} E_i, E_j \rangle &= \frac{1}{\lambda_i - \lambda_j} \langle (\nabla_{E_k} A) E_i, E_j \rangle \\ &= \frac{1}{\lambda_i - \lambda_j} \langle (\nabla_{E_j} A) E_i, E_k \rangle = 0 \qquad (\text{if } \lambda_k = \lambda_i \neq \lambda_j)\end{aligned}$$

and

$$\begin{aligned}\langle \nabla_{E_k} E_j, E_i \rangle &= \frac{1}{\lambda_j - \lambda_i} \langle (\nabla_{E_k} A) E_j, E_i \rangle \\ &= \frac{1}{\lambda_j - \lambda_i} \langle (\nabla_{E_j} A) E_i, E_k \rangle = 0 \qquad (\text{if } \lambda_k = \lambda_i \neq \lambda_j).\end{aligned}$$

Using these equations together with the Gauss and Codazzi equations, we obtain for all $i, j \in \{1, \ldots, n-1\}$ with $\lambda_i \neq \lambda_j$

$$\begin{aligned}\kappa + \lambda_i \lambda_j &= \langle R(E_i, E_j) E_j, E_i \rangle \\ &= \langle \nabla_{E_i} \nabla_{E_j} E_j, E_i \rangle - \langle \nabla_{E_j} \nabla_{E_i} E_j, E_i \rangle - \langle \nabla_{[E_i, E_j]} E_j, E_i \rangle \\ &= E_i \langle \nabla_{E_j} E_j, E_i \rangle - \langle \nabla_{E_j} E_j, \nabla_{E_i} E_i \rangle \\ &\quad - E_j \langle \nabla_{E_i} E_j, E_i \rangle + \langle \nabla_{E_i} E_j, \nabla_{E_j} E_i \rangle - \langle \nabla_{[E_i, E_j]} E_j, E_i \rangle \\ &= \langle \nabla_{E_i} E_j, \nabla_{E_j} E_i \rangle - \langle \nabla_{[E_i, E_j]} E_j, E_i \rangle \\ &= \langle \nabla_{E_i} E_j, \nabla_{E_j} E_i \rangle - \frac{1}{\lambda_j - \lambda_i} \langle (\nabla_{[E_i, E_j]} A) E_j, E_i \rangle \\ &= \langle \nabla_{E_i} E_j, \nabla_{E_j} E_i \rangle - \frac{1}{\lambda_j - \lambda_i} \langle (\nabla_{E_i} A) E_j, [E_i, E_j] \rangle \\ &= \langle \nabla_{E_i} E_j, \nabla_{E_j} E_i \rangle - \frac{1}{\lambda_j - \lambda_i} (\langle (\nabla_{E_i} A) E_j, \nabla_{E_i} E_j \rangle \\ &\quad - \langle (\nabla_{E_i} A) E_j, \nabla_{E_j} E_i \rangle) \\ &= \langle \nabla_{E_i} E_j, \nabla_{E_j} E_i \rangle - \frac{1}{\lambda_j - \lambda_i} (\langle (\nabla_{E_j} A) E_i, \nabla_{E_i} E_j \rangle \\ &\quad - \langle (\nabla_{E_i} A) E_j, \nabla_{E_j} E_i \rangle) \\ &= \langle \nabla_{E_i} E_j, \nabla_{E_j} E_i \rangle - \frac{1}{\lambda_j - \lambda_i} (\lambda_i - \lambda_j) \langle \nabla_{E_i} E_j, \nabla_{E_j} E_i \rangle \\ &= 2 \langle \nabla_{E_i} E_j, \nabla_{E_j} E_i \rangle \\ &= 2 \sum_{k=1}^{n-1} \langle \nabla_{E_i} E_j, E_k \rangle \langle \nabla_{E_j} E_i, E_k \rangle \\ &= 2 \sum_{\substack{k=1 \\ \lambda_k \neq \lambda_i, \lambda_j}}^{n-1} \frac{\langle (\nabla_{E_k} A) E_i, E_j \rangle^2}{(\lambda_j - \lambda_k)(\lambda_i - \lambda_k)} .\end{aligned}$$

This implies

$$\sum_{\substack{j=1\\ \lambda_j \neq \lambda_i}}^{n-1} \frac{\kappa + \lambda_i \lambda_j}{\lambda_i - \lambda_j} = 2 \sum_{\substack{j,k=1\\ \lambda_k \neq \lambda_i \neq \lambda_j \neq \lambda_k}}^{n-1} \frac{\langle (\nabla_{E_k} A) E_i, E_j \rangle^2}{(\lambda_i - \lambda_j)(\lambda_j - \lambda_k)(\lambda_i - \lambda_k)}$$
$$= - \sum_{\substack{k=1\\ \lambda_k \neq \lambda_i}}^{n-1} \frac{\kappa + \lambda_i \lambda_k}{\lambda_i - \lambda_k} ,$$

which induces Cartan's fundamental formula. □

From this formula one can easily deduce that the number of distinct principal curvatures of a hypersurface with constant principal curvatures in $\mathbb{R}^n$ or H^n is at most two. It is then easy to determine all isoparametric hypersurfaces in $\mathbb{R}^n$ and H^n. For $\mathbb{R}^n$ this classification is due to Levi Civita [130] for $n = 3$ and to Segre [199] in general, while for H^n it is due to E. Cartan [32].

An isoparametric hypersurface of $\mathbb{R}^n$ is one of the following:

(1) a geodesic hypersphere in $\mathbb{R}^n$, or

(2) an affine hyperplane in $\mathbb{R}^n$, or

(3) a tube around a k-dimensional affine subspace of $\mathbb{R}^n$ for some $1 \leq k \leq n-2$.

Similarly, an isoparametric hypersurface of H^n is one of the following:

(1) a geodesic hypersphere in H^n, or

(2) a horosphere in H^n, or

(3) a totally geodesic hyperbolic hyperplane H^{n-1} in H^n or an equidistant hypersurface to it, or

(4) a tube around a k-dimensional totally geodesic hyperbolic subspace H^k of H^n for some $1 \leq k \leq n-2$.

f) Isoparametric hypersurfaces of S^n

In a series of papers [32], [33], [34], [35], Cartan made an attempt to also classify the isoparametric hypersurfaces in the sphere S^n. He did not succeed, and, in fact, the full classification is still not known. In the following, we present some basic results regarding this.

As far as spheres are concerned, Cartan's fundamental formula does not provide sufficient information to determine the possible number of distinct principal curvatures. Only later Münzner [144] proved, using methods from algebraic topology, that the number g of distinct principal curvatures of an isoparametric hypersurface in S^n equals $1, 2, 3, 4$ or 6. This can also be deduced using Coxeter groups (see the example at page 148). Cartan classified isoparametric hypersurfaces with, at most, three distinct principal curvatures. They all turn out to be homogeneous. After the original proof of Cartan [33], alternative proofs were given using various approaches [109], [114], [62]. Surprisingly, for $g = 4$ there exist inhomogeneous isoparametric hypersurfaces. The first such examples were discovered by Ozeki and Takeuchi [184], and later Ferus, Karcher and Münzner [87] constructed new series of examples by using representations of Clifford algebras. Recently, Cecil, Chi and Jensen [43] proved that, with a few possible exceptions, all inhomogeneous isoparametric hypersurfaces of S^n with $g = 4$ are given by this construction. It was shown by Abresch [1] that the case $g = 6$ occurs only in S^7 and S^{13}, and Dorfmeister and Neher [73] proved that an isoparametric hypersurface in S^7 must be homogeneous. Now, as concerns the classification of homogeneous hypersurfaces, the following result by Hsiang and Lawson [102] settles the remaining cases $g = 4, 6$:

THEOREM 3.8.6
A hypersurface in S^n is homogeneous if and only if it is a principal orbit of the isotropy representation of a Riemannian symmetric space of rank two.

We can therefore read off the classification of homogeneous hypersurfaces in spheres from the list of compact, simply connected, Riemannian symmetric spaces. In detail, we get the following list of homogeneous hypersurfaces M in $S^n = SO(n+1)/SO(n)$:

$g = 1$: Then M is a geodesic hypersphere in S^n. A suitable subgroup of $SO(n+1)$ is the isotropy group $SO(n)$, and the corresponding Riemannian symmetric space of rank two is

$$(SO(2) \times SO(n+1))/SO(n) = S^1 \times S^n .$$

$g = 2$: M is a Riemannian product of two spheres, namely

$$S^k(r_1) \times S^{n-k-1}(r_2) \ , \ r_1^2 + r_2^2 = 1 \ , \ 0 < r_1, r_2 < 1 \ , \ 0 < k < n-1 \ .$$

A suitable subgroup of $SO(n+1)$ is $SO(k+1) \times SO(n-k)$, and the corresponding Riemannian symmetric space of rank two is

$$(SO(k+2) \times SO(n-k+1))/(SO(k+1) \times SO(n-k)) = S^{k+1} \times S^{n-k} .$$

$g = 3$: M is congruent to a tube around the Veronese embedding of $\mathbb{R}P^2$ into S^4, $\mathbb{C}P^2$ into S^7, $\mathbb{H}P^2$ into S^{13}, or $\mathbb{O}P^2$ into S^{25}. The corresponding Riemannian symmetric spaces of rank two are

$$SU(3)/SO(3) \ , \ SU(3) \ , \ SU(6)/Sp(3) \ , \ E_6/F_4 \ ,$$

respectively. These homogeneous hypersurfaces might also be characterized as the principal orbits of the natural actions of $SO(3)$, $SU(3)$, $Sp(3)$, F_4 on the unit sphere in the linear subspace of all traceless matrices in the real Jordan algebra of all

3×3 Hermitian matrices with coefficients in $\mathbb{R}, \mathbb{C}, \mathbb{H}, \mathbb{O}$, respectively. The singular orbits of these actions give the Veronese embeddings of the corresponding projective spaces.

$g = 4$: M is a principal orbit of the isotropy representation either of

$$Sp(2)\ ,\ SO(10)/U(5)\ ,\ E_6/Spin(10)U(1)\ ,$$

or of a two-plane Grassmannian

$$\begin{aligned} G_2^+(\mathbb{R}^{k+2}) &= SO(k+2)/SO(k) \times SO(2)\ (k \geq 3)\ , \\ G_2(\mathbb{C}^{k+2}) &= SU(k+2)/S(U(k) \times U(2))\ (k \geq 3)\ , \\ G_2(\mathbb{H}^{k+2}) &= Sp(k+2)/Sp(k) \times Sp(2)\ (k \geq 2)\ . \end{aligned}$$

The homogeneous hypersurfaces related to $G_2^+(\mathbb{R}^{k+2})$ are the principal orbits of the action of $SO(k) \times SO(2)$ on the unit sphere S^{2k-1} in $\mathrm{Mat}(k \times 2, \mathbb{R}) \approx \mathbb{R}^{2k}$ defined by $(A, B)X = AXB^{-1}$ with $A \in SO(k)$, $B \in SO(2)$ and $X \in \mathrm{Mat}(k \times 2, \mathbb{R})$. The homogeneous hypersurfaces related to the complex and quaternionic Grassmannians are obtained from the analogous actions of $S(U(k) \times U(2))$ and $Sp(k) \times Sp(2)$ on the unit sphere in $\mathrm{Mat}(k \times 2, \mathbb{C}) \approx \mathbb{C}^{2k} \approx \mathbb{R}^{4k}$ and $\mathrm{Mat}(k \times 2, \mathbb{H}) \approx \mathbb{H}^{2k} \approx \mathbb{R}^{8k}$, respectively.

The homogeneous hypersurfaces related to $Sp(2) \approx Spin(5)$ are the principal orbits of the adjoint representation of $Sp(2)$ on the unit sphere S^9 of its Lie algebra $\mathfrak{sp}(2) \approx \mathbb{R}^{10}$.

The unitary group $U(5)$ acts on $\mathbb{C}^5$ and hence on $\Lambda^2\mathbb{C}^5 \approx \mathbb{C}^{10} \approx \mathbb{R}^{20}$ in a natural way. The principal orbits of this action on the unit sphere S^{19} correspond to the principal orbits of the action of the isotropy representation of $SO(10)/U(5)$.

Denote by Δ^+ and Δ^- the two real half-spin representations of $Spin(10)$ on $\mathbb{R}^{32} \approx \mathbb{C}^{16}$, and by ξ the canonical representation of $U(1)$ on $\mathbb{C}^{16}$ given by multiplication by unit complex numbers.
Then the isotropy representation of $E_6/Spin(10)U(1)$ is equivalent to $\Delta^+ \otimes \xi^3 + \Delta^- \otimes \xi^{-3}$, and its principal orbits in the unit sphere $S^{31} \subset \mathbb{R}^{32}$ are homogeneous hypersurfaces.

$g = 6$: M is a principal orbit of the isotropy representation of $G_2/SO(4)$ or of the compact real Lie group G_2. The isomorphisms $Spin(4) \approx Sp(1) \times Sp(1)$ and $\mathbb{R}^8 \approx \mathbb{H}^2$ give rise to an action of $Spin(4)$ on $\mathbb{R}^8$ by means of $(\lambda, \mu) \cdot (z, v) = (\lambda z, v\mu^{-1})$, where $(\lambda, \mu) \in Sp(1) \times Sp(1)$ and $(z, v) \in \mathbb{H} \oplus \mathbb{H}$. The principal orbits of this action on the unit sphere S^7 are homogeneous hypersurfaces with six distinct principal curvatures. Miyaoka [138] proved that the orbits of this action are precisely the inverse images under the Hopf map $S^7 \to S^4$ of the $SO(3)$-orbits in S^4 as described in the case $g = 3$. The principal orbits in the unit sphere S^{13} of the Lie algebra $\mathfrak{g}_2 \approx \mathbb{R}^{14}$ of the adjoint representation of the Lie group G_2 are homogeneous hypersurfaces with six distinct principal curvatures, all of whose multiplicities are two.

3.9 Algebraically constant second fundamental form

We now introduce a class of submanifolds that generalize those that have been studied in this chapter. Indeed, a common feature of isoparametric hypersurfaces of space forms and of homogeneous submanifolds is that their second fundamental form at different points can be expressed by the same matrices.

DEFINITION 3.9.1 *Let M be a submanifold of a Riemannian manifold $\bar{M}$. We say that the second fundamental form α of M is algebraically constant, if for all $p, q \in M$ there exists a linear isometry $F_p : T_p\bar{M} \to T_q\bar{M}$ such that $F_pT_pM = T_qM$ and $\alpha_q(F_pX, F_pY) = F_p\alpha_p(X, Y)$ for all $X, Y \in T_pM$.*

Submanifolds with algebraically constant second fundamental form might be regarded as an extrinsic analogue of curvature-homogeneous manifolds (see [225]). Submanifolds of S^n with algebraically constant second fundamental form and nonzero parallel mean curvature vector fields have generically constant principal curvatures.

THEOREM 3.9.2 [61]
Let M be a locally irreducible full submanifold of S^n with nonzero parallel mean curvature vector field and algebraically constant second fundamental form. Then M has constant principal curvatures and, if the codimension of M is greater than one, it is an orbit of an s-representation.

This result follows from the more general Theorem 5.5.2, therefore we omit the proof here. Note that Theorem 3.9.2 implies that a homogeneous submanifold of S^n with nonzero parallel mean curvature vector field is an orbit of an s-representation [175]. Since every compact subgroup of $SO(n)$ has a minimal orbit in S^n [102], the assumption that the mean curvature vector field is nonzero cannot be removed.

In Section 3.7, we observed that a symmetric submanifold of a space form is uniquely determined by its second fundamental form at one point. The following improves that result (see also [62]).

THEOREM 3.9.3
Let M be a connected submanifold of a standard space form $\bar{M}^n(\kappa)$ with parallel mean curvature vector field and algebraically constant second fundamental form α. If α coincides algebraically with the second fundamental form of a symmetric submanifold N of $\bar{M}^n(\kappa)$, then α is parallel and M is isometrically congruent to an open part of N.

Note that, in the previous theorem, we also allow vanishing mean curvature.

If $\bar{M}^n(\kappa) = S^n$ and if N is irreducible, then M is minimal since, as we saw in Section 3.7, irreducible symmetric submanifolds of S^n are minimal. Note also that Theorem 3.9.3 generalizes a rigidity theorem for immersions with parallel second fundamental form due to Reckziegel [192]: if two immersions with parallel second fundamental form at some point have the same tangent space and second fundamental form, then they locally coincide.

PROOF Since M has algebraically the same second fundamental form as the symmetric submanifold N, we have

$$\hat{R}(x, y) \cdot \alpha = 0 \ , \tag{3.14}$$

where $\hat{R}(x, y)$ acts as a derivation on TM via the Riemannian curvature tensor $R(x, y)$ of M and on νM via the normal curvature tensor $R^\perp(x, y)$. Submanifolds satisfying (3.14) are also known as semisymmetric submanifolds (see, for instance, [137]). Note that, due to the Gauss and Ricci equations, both $R(x, y)$ and $R^\perp(x, y)$ are determined by α. From the Codazzi equation, the Ricci formula

$$\nabla^2_{xy}\alpha - \nabla^2_{yx}\alpha = -\hat{R}(x, y) \cdot \alpha$$

and (3.14), it follows that $\nabla^2_{xy}\alpha$ is symmetric in all its 4 entries. By [55, formula (3.12)] (cf. also Remark 2.1.3) we have

$$\frac{1}{2}\Delta||\alpha||^2 = ||\nabla^\perp \alpha||^2 + \langle \alpha, \Delta\alpha \rangle \ ,$$

where Δ is the Laplace-Beltrami operator, and the norm and inner product is the usual one for tensor fields. Since α is algebraically constant, it has constant length and hence

$$||\nabla^\perp \alpha||^2 + \langle \alpha, \Delta\alpha \rangle = 0 \ .$$

Thus it suffices to show that $\Delta\alpha = 0$. We choose an orthonormal basis $e_1, \ldots, e_m$ of T_pM, $p \in M$. Then, using the symmetry of $\nabla^2\alpha$, we have

$$(\Delta\alpha)(x, y) = \sum_{i=1}^{m} (\nabla^2_{e_ie_i}\alpha)(x, y) = \sum_{i=1}^{m} (\nabla^2_{xy}\alpha)(e_i, e_i) = m\nabla^2_{xy}H = 0$$

for all $x, y \in T_pM$, since the mean curvature vector field H of M is parallel. This concludes the proof. □

3.10 Exercises

Exercise 3.10.1 Consider the symmetric pair $(SL(n,\mathbb{R}), SO(n)), n \geq 2$. The corresponding Cartan decomposition is

$$\mathfrak{sl}(n,\mathbb{R}) = \mathfrak{so}(n) \oplus \mathcal{S}_n \ ,$$

where $\mathcal{S}_n$ is the real vector space of all symmetric $n \times n$-matrices with real coefficients and trace zero. Show that one can identify the tangent space of $M = SL(n,\mathbb{R})/SO(n)$ at $o = eSO(n)$ with $\mathcal{S}_n$ and that the isotropy representation of $SL(n,\mathbb{R})/SO(n)$ is conjugation on $\mathcal{S}_n$ by elements in $SO(n)$. Prove that this action is orthogonal with respect to the inner product $\langle X, Y \rangle = \mathrm{tr}(XY)$ on $\mathcal{S}_n$. The orbits are the standard embeddings of the real flag manifolds. A special case is the Veronese embedding of the real projective space $\mathbb{R}P^{n-1}$. Compute the shape operator of an orbit for $n = 3$.

Exercise 3.10.2 Let G be a closed subgroup of the isometry group of a Riemannian manifold M, $p \in M$ and $v \in \nu_p(G \cdot p)$ a point on a principal orbit of the slice representation at p. Then there exists a real number $\epsilon > 0$ such that $G \cdot \exp_p(tv)$ is a principal orbit of the G-action for all $t \in (0, \epsilon)$. Conclude that the cohomogeneity of the G-action on M is equal to the cohomogeneity of the slice representation at any point $p \in M$.

Exercise 3.10.3 Let G be a compact subgroup of the isometry group of a Riemannian manifold M and $G \cdot p$ be a singular orbit of the G-action on M. Let $V \subset \nu_p(G \cdot p)$ be the fixed point set of the slice representation at p. Then the dimension of V is strictly smaller than the cohomogeneity of the G-action of M. *Hint:* If $\dim G \cdot p$ is not maximal, then the cohomogeneity of the slice representation is greater than $\dim V$ (see Exercise 3.10.2).

Exercise 3.10.4 Assume that a connected Lie group G acts locally polar on a Riemannian manifold M. Let $q \in M$ and q_k be a sequence of points on M that lie on principal orbits and converge toward q. Let Σ_k be a local section of the G-action through q_k. Assume that $T_{q_k}\Sigma_k$ converges to some subspace V of T_qM. Then V is a section for the slice representation at q. Conclude that the slice representation at q is polar.

Exercise 3.10.5 Let G be a closed subgroup of $SO(n)$. Then the action of G is polar on $\mathbb{R}^n$ if and only if it is polar on $S^{n-1} \subset \mathbb{R}^n$.

Exercise 3.10.6 Prove that the action of $S^1 = \{e^{i\theta} \mid \theta \in \mathbb{R}\}$ on $\mathbb{C} \times \mathbb{C}$ defined by $e^{i\theta} \cdot (z_1, z_2) := (e^{i\theta} z_1, e^{i\theta} z_2)$ is not polar. Observe that by Exercise 3.10.5 the action is not polar on the sphere S^3 either.

Exercise 3.10.7 Assume that the Lie subgroup G of $SO(n)$ acts on $\mathbb{R}^n \setminus \{0\}$ without fixed points. Show that if G has a principal orbit that is not full in $\mathbb{R}^n$, then the action cannot be polar. Deduce from this another proof of Exercise 3.10.6.

Exercise 3.10.8 Prove that the first normal space at any point of an orbit of an irreducible polar representation on $\mathbb{R}^n$ coincides with the normal space at that point.

Hint: The first normal space is $\nabla^\perp$-parallel.

Exercise 3.10.9 Let G be a compact Lie group that acts polarly on a complete Riemannian manifold M, and let Σ be a section for the G-action. Prove that the set Σ_r of points in Σ that lie on a principal orbit is open and dense in Σ. *Hints:* (i) If Σ_r is not dense in Σ, then there exists an open subset Ω of Σ such that $\dim(G \cdot p)$ is constant on Ω and smaller than the cohomogeneity of the G-action. (ii) Let $q \in \Sigma$ be such that $\dim(G \cdot q) = \dim(G \cdot p)$ for any q in an open neighbourhood of p in Σ. Then $T_p\Sigma$ is pointwise fixed by the slice representation of p (same ideas as used to solve Exercise 3.10.2).

Exercise 3.10.10 Let M be a submanifold of a Riemannian homogeneous space $\bar{M}$. Prove that M is totally geodesic in $\bar{M}$ if and only if the orthogonal projection onto TM of any Killing vector field on $\bar{M}$ is a Killing vector field on M.

Exercise 3.10.11 Let G be a closed subgroup of the isometry group of a Riemannian manifold M and $p \in M$. Prove that $S = \exp_p(\nu_p(G \cdot p))$ intersects each orbit of the G-action on M.

Exercise 3.10.12 Let G be a closed subgroup of the isometry group of a Riemannian manifold M and assume that G acts with cohomogeneity one on M. If every geodesic in M is a closed submanifold, then G acts polarly on M.

Exercise 3.10.13 Let M be a full irreducible symmetric submanifold of $\mathbb{R}^n$. Prove that M does not have a parallel nonumbilical local normal vector field.

Exercise 3.10.14 Prove that any equivariant vector field on a principal orbit of a polar representation is parallel with respect to the normal connection. Deduce that a principal orbit of a polar representation is isoparametric.

Exercise 3.10.15 Let X be a Killing vector field that is orthogonal to all sections of a polar action of G on a Riemannian manifold M. Then the one-parameter group of isometries generated by X preserves any G-orbit.

Exercise 3.10.16 Let $M = G/K$ be a simply connected semisimple Riemannian symmetric space with $G = I^o(M)$, $p \in M$ and $K = G_p$. Let $\mathfrak{g} = \mathfrak{k} \oplus \mathfrak{p}$ be the corresponding Cartan decomposition of $\mathfrak{g}$ and identify the isotropy representation of G/K with $\mathrm{Ad} : K \to SO(\mathfrak{p})$ via the isomorphism $T_pM \cong \mathfrak{p}$.

Let $M = K \cdot \eta \cong K/K_0$ be a symmetric R-space regarded as an immersed submanifold $f : M \to \mathbb{R}^n \cong \mathfrak{p}$, $k \cdot \eta \mapsto \mathrm{Ad}(k)\eta$. Let $\mathfrak{k} = \mathfrak{k}_0 \oplus \mathfrak{m}$ be the reductive decomposition of $\mathfrak{k}$ given by $\mathfrak{m} = \mathfrak{k}_0^\perp$ with respect to a K-invariant inner product on $\mathfrak{k}$. Prove that for any $X, Y, Z \in \mathfrak{m} \cong T_\eta M$,

(a) $f_* X = \mathrm{ad}(X)\eta$,

(b) $\alpha(X, Y) = \mathrm{ad}(X)\mathrm{ad}(Y)\eta$,

(c) $(\nabla^\perp_Z \alpha)(X, Y) = (\mathrm{ad}(Z)\mathrm{ad}(X)\mathrm{ad}(Y)\eta)^\perp = 0$, where $^\perp$ denotes the normal component.

(See e.g. [112].) Note that this yields a new proof of Proposition 3.7.7.

Exercise 3.10.17 (Existence of principal orbits for isometric actions) Let M be a connected complete Riemannian manifold and H a connected closed subgroup of $I(M)$. Let $p \in M$ and denote by $\exp^\perp : \nu(H \cdot p) \to M$ the normal exponential map of the orbit $H \cdot p$ defined by $\exp^\perp(\xi_q) = \exp_q(\xi_q)$ for $\xi_q \in \nu_q(H \cdot p)$, $q \in H \cdot p$. Let $\varepsilon > 0$ be sufficiently small so that $\exp^\perp$ is a diffeomorphism from V_ε onto an open subset of M that contains p, where

$$V_\varepsilon = \{\xi_q \in \nu_q(H \cdot p) \mid ||\xi_q|| < \varepsilon \, , \, q \in B_\varepsilon(p) \cap H \cdot p\}$$

and $B_\varepsilon(p)$ is the open ε-ball in M centered at p. Prove the following statements:

(a) Let $q = \exp^\perp(\xi_p)$, where $\xi_p \in \nu_p(H \cdot p)$ and $||\xi_q|| < \varepsilon/2$. Then the isotropy subgroup H_q is contained in H_p. *Hint:* If $h \in H_q$ then $hp \in B_\varepsilon(p) \cap H \cdot p$ and $q = \exp^\perp(h_*(\xi_p))$.

(b) If $q \in \exp^\perp(V_{\varepsilon/2})$, then there exists $\tilde{q} \in H \cdot q$ such that $\tilde{q} = \exp^\perp(\xi_p)$ with $||\xi_p|| < \varepsilon/2$.

(c) Using (a) and (b) and the compactness of the isotropy subgroups, prove that there exists an open and dense subset $\mathcal{O}$ of M where the conjugacy class of the isotropy subgroup H_q is locally constant. Conclude that if $q \in \mathcal{O}$, then $h_{*q}|\nu_q(H \cdot q) = Id$ for all $h \in H_q$.

(d) Prove that for any $q \in M$ the subset $\exp^\perp(\nu_q(H \cdot q)) \subset M$ intersects any other orbit of H. *Hint:* Let $r = hp \in H \cdot p$ and $s \in H \cdot q$ be such that the distance

$d(r, s)$ is minimal. Then the minimizing geodesic $\gamma : [0, 1] \to M$ from r to s is perpendicular to both orbits. Then $h^{-1} \circ \gamma$ is geodesic starting at p, perpendicular to $T_p(H \cdot p)$, and which intersects $H \cdot q$.

(e) Conclude that the isotropy subgroups H_q, $q \in \mathcal{O}$, belong all to the same conjugacy class.

Chapter 4

The Normal Holonomy Theorem

A connection on a Riemannian manifold M can be interpreted as a tool for comparing different tangent spaces by means of parallel transport. In general, parallel transport depends upon the curve that is chosen for joining two given points. This dependence is measured by the so-called holonomy group, that is, the group of linear isometries of a tangent space T_pM generated by all parallel transports along loops based at p. Holonomy plays an important role in Riemannian geometry, in particular in the context of special structures on manifolds, for example, Kähler, hyperkähler, or quaternionic Kähler structures. Holonomy is a concept that can be defined for any connection on a vector bundle. In this chapter, we will deal with the holonomy group of the normal connection of a submanifold, the so-called *normal holonomy group* of a submanifold. The purpose of this chapter is to explain how the theory of holonomy can be used to study submanifold geometry.

In Section 4.1, we recall some important facts on holonomy. There are many analogies with the holonomy of a Riemannian manifold, due to the *a priori* surprising fact that normal holonomy groups look like non-exceptional Riemannian holonomy groups. Furthermore, we introduce higher order mean curvature tensors, and we characterize submanifolds whose higher order mean curvature tensors are parallel with respect to the normal connection as submanifolds with constant principal curvatures. This class of submanifolds is very important in the context of normal holonomy and includes orbits of s-representations (which were introduced and discussed in the previous chapter). We show in Theorem 4.1.7 that the normal holonomy of an orbit of an s-representation is equal to the slice representation, that is, the effectivized action of the isotropy group on the normal space.

In Section 4.2 we explain the Normal Holonomy Theorem 4.2.1 [173], which asserts that the nontrivial part of the normal holonomy action on the normal space is an s-representation. The Normal Holonomy Theorem is some sort of extrinsic analogue to de Rham's Decomposition Theorem and Berger's Theorem on Riemannian holonomy.

In Section 4.3, we present a proof of the Normal Holonomy Theorem based on holonomy systems as introduced by Simons.

One of the main consequences of the Normal Holonomy Theorem is that orbits of s-representations play a role in submanifold geometry that is similar to that of Riemannian symmetric spaces in Riemannian geometry. This is illustrated in Section 4.4, where we define some important tools for the study of submanifolds: focalization and construction of holonomy tubes. These tools are very suitable for studying submanifolds that satisfy simple geometric conditions, like submanifolds admitting a parallel normal vector field whose shape operator has constant eigenvalues, isoparametric submanifolds, submanifolds with constant principal curvatures, et cetera.

We finish this chapter in Section 4.5 with some further remarks about normal holonomy.

4.1 Normal holonomy

a) General facts

We first present the definition of the normal holonomy group of a submanifold, even though it is along the same lines as in the Riemannian case (cf. Appendix, Section A.1). In fact, one can give a general definition of holonomy group for any connection on a vector bundle. For general facts about parallel transport and holonomy see [117] (cf. Appendix).

Let M be a submanifold of a standard space form $\bar{M}^n(\kappa)$, and denote, as usual, by νM the normal bundle of M and by $\nabla^\perp$ the induced normal connection. Let $p, q \in M$ and $\gamma : [0,1] \to M$ be a piecewise differentiable curve in M with $\gamma(0) = p$ and $\gamma(1) = q$. Then the $\nabla^\perp$-parallel transport along γ induces a linear isometry

$$\tau_\gamma^\perp : \nu_p M \to \nu_q M \ .$$

It is easy to see that parallel transport does not depend on the parametrization of the curve γ.

We now fix a point $p \in M$ and denote by $\Omega_p M$ the set of all piecewise differentiable loops based at p. Recall that a loop based at p is a curve $\gamma : [0,1] \to M$ with $\gamma(0) = p = \gamma(1)$. We denote by $\gamma * \gamma'$ the composition of the loops γ and γ' given by $\gamma * \gamma'(t) = \gamma(2t)$ for $0 \le t \le 1/2$ and $\gamma * \gamma'(t) = \gamma'(2t-1)$ for $1/2 \le t \le 1$. Then we have a map

$$\tau^\perp : \Omega_p M \to O(\nu_p M) \ , \ \gamma \mapsto \tau_\gamma^\perp \ ,$$

which satisfies $\tau_{\gamma * \gamma'}^\perp = \tau_{\gamma'}^\perp \tau_\gamma^\perp$ and $\tau_{\tilde{\gamma}}^\perp = (\tau_\gamma^\perp)^{-1}$, where $\tilde{\gamma} : [0,1] \to M$ is given by $\tilde{\gamma}(t) = \gamma(1-t)$. The image $\tau^\perp(\Omega_p M)$ is a subgroup of $O(\nu_p M)$, the so-called *normal holonomy group of M at p*. We denote this normal holonomy group by Φ_p. If M is connected, then all normal holonomy groups are conjugate to each other. Indeed, if γ is a piecewise differentiable curve from p to q, then $\Phi_q = \tau_\gamma \Phi_p (\tau_\gamma)^{-1}$. In this situation, we will often omit the base point and refer to the "normal holonomy group" Φ.

If we replace $\Omega_p M$ by the set $\Omega_p^* M$ of null homotopic piecewise differentiable loops in M based at p, then the resulting subgroup of $O(\nu_p M)$ is called the *restricted normal holonomy group of M at p* and denoted by Φ_p^*. Note that Φ_p^* is a normal subgroup of Φ_p and Φ_p/Φ_p^* is countable. One can show that Φ_p^* is an arcwise connected subgroup of $O(\nu_p M)$ (see [117, vol. I, Chapter II, Theorem 4.2], cf. also [22, page 289]). As a consequence of a result by Yamabe [245], which states that every arcwise connected subgroup of a Lie group is a Lie subgroup, we get that *Φ_p^* is a Lie subgroup of $O(\nu_p M)$* (cf. also [117, vol I, Appendix 4, page 275] for a an easier proof, using that Φ_p^* is C^1-arcwise connected in this case). This implies that Φ_p^* is the identity component of Φ_p (see [22, Corollary 10.48, page 289]). In particular, Φ_p^* and Φ_p have the same Lie algebra, which is called the *normal holonomy algebra* of M at p. Of course, if M is simply connected, then $\Phi_p = \Phi_p^*$.

Since we are mainly concerned with the local geometry of submanifolds, we will mainly use the restricted normal holonomy group throughout these notes. Another reason is that Φ_p is, in general, not a closed subgroup of $O(\nu_p M)$, whereas Φ_p^* is closed in $O(\nu_p M)$ and, in particular, compact, as we will prove in Theorem 4.2.1. The normal holonomy group will always be regarded as a Lie subgroup of the orthogonal group $O(\nu_p M)$, so Φ_p will act in a natural way on $\nu_p M$.

For a submanifold, we can also consider the combined holonomy

$$\Phi_p^c = \{(\tau_\gamma, \tau_\gamma^\perp) \mid \gamma \text{ is a loop in } M \text{ based at } p\} \subset \mathrm{Hol}_p(M) \times \Phi_p ,$$

where $\mathrm{Hol}_p(M)$ denotes the holonomy group of M at p and τ_γ is the parallel transport in M along γ. Note that, in general, Φ_p^c does not coincide with $\mathrm{Hol}_p M \times \Phi_p$. There is a natural action of Φ_p^c on tensor fields with tangent and normal variables. An important example is the action on the normal curvature tensor $R_p^\perp$ given by

$$(\bar\tau_\gamma \cdot R_p^\perp)(x, y)\xi = (\tau_\gamma^\perp)^{-1}(R_p^\perp)(\tau_\gamma x \, , \tau_\gamma y)\tau_\gamma^\perp \xi \, ,$$

for all $x, y \in T_p M$ and $\xi \in \nu_p M$.

REMARK 4.1.1 More generally, if γ is a path from p to another point $q \in M$, then parallel transport $\bar\tau_\gamma$ along γ (with respect to Levi Civita and normal connection of M) acts on $R_p^\perp$ as described above. □

b) Higher order mean curvatures

The existence of parallel tensor fields has often strong implications on the geometry of a manifold. In Riemannian geometry, a basic example of this situation is a classical result by Cartan: If the Riemannian curvature tensor of a Riemannian manifold M is parallel, then M is locally symmetric. Symmetric spaces play a prominent role in Riemannian geometry and are closely related to holonomy, as we will soon see.

In submanifold geometry, it is therefore natural to look for tensors that are parallel with respect to the normal connection. As we observed in Chapter 2, the second fundamental form, or equivalently the shape operator, is in some sense an analogue to the Riemannian curvature tensor. From the second fundamental form, one can derive both the Riemannian curvature tensor of the submanifold, via the Gauss equation, and the normal curvature tensor, via the Ricci equation. We will now introduce some symmetric tensor fields on the normal bundle of a submanifold, the so-called *higher order mean curvatures.*

Let M be an m-dimensional submanifold of a standard space form $\bar{M}^n(\kappa)$. The mean curvature $H_k(\xi)$ of order $k \in \{1, \dots, m\}$ in direction $\xi \in \nu M$ is defined as the k-th elementary symmetric function of the eigenvalues of A_ξ divided by the constant $\binom{m}{k}$. In terms of the principal curvatures $\lambda_1, \dots, \lambda_m$ of M with respect to ξ, counted with multiplicities, we have

$$H_k(\xi) = \frac{k!(m-k)!}{m!} \sum_{i_1 < \dots < i_k} \lambda_{i_1} \cdot \ldots \cdot \lambda_{i_k} .$$

Note that $H_1(\xi) = \langle H, \xi \rangle$, where H is the mean curvature vector field of M. We denote by $h_k(\xi_1, \dots, \xi_k)$ the real-valued symmetric tensor field on νM that is obtained by polarization of $H_k(\xi)$, that is,

$$h_k(\xi_1, \dots, \xi_k) = \frac{1}{k!} \sum_{\rho=1}^{k} (-1)^{k-\rho} \sum_{1 \le i_1 < \dots < i_\rho \le k} H_k(\xi_{i_1} + \dots + \xi_{i_\rho}) .$$

For example, $m(m-1)h_2(\xi_1, \xi_2) = \mathrm{tr} A_{\xi_1} \mathrm{tr} A_{\xi_2} - \mathrm{tr}(A_{\xi_1} A_{\xi_2})$.

Suppose that $h_1, \dots, h_m$, or equivalently $H_1, \dots, H_m$, are invariant under $\nabla^\perp$-parallel transport. Then $H_k(\xi(t))$ is constant for any parallel normal vector field $\xi(t)$ along any piecewise differentiable curve in M. Since the elementary symmetric functions of the eigenvalues of $A_{\xi(t)}$ are the coefficients of the characteristic polynomial of $A_{\xi(t)}$, this polynomial does not depend on t. Thus $A_{\xi(t)}$ has constant eigenvalues. Conversely, it is easy to see that if $A_{\xi(t)}$ has constant eigenvalues, then $h_1, \dots, h_m$, or equivalently $H_1, \dots, H_m$, are $\nabla^\perp$-parallel. A submanifold M with this property is called a *submanifold with constant principal curvatures.*

This class of submanifolds can be regarded as the analogue in submanifold geometry of locally symmetric spaces in Riemannian geometry, for reasons that we will discuss later. Important examples of submanifolds with constant principal curvatures are given by the *orbits of s-representations*, which play the same role in submanifold geometry as symmetric spaces play in Riemannian geometry, as we will illustrate later. It follows readily from our definition that a submanifold with constant principal curvatures has parallel mean curvature vector field.

An important subclass of the class of submanifolds with constant principal curvatures is given by those with flat normal bundle, which are called *isoparametric submanifolds.* These are somehow "generic" among submanifolds with constant principal curvatures. Indeed Heintze, Olmos and Thorbergsson [96] gave a complete characterization of submanifolds of standard space forms with constant principal curvatures as being either isoparametric or a focal manifold of an isoparametric submanifold (see Section 5.3).

REMARK 4.1.2 As we have seen, one can characterize a submanifold with constant principal curvatures geometrically by the property that, for any parallel normal vector field $\xi(t)$ along any piecewise differentiable curve, the shape operator $A_{\xi(t)}$ has constant eigenvalues. The advantage of the above definition in terms of higher order mean curvatures is that it shows that the constancy of the principal curvatures is a tensorial property. □

c) An illustration of normal holonomy: classification of surfaces with constant principal curvatures

We now illustrate the use of normal holonomy in a simple example: we classify surfaces in standard space forms with constant principal curvatures (see [205, Section 6]). As a first step, we show that we can reduce the problem to full surfaces. We will then analyze the possible cases according to the codimension. Indeed, an important general property of submanifolds with constant principal curvatures, which we will make use of now and in the sequel, is that one can always reduce their codimension to the dimension of the first normal space. This follows from the Theorem 2.5.1 on the reduction of codimension and the next

LEMMA 4.1.3
Let M be a submanifold of $\bar{M}^n(\kappa)$ with constant principal curvatures. Then, for each $p \in M$, the first normal space $\mathcal{N}_p^1$ is invariant under $\nabla^\perp$-parallel transport.

PROOF As we remarked in Section 2.5, the orthogonal complement $(\mathcal{N}_p^1)^\perp$ of $\mathcal{N}_p^1$ in $\nu_p M$ is given by $(\mathcal{N}_p^1)^\perp = \{\xi \in \nu_p M \mid A_\xi \equiv 0\}$. It follows immediately from the definition of submanifolds with constant principal curvatures that $(\mathcal{N}_p^1)^\perp$ is invariant under $\nabla^\perp$-parallel transport. But this implies that $\mathcal{N}_p^1$ is invariant under $\nabla^\perp$-parallel transport as well. □

In particular, we get that the codimension of a full submanifold with constant principal curvatures cannot exceed the maximal number $m(m+1)/2$ of linearly independent symmetric $m \times m$-matrices. In the case of surfaces, this shows that the codimension h of a surface M^2 in $\bar{M}^{2+h}(\kappa)$ with constant principal curvatures is, at most, three. We now investigate the three possible cases for the codimension h.

1. The case $h = 1$. In this case, the normal bundle νM has rank one and hence is flat. Thus M is an isoparametric hypersurface of $\bar{M}^3(\kappa)$. Then, by Exercise 4.6.12, M has parallel second fundamental form. Moreover, M is either totally umbilical or it is locally extrinsically reducible. In the latter case, M is locally an extrinsic product of two one-dimensional totally umbilical submanifolds with parallel mean curvature vector field (see for instance the classification of symmetric submanifolds in Section 3.7, part e).

2. The case $h = 2$. Here we will make use of the restricted normal holonomy group Φ^* of the surface. Since Φ^* is a connected Lie subgroup of $SO(2)$, we have two possibilities: Φ^* is either trivial or equal to $SO(2)$.

If Φ^*_p is trivial, then νM is flat and M is isoparametric with two distinct common eigenvalue functions. Note that M cannot be totally umbilical since the codimension of M is bigger than one. Let n_1 and n_2 be the curvature normals (see Exercise 4.6.9), which, by Exercise 4.6.11, span the first normal space. Then n_1 and n_2 are linearly independent, and locally there exists a parallel unit normal vector field ξ such that $\langle n_1, \xi\rangle = \langle n_2, \xi\rangle$. Therefore, A_ξ is a constant multiple of the identity, that is, ξ is an umbilical section. Since M is full and, by Theorem 2.5.1 on the reduction of codimension, ξ must be a multiple of the mean curvature vector field H, and hence M is pseudoumbilical. By Theorem 2.6.3, M is a minimal submanifold of some extrinsic sphere $\bar{M}^3(\kappa + \|H\|^2)$ in $\bar{M}^4(\kappa)$, and hence the problem is reduced to the previous case $h = 1$.

If $\Phi^*_p = SO(2)$, then Φ^*_p is transitive on the unit sphere in $\nu_p M$ for all $p \in M$. This implies that there exist constants $\bar{H}_k$ such that $H_k(\xi) = \bar{H}_k$, $k = 1, 2$, for all unit vectors $\xi \in \nu_p M$. In particular, $\bar{H}_1 = H_1(\xi) = \langle H, \xi\rangle = 0$ for ξ orthogonal to H. Thus $H = 0$, that is, M is a minimal submanifold. Moreover, as a reformulation of the condition on H_2, for any normal vectors ξ and η we have $\langle A_\xi, A_\eta\rangle = \mathrm{tr}(A_\xi A_\eta) = \mu^2\langle \xi, \eta\rangle$ for some $\mu \geq 0$. Now, either M is totally geodesic or $\mu > 0$. In the latter case, each shape operator A_ξ with respect to a unit normal vector field ξ has two distinct eigenvalues $\pm\mu$. We can then choose local orthonormal frame fields e_1, e_2 of TM and ξ_1, ξ_2 of νM such that A_{ξ_1} is represented by the matrix $\begin{pmatrix} \mu & 0 \\ 0 & -\mu \end{pmatrix}$. Suppose that A_{ξ_2} is represented by the matrix $\begin{pmatrix} b & d \\ d & -b \end{pmatrix}$. Since $\langle A_{\xi_1}, A_{\xi_2}\rangle = 0$ we get $b = 0$, and since $\langle A_{\xi_1}, A_{\xi_1}\rangle = \langle A_{\xi_2}, A_{\xi_2}\rangle$ we get $d = \pm\mu$. Without loss of generality we may assume that $d = \mu$, so that A_{ξ_2} has the form $\begin{pmatrix} 0 & \mu \\ \mu & 0 \end{pmatrix}$. In particular, from the equations of Gauss and Ricci we have that the sectional curvature of M is equal to $K = \kappa - 2\mu^2$, while the normal curvature $K^\perp = \langle R^\perp(e_1, e_2)\xi_1, \xi_2\rangle = 2\mu^2$. Next, using the Codazzi equation, we can see that $0 = \Delta \log \mu = 2K - K^\perp$, where Δ is the Laplace-Beltrami operator on M acting on functions by $\Delta f = \sum_i \nabla^2_{e_i e_i} f = \sum_i e_i e_i f - \sum_i \nabla_{e_i} e_i f$ (this is left as an exercise, cf. also [55, formula 3.5]). So $2\mu^2 = 2\kappa - 4\mu^2$, that is, $\kappa = 3\mu^2 > 0$. Hence M has the same second fundamental form as the Veronese surface $S^2(\kappa/\sqrt{3}) \to S^4(\kappa)$ (cf. Section 3.3). We now apply Theorem 3.9.3, which shows that M coincides locally with the Veronese surface $S^2(\kappa/\sqrt{3}) \to S^4(\kappa)$.

3. The case $h = 3$. If M is minimal, then the first normal space has dimension two, namely the dimension of the space of traceless 2×2-symmetric matrices. By assumption it must be equal to the codimension. It follows that M cannot be a minimal submanifold. Moreover, $\xi_1 = H/\|H\|$ is a global parallel unit normal vector field, and A_{ξ_1} commutes with all shape operators by the Ricci equation. If A_{ξ_1} has two distinct eigenvalues, then all shape operators are simultaneously diagonalizable and M is isoparametric, since it has flat normal bundle. This is a contradiction, since the number of distinct principal curvatures must be bigger or equal than the dimension of the first normal space (cf. Exercise 4.6.12). So A_{ξ_1} is a constant multiple of the identity, that is, M is pseudoumbilical. By Theorem 2.6.3, M is a minimal submanifold of some extrinsic sphere $\bar{M}^4(\kappa + ||H||^2)$ in $\bar{M}^5(\kappa)$ and $M \to \bar{M}^4(\kappa + ||H||^2)$ has constant principal curvatures. This reduces the problem to the previous case.

Altogether, we have now proved (cf. [205, Theorem 3 and Corollary 3]):

THEOREM 4.1.4
Let M be a full surface of a standard space form $\bar{M}^{2+h}(\kappa)$ with constant principal curvatures. Then $h \leq 3$, M has parallel second fundamental form, and M is locally one of the following surfaces:

(i) an isoparametric surface in $\bar{M}^3(\kappa')$ $(\kappa' \geq \kappa)$ that is either totally umbilical or locally an extrinsic product of two one-dimensional totally umbilical submanifolds with parallel mean curvature vector field;

(ii) the Veronese surface in a 4-sphere or the composition of the Veronese surface with a totally umbilical embedding of the 4-sphere in $\bar{M}^5(\kappa)$.

d) The geometry of an orbit of an s-representation

We have already shown in Section 3.6 that every principal orbit of an s-representation is isoparametric. We now prove that, more generally, every orbit of an s-representation is a submanifold with constant principal curvatures.

Let (G, K) be a Riemannian symmetric pair of noncompact type, $\mathfrak{g} = \mathfrak{k} \oplus \mathfrak{p}$ the corresponding Cartan decomposition of $\mathfrak{g}$, and $(\cdot, \cdot)$ the usual $\mathrm{Ad}(K)$-invariant inner product on $\mathfrak{g}$ that is induced from the Killing form and the Cartan involution of $\mathfrak{g}$. Let $0 \neq A_0 \in \mathfrak{p}$ and $M = \mathrm{Ad}(K) \cdot A_0$ the corresponding adjoint orbit, which is a real flag manifold. The isotropy subalgebra at A_0 is

$$\mathfrak{k}_{A_0} = \{Y \in \mathfrak{k} \mid [Y, A_0] = 0\} .$$

Let $\mathfrak{m}$ be the orthogonal complement with respect to $(\cdot, \cdot)$ of $\mathfrak{k}_{A_0}$ in $\mathfrak{k}$. Then $\mathfrak{k} = \mathfrak{k}_{A_0} \oplus \mathfrak{m}$ is a reductive decomposition of $\mathfrak{k}$. Recall that the tangent and normal space of M at A_0 are given by

$$T_{A_0}M = [\mathfrak{m}, A_0] \, , \; \nu_{A_0}M = \{\xi \in \mathfrak{p} \mid [\xi, A_0] = 0\} .$$

LEMMA 4.1.5
We have $[\mathfrak{m}, \nu_{A_0}M] \subset T_{A_0}M$.

PROOF For $Y \in \mathfrak{m}$ and $\xi, \zeta \in \nu_{A_0}M$ we have $[Y, \xi] \in \mathfrak{p}$ and $([Y, \xi], \zeta) = (Y, [\xi, \zeta])$. The Jacobi identity implies $[[\xi, \zeta], A_0] = [\xi, [\zeta, A_0]] - [\zeta, [\xi, A_0]] = 0$. Thus $[\xi, \zeta] \in \mathfrak{k}_{A_0}$, and hence $(Y, [\xi, \zeta]) = 0$. $\square$

Let γ be a piecewise differentiable curve in M with $\gamma(0) = A_0$. There exists a unique piecewise differentiable curve $k(t) \in K$ such that $k(0) = e$, $k^{-1}(t)k'(t) \in \mathfrak{m}$ and $\gamma(t) = \mathrm{Ad}(k(t))A_0$. Then the normal vector field $\hat{\xi}(t) = \mathrm{Ad}(k(t))\xi$ satisfies

$$\frac{d}{dt}\hat{\xi}(t) = \frac{d}{dt}\mathrm{Ad}(k(t))\xi = \mathrm{Ad}(k(t)[k^{-1}(t)k'(t), \xi] \in \mathrm{Ad}(k(t))[\mathfrak{m}, \nu_{A_0}M] ,$$

which is contained in $\mathrm{Ad}(k(t))T_{A_0}M = T_{\gamma(t)}M$ by Lemma 4.1.5. This shows that $\hat{\xi}$ is the $\nabla^{\perp}$-parallel transport of ξ along γ. Since the shape operators A_ξ and $A_{\mathrm{Ad}(k(t))\xi}$ are conjugate to each other (see, for instance, (3.4)), it follows that the principal curvatures of M with respect to $\hat{\xi}$ are constant along γ. Thus, we have proved

PROPOSITION 4.1.6
Every orbit of an s-representation is a submanifold with constant principal curvatures.

e) The normal holonomy of an orbit of an s-representation

It is well known that the holonomy representation of an irreducible Riemannian symmetric space coincides with the isotropy representation (see Section A.1). We now discuss an analogue for submanifold geometry that involves the orbits of s-representations and is due to Heintze and Olmos [95].

THEOREM 4.1.7
Let $N = G/K$ be a symmetric space of noncompact type with $G = I^o(N)$, $K = G_p$ and $p \in N$. Consider $0 \neq A_0 \in T_pN$ and the orbit $M = \mathrm{Ad}(K)\cdot A_0 \subset T_pN$ of the isotropy representation of G/K. Assume that M is full. Then the normal holonomy representation of M at A_0 is equal to the effectivized slice representation of A_0, that is, the effectivized action of the isotropy group K_{A_0} on the normal space $\nu_{A_0}M$.

PROOF In Section 6.2, page 198, we will give an alternative proof using a description of normal holonomy of orbits in terms of projection of Killing vector fields (Theorem 6.2.7), and Lemma 4.1.5, which implies that normal Killing vector fields induced by elements in the orthogonal complement of the isotropy algebra vanish.

Here, we give the same proof as in [95], and we use the restricted root space decomposition associated with the symmetric space $N = G/K$, following the notation of Section 3.2. Let $\mathfrak{g}$ and $\mathfrak{k}$ be the Lie algebra of G and K, respectively, and $\mathfrak{g} = \mathfrak{k} \oplus \mathfrak{p}$ the Cartan decomposition of $\mathfrak{g}$ induced from $\mathfrak{k}$. Recall that we identify T_pN with $\mathfrak{p}$ and the isotropy representation with the adjoint representation $\mathrm{Ad}: K \to O(\mathfrak{p})$. The orbit $M = K \cdot A_0$ is diffeomorphic to K/K_{A_0} via $f : K/K_{A_0} \to M$, $kK_{A_0} \mapsto \mathrm{Ad}(k)A_0$. Note that K is connected since $N = G/K$ is simply connected and $G = I^o(N)$ is connected.

In Section 3.2, we related tangent and normal spaces of M to the restricted root space decomposition with respect to a maximal Abelian subspace of $\mathfrak{p}$ containing A_0. The tangent and normal spaces of M are respectively given by

$$T_{A_0}M = \mathfrak{p}_+ = \textstyle\sum_{\lambda\in\Delta_+} \mathfrak{p}_\lambda = ad(A_0)\mathfrak{k}_+ = [A_0, \mathfrak{k}_+] \ ,$$
$$\nu_{A_0}M = \mathfrak{p}_{A_0} = \{Z \in \mathfrak{p} \mid [Z, A_0] = 0\} \ .$$

An important observation now is that $M = K/K_0$ can be endowed with the normal homogeneous metric induced by the negative of the Killing form of $\mathfrak{g}$ restricted to $\mathfrak{k}$. Since this metric is naturally reductive, the curves in M determined by one-parameter subgroups,

$$c(t) = f\left((k\mathrm{Exp}tX)K_{A_0}\right) = \mathrm{Ad}(k\mathrm{Exp}tX)A_0 \ , \ X \in \mathfrak{k}_+$$

are geodesics in M. In general, however, this normal homogeneous metric on K/K_0 does not coincide (even up to a constant factor) with the induced metric on the immersed submanifold M. This actually happens if and only if M is a symmetric submanifold of $\mathfrak{p}$ (cf. [95]).

Next, we consider the ideal $\mathfrak{I}$ spanned by $\mathfrak{k}_+$, which coincides with $[\mathfrak{k}_+, \mathfrak{k}_+] + \mathfrak{k}_+$ (exercise). We shall need the following lemma that combines algebraic properties of the above decomposition with the geometric assumption that M is full in $\mathbb{R}^n \cong \mathfrak{p}$.

LEMMA 4.1.8
The ideal $\mathfrak{I}$ spanned by $\mathfrak{k}_+$ is equal to $\mathfrak{k}$.

PROOF Let $\mathfrak{I}^\perp$ be the orthogonally complementary ideal of $\mathfrak{I}$ in $\mathfrak{k}$ (here $\mathfrak{k}$ is compact). We will see that $[\mathfrak{I}^\perp, v] = 0$ for all $v \in T_{A_0}M = [\mathfrak{k}_+, A_0]$. Indeed, by the Jacobi identity, $[\mathfrak{I}^\perp, [\mathfrak{k}_+, A_0]] = 0$, since $[\mathfrak{I}^\perp, \mathfrak{k}_+] = 0$ and $\mathfrak{I}^\perp \subseteq \mathfrak{k}_0$ (so $[\mathfrak{I}^\perp, A_0] = 0$). This implies that, for any $t_0 \in \mathbb{R}$ and $X \in \mathfrak{I}^\perp$, $\mathrm{Ad}(\mathrm{Exp}\, t_0X)$ determines an isometry of $M = \mathrm{Ad}(K)A_0$ such that $\mathrm{Ad}(\mathrm{Exp}\, t_0X)A_0 = A_0$ and $(\mathrm{Ad}(\mathrm{Exp}\, t_0X))_{*A_0} = \mathrm{id}$. Thus, $\mathrm{Ad}(\mathrm{Exp}\, t_0X)$ is the identity on M. Since the fixed points in $\mathfrak{p}$ of $\mathrm{Ad}(\mathrm{Exp}\, t_0X)$ form a linear subspace and M is full (and contained in a sphere), it follows that $\mathrm{Ad}(\mathrm{Exp}\, t_0X)$ is the identity on $\mathfrak{p}$ for each t_0. Then, since (G, K) is effective, $\mathrm{Exp}\, tX = e$ and $X = 0$. ∎

The parallel transport in the normal bundle of M along the geodesics of K/K_0 given by $c(t) = \mathrm{Ad}(k\mathrm{Exp}tX)A_0$ is determined by the action of the corresponding one-parameter subgroup. Explicitly, for any $\xi \in \mathfrak{p}_{A_0} = \nu_{A_0}M$ and $X \in \mathfrak{k}_+$ the vector field $\xi(t) = \mathrm{Ad}(k\mathrm{Exp}tX)\xi$ is $\nabla^\perp$-parallel along $c(t) = \mathrm{Ad}(k\mathrm{Exp}tX)A_0$. This can be seen using the same arguments as in the proof of Proposition 4.1.6. So parallel transport along any broken geodesic is given by the differential of the action at some point in K. Any broken geodesic starting at eK_{A_0} can be written as

$$\mathrm{Exp}x_1 \dots \mathrm{Exp}x_{i-1}\mathrm{Exp}(t - t_{i-1})X_i \ , t \in [t_{i-1}, t_i] \ , \ i = 1, \dots, r \ ,$$

where $X_i \in \mathfrak{k}_+$ and $x_i = (t_i - t_{i-1})X_i$. Then

$$I = \{\mathrm{Exp}x_1 \dots \mathrm{Exp}x_r \mid r \in \mathbb{N}, x_i \in \mathfrak{k}_+\}$$

is a Lie subgroup of K, since it is arcwise connected. Moreover, the Lie algebra of I coincides with $\mathfrak{I}$ (exercise). By Lemma 4.1.8 we get $I = K$, since both I and K are connected. Since any curve in M can be approximated by broken geodesics in M (and K_0 is closed), it follows that K_{A_0} acts on $\nu_{A_0}M$ as the normal holonomy group. □

f) Normal holonomy and normal curvature tensor

An important fact about holonomy groups is that the holonomy algebra is related to the curvature tensor. We briefly discuss this fact for the normal holonomy group, but everything in this paragraph holds in full generality for the holonomy of any metric connection on a vector bundle.

Example 4.1
We first examine a special case to get a feeling for this relation.

Suppose that the normal bundle is flat, that is, $R^\perp = 0$. Then it is not hard to show that any normal vector has the same $\nabla^\perp$-parallel transport along homotopic curves (with fixed endpoints). This clearly implies that *if $R^\perp = 0$ then the restricted normal holonomy group is trivial*, or equivalently, that the normal holonomy algebra $L(\Phi_p)$ is trivial. Thus the normal curvature tensor can be regarded as an obstruction for the restricted holonomy group to be trivial.

To see this more precisely, let us consider homotopic paths γ_0 and γ_1, both starting at p and ending at q. Suppose $H(t,s) = \gamma_s(t)$ is a piecewise smooth homotopy. Fix s, take a normal vector $\xi \in \nu_p M$, and let $t \to \xi_{t,s}$ be its $\nabla^\perp$-parallel transport along γ_s. We will show that $\xi_{1,s}$ does not depend on s. Since $\xi_{t,s}$ is $\nabla^\perp$-parallel along γ_s, $\frac{D^\perp}{\partial t}\xi_{t,s} = 0$ and $\frac{D^\perp}{ds}\xi_{0,s} = 0$, because $\xi_{0,s} = \xi$. Thus $\frac{D^\perp}{\partial s}\frac{D^\perp}{\partial t}\xi_{t,s} = 0$, and since $R^\perp = 0$ this yields $\frac{D^\perp}{\partial t}\frac{D^\perp}{\partial s}\xi_{t,s} = 0$. So $\frac{D^\perp}{\partial s}\xi_{t,s}$ is $\nabla^\perp$-parallel along γ_s. Now, the value at $t = 0$ of the parallel vector field $\frac{D^\perp}{\partial s}\xi_{t,s}$ is $\frac{D^\perp}{ds}\xi_{0,s} = \frac{d}{ds}\xi = 0$. Thus $\frac{D^\perp}{\partial s}\xi_{t,s} = 0$. In particular $\xi_{1,s}$ is constant (observe that $\frac{D^\perp}{ds}\xi_{1,s} = \frac{d}{ds}\xi_{1,s}$). □

In general, one can show that the normal curvature tensor endomorphisms always belong to the normal holonomy algebra (see [22, 10.52, p. 290]). This can be seen as a consequence of the following simple formula relating the curvature tensor with parallel transport. Let $u, v \in T_pM$, $\xi \in \nu_p M$. Construct a parametrized surface $f(s,t)$ on M such that $f(0,0) = p$, $f_s(0,0) = u$ and $f_t(0,0) = v$.

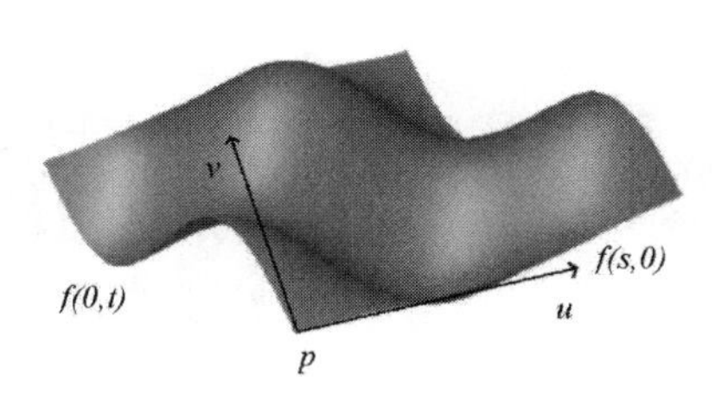

FIGURE 4.1: A parametrized surface $f(s,t)$ on M.

We now move along coordinate lines from $f(0,0)$ to $f(t,0)$, then to $f(t,t)$, then backward to $f(0,t)$ and finally to $f(0,0)$ again. In this way, a loop γ_t is defined. Let $\tau^{\perp}_{\gamma_t}\xi$ be the $\nabla^{\perp}$-parallel displacement of ξ along this loop. Then, by Exercise 4.6.2, $\frac{d}{dt}_{|t=0}\tau^{\perp}_{\gamma_t}\xi = 0$ and

$$R^{\perp}(u,v)\xi = -\frac{1}{2}\frac{d^2}{dt^2}_{|t=0}\tau^{\perp}_{\gamma_t}\xi \ .$$

As a linear operator, or more precisely as an element of $\mathfrak{so}(\nu_p M)$,

$$R^{\perp}(u,v) = -\frac{1}{2}\frac{d^2}{dt^2}_{|t=0}\tau^{\perp}_{\gamma_t} \ .$$

Note that $\tau^{\perp}_{\gamma_t}$ is a curve in the normal holonomy group Φ_p that is the identity at $t = 0$ and it has zero velocity. Hence, the second derivative $\frac{d^2}{dt^2}_{|t=0}\tau^{\perp}_{\gamma_t} = -2\,R^{\perp}(u,v)$ makes sense and belongs to the Lie algebra of the normal holonomy group $L(\Phi_p)$. Moreover, if γ is a curve from p to q and $x, y \in T_qM$, then $(\tau^{\perp}_{\gamma})^{-1}R^{\perp}(x,y)\tau^{\perp}_{\gamma} \in L(\Phi_p)$.

In a suitable sense, the holonomy algebra is spanned by the curvature tensors produced in this way. This is due to a remarkable result of Ambrose and Singer (which holds for the holonomy of any linear connection on a vector bundle; see, for instance, [117] or [22], [189]).

THEOREM 4.1.9 Ambrose-Singer Holonomy Theorem
Let M be a connected submanifold of a standard space form $\bar{M}^n(\kappa)$. The normal holonomy algebra $L(\Phi_p)$ is the subalgebra of $\mathfrak{so}(\nu_p M)$ generated by the endomorphisms

$$\tau^{\perp\,-1}_{\gamma}R^{\perp}(x,y)\tau^{\perp}_{\gamma} \ ,$$

where $\tau^{\perp}_{\gamma}$ is the $\nabla^{\perp}$-parallel transport on M along a piecewise differentiable curve $\gamma : [0,1] \to M$ starting from p and $x, y \in T_{\gamma(1)}M$.

For a proof of this theorem see Exercise 8.6.4.

REMARK 4.1.10 The Ambrose-Singer Holonomy Theorem implies that the Lie algebra of Φ_p can be also recovered as linear span of

$$\{\tau^{\perp\,-1}_{\gamma}R^{\perp}(\tau_{\gamma}x, \tau_{\gamma}y)\tau^{\perp}_{\gamma}\}$$

where τ_γ is the tangential parallel transport on M along a piecewise differentiable curve $\gamma : [0,1] \to M$ starting from p and $x, y \in T_pM$ (actually, the tangential connection is irrelevant). $\Box$

For the local study of the geometry of submanifolds we will use the *local normal holonomy group*. The local normal holonomy group Φ_p^{loc} at p is defined as intersection of all normal holonomy groups $\Phi_p^*(U)$, where U runs through all open neighbourhoods of p. Observe that there always exists an open neighbourhood V of p such that the normal holonomy group of V at p coincides with Φ_p^{loc}, and the same is true for smaller neighbourhoods of p, so we can assume that V is diffeomorphic to an open ball. If the dimension of Φ_p^{loc} is constant, then $\Phi_p^{loc} = \Phi_p^*$.

A property that will be useful in the sequel is the following (see [82, Appendix] for a proof).

PROPOSITION 4.1.11
If $\varepsilon > 0$ is sufficiently small, then parallel transport along loops of length at most ε contains a neighbourhood of the identity of Φ_p^{loc} and belongs to Φ_p^{loc}.

4.2 The Normal Holonomy Theorem

We begin with some motivating facts about holonomy of Riemannian manifolds. The two fundamental results regarding the restricted holonomy group of a Riemannian manifold are the *de Rham Decomposition Theorem* and the *Berger Holonomy Theorem*. Both have local and global versions, and we will refer only to the first one. De Rham's Theorem asserts that a Riemannian manifold M is locally irreducible around p if and only if its local holonomy group acts irreducibly on T_pM. Berger's Theorem says that if M is irreducible around p and not locally symmetric, then the restricted holonomy group $\mathrm{Hol}_p^o(M)$ acts transitively on the unit sphere in T_pM. In particular, one has the following property: *for each $p \in M$ there exist a unique (up to order) orthogonal decomposition $T_pM = V_0 \oplus \ldots \oplus V_k$ of T_pM into $\mathrm{Hol}_p^o(M)$-invariant subspaces $V_1, \ldots, V_k$ and normal subgroups $G_0, \ldots, G_k$ of $\mathrm{Hol}_p^o(M)$ such that*

(i) $\mathrm{Hol}_p^o(M) = G_0 \times \ldots \times G_k$ *(direct product).*

(ii) G_i *acts trivially on* V_j *if* $i \neq j$.

(iii) $G_0 = \{1\}$ *and, if* $i \geq 1$, G_i *acts transitively on the unit sphere in* T_pM *or it acts irreducibly on* V_i *as the isotropy representation of a simple Riemannian symmetric space.*

We call this result the *algebraic de Rham-Berger Theorem.*

A key property for dealing with holonomy groups of Riemannian manifolds is that the holonomy algebra is generated by algebraic curvature tensors, that is, linear tensors satisfying all the algebraic identities of a curvature tensor. More precisely, the holonomy algebra of a Riemannian manifold is generated by endomorphisms of the form

$$\gamma^*(R)(x,y) = (\tau_\gamma)^{-1} R(\tau_\gamma x, \tau_\gamma y)\tau_\gamma$$

where $x, y \in T_pM$, $\gamma : [0,1] \to M$ is a piecewise differentiable curve starting at p, and τ_γ denotes parallel transport along γ.

It is surprising that, for the normal connection of a submanifold of a standard space form, the algebraic de Rham-Berger Theorem holds in a simpler version.

THEOREM 4.2.1 Normal Holonomy Theorem [173]
Let M be a connected submanifold of a standard space form $\bar{M}^n(\kappa)$. Let $p \in M$ and let Φ^* be the restricted normal holonomy group at p. Then Φ^* is compact, there exists a unique (up to order) orthogonal decomposition $\nu_pM = V_0 \oplus \ldots \oplus V_k$ of the normal space ν_pM into Φ^*-invariant subspaces and there exist normal subgroups $\Phi_0, \ldots, \Phi_k$ of Φ^* such that

(i) $\Phi^* = \Phi_0 \times \ldots \times \Phi_k$ *(direct product).*

(ii) Φ_i *acts trivially on* V_j *if* $i \neq j$.

(iii) $\Phi_0 = \{1\}$ *and, if* $i \geq 1$, Φ_i *acts irreducibly on* V_i *as the isotropy representation of an irreducible Riemannian symmetric space.*

This result is an important tool for the study of the geometry of submanifolds of space forms. Indeed, in Section 4.4, we will construct so-called holonomy tubes, many properties of which rely on the Normal Holonomy Theorem. Holonomy tubes have many applications, as we will see later.

We first sketch the proof by listing the ingredients of the Normal Holonomy Theorem recipe, before we provide the arguments in full detail. We will define a tensor field

$$\mathcal{R}^\perp : \otimes^3 \nu M \to \nu M$$

that contains the same geometric information as the normal curvature tensor $R^\perp$, yet bears the algebraic properties of a Riemannian curvature tensor, that is, it is an algebraic curvature tensor (see, for instance, Section A.1). A very important property of $\mathcal{R}^\perp$ is that its scalar curvature is nonpositive and vanishes if and only if $\mathcal{R}^\perp$ vanishes. This implies that normal holonomy groups look like nonexceptional Riemannian holonomy groups.

The argument follows some ideas of Cartan and the methods used by Simons [200] in his proof of Berger's Theorem. One defines a holonomy system, that is, a triple $[V, R, G]$, where V is a Euclidean vector space, R an algebraic curvature tensor on V and G a compact Lie group acting effectively on V by isometries, such that R_{xy} belongs to the Lie algebra of G for all $x, y \in V$. Some reduction results allow us to concentrate on irreducible holonomy actions. Roughly speaking, we will take $G = \Phi_p^*$, $V = \nu_p M$ and $R = \mathcal{R}_p^\perp$. Note that, since a connected Lie subgroup of the orthogonal group acting irreducibly on a vector space is compact, one gets that Φ^* is compact.

A prominent role among holonomy systems is played by the so-called symmetric holonomy systems, which are strictly related to symmetric spaces and s-representations.

4.3 Proof of the Normal Holonomy Theorem

a) Holonomy systems

Let us consider an n-dimensional Euclidean vector space $(V, \langle \cdot, \cdot \rangle)$ and the real vector space $\mathcal{P}$ of all tensors of type $(1,3)$ on V. We identify such a tensor with a bilinear map $P : V \times V \to \mathrm{End}(V)$, $(x, y) \mapsto P_{x,y}$. Then the group $O(n)$ of isometries of V acts on $\mathcal{P}$ by

$$(g \cdot P)_{x,y} = g P_{g^{-1}(x), g^{-1}(y)} g^{-1} \ , \ g \in O(n) \ , x, y \in V \ .$$

By differentiation, we get an action of $\mathfrak{so}(n)$ on $\mathcal{P}$ by

$$(A \cdot P)_{x,y} = -P_{Ax,y} - P_{x,Ay} - [P_{x,y}, A] \ , \ A \in \mathfrak{so}(n) \ , \ x, y \in V \ .$$

Next, we recall the following definition (cf. Section A.1).

DEFINITION 4.3.1 *A tensor $R \in \mathcal{P}$ is called an algebraic curvature tensor if*

(1) $R_{xy} = -R_{yx}$

(2) $\langle R_{xy} z, w \rangle = -\langle R_{xy} w, z \rangle$

(3) $\langle R_{xy} z, w \rangle = \langle R_{zw} x, y \rangle$

(4) $R_{xy} z + R_{yz} x + R_{zx} y = 0$ *(algebraic or first Bianchi identity)*

Note that every linear combination of algebraic curvature tensors is again an algebraic curvature tensor. If R is an algebraic curvature tensor, $g \in SO(n)$ and $A \in \mathfrak{so}(n)$, then $g \cdot R$ and $A \cdot R$ are algebraic curvature tensors as well. Thus, the real vector space of all algebraic curvature tensors on V is an $SO(n)$-module.

Associated to every algebraic curvature tensor R is its *scalar curvature* (cf. Appendix)

$$k(R) = 2 \sum_{i<j} \langle R_{e_i e_j} e_j, e_i \rangle \ ,$$

where $e_1, \ldots, e_n$ is an orthonormal basis of V. Note that $k(g \cdot R) = k(R)$ holds for all $g \in O(n)$.

DEFINITION 4.3.2 *Let R be an algebraic curvature tensor on V. A compact subgroup G of $O(n)$ is called a holonomy group of R if $R_{xy} \in \mathfrak{g}$ for all $x, y \in V$, where $\mathfrak{g}$ denotes the Lie algebra of G.*

If $R_{xy} \in \mathfrak{g}$ holds for all $x, y \in V$, then $(g \cdot R)_{xy} \in \mathfrak{g}$ and $(A \cdot R)_{xy} \in \mathfrak{g}$ holds for all $g \in G$, $A \in \mathfrak{g}$ and $x, y \in V$. Thus, if G is a holonomy group for R, it is also a holonomy group for $g \cdot R$ and $A \cdot R$ for all $g \in G$ and $A \in \mathfrak{g}$.

DEFINITION 4.3.3 *A triple $S = [V, R, G]$, where V is an Euclidean vector space, R an algebraic curvature tensor on V and G a connected holonomy group of R is called a holonomy system.*

The definition of holonomy systems is motivated by the fact that, on a Riemannian manifold M with Riemannian curvature tensor R, the elements $R(x, y)$, $x, y \in T_pM$, lie in the holonomy algebra at p. Thus, *if M is a Riemannian manifold, then $[T_pM, R_p, \mathrm{Hol}_p^o(M)]$ is a holonomy system for each $p \in M$.*

b) Symmetric holonomy systems and holonomy of symmetric spaces

We now turn to a fundamental class of holonomy systems that are strictly related to symmetric spaces.

DEFINITION 4.3.4 *A holonomy system $S = [V, R, G]$ is called symmetric if $g \cdot R = R$ for all $g \in G$, or equivalently, if $A \cdot R = 0$ for all $A \in \mathfrak{g}$.*

In other words, a holonomy system is symmetric if its algebraic curvature tensor is G-invariant. The definition of symmetric holonomy systems is motivated by Cartan's theory of symmetric spaces. Indeed, given a symmetric holonomy system $S = [V, R, G]$, one can carry out the following construction due to Cartan:

Consider the real vector space $\mathfrak{L} := \mathfrak{g} \oplus V$ and define on it a bilinear skewsymmetric map $[\cdot, \cdot] : \mathfrak{L} \times \mathfrak{L} \to \mathfrak{L}$ by

$$[A, B] = [A, B]_{\mathfrak{g}} \ , \ [x, y] = R_{xy} \ , \ [A, x] = Ax \ , \ A, B \in \mathfrak{g} \ , \ x, y \in V \ ,$$

where $[\cdot, \cdot]_{\mathfrak{g}}$ denotes the Lie algebra structure on $\mathfrak{g}$. It turns out that $[\cdot, \cdot]$ defines a Lie algebra structure on $\mathfrak{L}$, that is, it satisfies the Jacobi identity. To verify this, note first that the only nontrivial case occurs when two elements are in V and one element is in $\mathfrak{g}$. Indeed, if all three elements are in V, the Jacobi identity is just the algebraic Bianchi identity for R, and if $A, B \in \mathfrak{g}$ and $x \in V$ we have

$$[[A, B], x] + [[B, x], A] + [[x, A], B] = (AB - BA)x - A(Bx) + B(Ax) = 0 \ .$$

For $A \in \mathfrak{g}$ and $x, y \in V$, the definition of symmetric holonomy systems implies

$$\begin{aligned}&[A,[x,y]] + [y,[A,x]] + [x,[y,A]] \\ &= [A, R_{x,y}] - R_{y,Ax} - R_{x,Ay} = (A \cdot R)_{xy} = 0 \,.\end{aligned}$$

From the very definition, it follows that the *Cartan relations*

$$[\mathfrak{g},\mathfrak{g}] \subset \mathfrak{g} \,,\; [\mathfrak{g}, V] \subset V \,,\; [V,V] \subset \mathfrak{g}$$

hold. Then $\mathfrak{L}$ corresponds to a Riemannian symmetric space M, whose tangent space at a point p can be identified with V, whose curvature tensor at p is R and whose holonomy algebra at p is $\mathfrak{g}$. To construct M explicitly, consider the involutive automorphism

$$\sigma : \mathfrak{L} = \mathfrak{g} \oplus V \to \mathfrak{L} = \mathfrak{g} \oplus V \,,\; A + x \mapsto A - x \,.$$

The pair $(\mathfrak{L}, \sigma)$ is an orthogonal symmetric Lie algebra, since G is compact and hence $\mathfrak{g}$ is a compact Lie algebra. If L is a simply connected Lie group with Lie algebra $\mathfrak{L}$ and G is the connected closed subgroup of L with Lie algebra $\mathfrak{g}$, then (L, G) is a Riemannian symmetric pair and $M = L/G$ is a simply connected Riemannian symmetric space (cf. [99, page 213]).

Let $\pi : L \to M$ be the canonical projection and $p = \pi(G) = eG \in M$. Let R_p be the Riemannian curvature tensor of M at p. Then

$$R_p(x,y)z = -[[x,y],z] \,,\; x,y,z \in T_pM \,.$$

Using the definition of $[\cdot,\cdot]$ on $\mathfrak{L}$, we have

$$R_p(x,y)z = -[[x,y],z] = -[-R_{xy}, z] = R_{xy}z \,,$$

and hence $R_p = R$. Note also that, by the Ambrose Singer Holonomy Theorem, the holonomy algebra of M at p is generated by the endomorphisms $\tau_\gamma^{-1} R_{\tau_\gamma x, \tau_\gamma y} \tau_\gamma$, where γ is any piecewise differentiable curve in M with $\gamma(0) = p$ and τ_γ denotes parallel transport along γ. Since S is a symmetric holonomy system, the holonomy algebra coincides with

$$\mathrm{span}\{R_{xy} \mid x, y \in T_pM\} = \mathrm{im}(R) = [V,V] \,.$$

Hence *a symmetric holonomy system $S = [V, R, G]$ with $R \neq 0$ corresponds to a simply connected symmetric space $M = L/G$ in such a way that R identifies with the Riemannian curvature tensor of M at $p = eG$.*

Note that the restriction of the Killing form $B^{\mathfrak{L}}$ of $\mathfrak{L}$ to $\mathfrak{g}$ is given by

$$\begin{aligned}B^{\mathfrak{L}}(A,B) &= \mathrm{tr}^{\mathfrak{g}}(\mathrm{ad}(A) \circ \mathrm{ad}(B)) + \mathrm{tr}^V(\mathrm{ad}(A) \circ \mathrm{ad}(B)) \\ &= B^{\mathfrak{g}}(A,B) + \frac{1}{n-2} B^{\mathfrak{so}(n)}(A,B)\end{aligned}$$

for all $A, B \in \mathfrak{g}$.

Now suppose that the action of G on V is irreducible. In this case, one says that the holonomy system is *irreducible*. This obviously implies that M is irreducible, since by the de Rham Decomposition Theorem (Appendix, page 290) this is equivalent to the holonomy acting irreducibly on $T_pM = V$.

Without loss of generality we can also assume that G acts faithfully on V (so $\mathfrak{L}$ is effective, that is, $\mathfrak{g}$ does not contain a nontrivial ideal of $\mathfrak{L}$). Using the lemma of Schur, one can show that $\mathfrak{L}$, and thus also L, are *semisimple*, i.e. $B^{\mathfrak{L}}$ is nondegenerate. Indeed, any effective orthogonal symmetric algebra $\mathfrak{L} = \mathfrak{g} \oplus V$ with $\mathfrak{g}$ acting irreducibly on V and $[V, V] \neq 0$ is the direct sum of, at most, two simple ideals [117, Chapter 11, Proposition 7.5]. Thus, the restriction of $B^{\mathfrak{L}}$ to V, which is also nondegenerate, must equal a nonzero constant multiple $1/c$ of the inner product on V (via the identification of V and T_pM), since both bilinear forms are G-invariant. Hence,

$$\langle R_{xy}z, w\rangle = cB^{\mathfrak{L}}([[x, y], z], w) = cB^{\mathfrak{L}}([x, y], [z, w]) \,. \tag{4.1}$$

Thus, we obtain the following result, due to Kostant, which allows us to read off the curvature tensor of the irreducible symmetric space corresponding to the orthogonal symmetric algebra $\mathfrak{L} = \mathfrak{g} \oplus V$ from the Lie algebras of $\mathfrak{g}$ and V alone (see also [196]).

THEOREM 4.3.5

The curvature tensor of the irreducible Riemannian symmetric space M corresponding to the orthogonal symmetric algebra $\mathfrak{L} = \mathfrak{g} \oplus V$ has the form

$$\langle R_{xy}z, w\rangle = cB^{\mathfrak{L}}([x, y], [z, w]) = c\left(B^{\mathfrak{g}} + \frac{1}{n-2}B^{\mathfrak{so}(n)}\right)([x, y], [z, w]) \;, \tag{4.2}$$

and M is Einstein with nonzero scalar curvature.

REMARK 4.3.6 This result implies that, if $S = [V, R, G]$ and $S' = [V, R', G]$ are two irreducible symmetric holonomy systems with $\dim V \geq 2$, then $R' = \lambda R$ for some constant λ. □

Observe now that $[V, V] = \mathfrak{hol} = \mathfrak{g}$, where $\mathfrak{hol}$ is the holonomy algebra. Indeed, since we are assuming that $\mathfrak{L}$ is effective, so that $\mathfrak{L}$ is semisimple, the orthogonal complement of $[V, V] \oplus V$ with respect to $B^{\mathfrak{L}}$ is an ideal of $\mathfrak{L}$ contained in $\mathfrak{g}$, and hence it vanishes. So:

The holonomy algebra of M coincides with $\mathfrak{g}$ and *an irreducible symmetric holonomy system $S = [V, R, G]$ corresponds to an s-representation with*

$$G = \text{holonomy group} = \text{isotropy group of the symmetric space} .$$

The following remarkable result due to Simons [200] is the essential tool in Simons' proof of Berger's Theorem. It enables one to prove that an irreducible Riemannian manifold with nontransitive holonomy group is locally symmetric. Nevertheless, we choose not to provide the long technical proof, which is not needed for proving the Normal Holonomy Theorem.

THEOREM 4.3.7 (J. Simons)
Let $[V, R, G]$ be an irreducible holonomy system. If G is not transitive on the unit sphere in V, then $[V, R, G]$ is symmetric.

REMARK 4.3.8 If M is a locally irreducible Riemannian manifold and $p \in M$ satisfies $R_p \neq 0$, then $[T_pM, \mathrm{Hol}_p^o(M), R]$ is an irreducible holonomy system. From the above theorem and standard facts, Simons provided a conceptual proof of the following result (which was obtained 7 years earlier, using classification results, by Berger [10] and is therefore known as *Berger's Theorem*): *Let M be a Riemannian manifold that is locally irreducible at some point p. If the holonomy group is not transitive on the unit sphere in T_pM, then M is locally symmetric.*

The (restricted) holonomy group of a locally irreducible Riemannian manifold is called *nonexceptional* if there exists a symmetric space with an isomorphic holonomy group. By Berger's Theorem, exceptional holonomies are always transitive on the unit sphere of the tangent space. If the scalar curvature of M is not identically zero, then the holonomy is nonexceptional (this is a consequence of Lemma 4.3.14 that will be stated later). $\square$

From Theorem 4.3.7 we deduce the following important property of s-representations.

PROPOSITION 4.3.9
Let K and K' be two irreducible s-representations on V, $\dim V \geq 2$, that are not transitive on the unit sphere in V. If K and K' have the same orbits, then $K = K'$ and the s-representations coincide.

PROOF The group $\tilde{K}$ generated by K and K' is not transitive on the unit sphere in V, since it has the same orbits as K and K'. Let R and R' be the curvature tensors corresponding to the s-representations of K and K'. Then $[V, R, \tilde{K}]$ is an irreducible nontransitive holonomy system. By Theorem 4.3.7, $[V, R, \tilde{K}]$ is symmetric and hence the Lie algebra of $\tilde{K}$ is generated by R. Then $K = \tilde{K}$ and, by Remark 4.3.6, $R' = \lambda R$. $\square$

c) Normal curvature tensor and proof of the Normal Holonomy Theorem

Let M be a submanifold of a standard space form $\bar{M}^n(\kappa)$. We start with the normal curvature tensor $R^\perp$ (at a point $p \in M$) in order to construct a holonomy system. For each $x, y \in T_pM$ the endomorphism $R^\perp(x, y)$ lies in the normal holonomy algebra. But it follows from the very definition that $R^\perp$ is not an algebraic curvature tensor. To construct a tensor field of type $(1, 3)$ on the normal bundle of

M we consider $R^\perp$ as a homomorphism

$$R^\perp : \Lambda^2 T_p M \to \Lambda^2 \nu_p M \ .$$

The homomorphism $R^\perp$ composed with its adjoint homomorphism $R^{\perp *}$ gives rise to an endomorphism

$$\mathcal{R}^\perp = R^\perp \circ R^{\perp *} : \Lambda^2 \nu_p M \to \Lambda^2 \nu_p M \ ,$$

which can be identified with a $(1,3)$-tensor on $\nu_p M$.

The Ricci equation $\langle R^\perp(x,y)\xi, \eta\rangle = \langle [A_\xi, A_\eta]x, y\rangle$ implies

$$R^{\perp *}(\xi \wedge \eta) = [A_\xi, A_\eta] \ ,$$

hence

$$\begin{aligned} \langle \mathcal{R}^\perp(\xi_1,\xi_2)\xi_3,\xi_4\rangle &= \langle R^{\perp *}(\xi_1 \wedge \xi_2), R^{\perp *}(\xi_3 \wedge \xi_4)\rangle \\ &= -\mathrm{tr}([A_{\xi_1}, A_{\xi_2}] \circ [A_{\xi_3}, A_{\xi_4}]) \ , \end{aligned} \tag{4.3}$$

since the inner product on $\Lambda^2 T_p M$ is given by $\langle A, B\rangle = -\mathrm{tr}(AB)$. Using (4.3) we get

LEMMA 4.3.10 [173]
$\mathcal{R}^\perp$ is an algebraic curvature tensor on $\nu_p M$.

PROOF We must verify that conditions (1)-(4) in Definition 4.3.1 hold. Now (1)-(3) are clear from (4.3). To verify the algebraic Bianchi identity (4), easy computations transform (4.3) into the form

$$\langle \mathcal{R}^\perp(\xi_1,\xi_2)\xi_3,\xi_4\rangle = 2\mathrm{tr}(A_{\xi_1}A_{\xi_2}A_{\xi_3}A_{\xi_4}) - 2\mathrm{tr}(A_{\xi_3}A_{\xi_1}A_{\xi_2}A_{\xi_4}) \ . \tag{4.4}$$

Cyclic sum over the indices $1, 2, 3$ yields the algebraic Bianchi identity. ☐

Note that (4.3) implies that $\mathcal{R}^\perp$ has nonpositive scalar curvature and its scalar curvature vanishes if and only if $\mathcal{R}^\perp$ is identically zero.

REMARK 4.3.11 The existence of this algebraic curvature tensor with nonpositive scalar curvature is of great importance for the structure theory of normal holonomy. It implies that normal holonomy groups behave like nonexceptional holonomy groups of Riemannian manifolds (see Remark 4.3.8) and simplifies the proof of the Normal Holonomy Theorem, compared with the one of Berger's Theorem. ☐

Using again the Ricci equation [173], an alternative expression for $\mathcal{R}^\perp$ is given by

$$\mathcal{R}^\perp(\xi_1,\xi_2)\xi_3 = 2\sum_{i=1}^{m}(R^\perp)(A_{\xi_1}e_i, A_{\xi_2}e_i)\xi_3 \ , \tag{4.5}$$

where $e_1, \ldots, e_m$ is an orthonormal basis of T_pM. Moreover, since $\ker R^\perp = (\mathrm{im} R^{\perp *})^\perp$, the image of $\mathcal{R}^\perp$ is the same as that of $R^\perp$ (cf. [173]). The Ambrose Singer Holonomy Theorem and the previous lemma therefore imply

LEMMA 4.3.12

Let M be a submanifold of a standard space form $\bar{M}^n(\kappa)$ and $p \in M$. Then the Lie algebra of the normal holonomy group is spanned by the tensors of the form $\gamma^* \mathcal{R}^\perp(\xi, \eta) = \tau_\gamma^{\perp -1} \mathcal{R}^\perp(\tau_\gamma^\perp \xi, \tau_\gamma^\perp \eta) \tau_\gamma^\perp$, where γ is any piecewise differentiable curve in M with p as endpoint.

We denote by $\mathcal{S}$ the real vector space of tensors on $\nu_p M$ that is spanned by all $\gamma^* \mathcal{R}^\perp$, where γ runs through all piecewise differentiable curves in M with p as endpoint. Since $\mathcal{R}^\perp$ is an algebraic curvature tensor, any $R \in \mathcal{S}$ has the algebraic properties of a curvature tensor.

Next, we define a holonomy system. At this stage, it is still not clear that Φ_p^* is compact, but this obstacle is avoided by splitting algebraically both $\nu_p M$ and the Lie algebra of Φ_p^* using the algebraic properties of the tensors in $\mathcal{S}$. What we achieve is an "algebraic" de Rham decomposition.

Explicitly, we decompose $\nu_p M$ into Φ_p^*-invariant orthogonal subspaces,

$$\nu_p M = V_0 \oplus \ldots \oplus V_k ,$$

where Φ_p^* acts trivially on V_0 and irreducibly on V_i for all $i \geq 1$. If ξ_i denotes the orthogonal projection of $\xi \in \nu_p M$ onto the factor V_i, the following holds for any $R \in \mathcal{S}$:

(a) $R(\xi_i, \xi_j) = 0$ if $i \neq j$

(b) $R(\xi, \eta) = \sum_i R(\xi_i, \eta_i)$

(c) $R(\xi_i, \eta_i) V_j = \{0\}$, if $i \neq j$

(d) $R(\xi_i, \eta_i) V_i \subseteq V_i$

These statements are all obtained from the properties of algebraic curvature tensors. As for (a), if $i \neq j$ and $\eta, \eta' \in \nu_p M$, then

$$\langle R(\xi_i, \xi_j)\eta, \eta' \rangle = \langle R(\eta, \eta')\xi_i, \xi_j \rangle = 0 ,$$

because $R(\eta, \eta') \in \mathfrak{g}$ and $\mathfrak{g}$ leaves V_i invariant. This readily implies (b). For (c), if $\zeta_j \in V_j$, the algebraic Bianchi identity together with (a) give

$$R(\xi_i, \eta_i)\zeta_j = -R(\eta_i, \zeta_j)\xi_i - R(\zeta_j, \xi_i)\eta_i = 0 .$$

Part (d) is clear.

We denote by $\mathfrak{g}_i$ is the real vector space spanned by

$$R(\xi_i, \eta_i) \ , \ \xi_i, \eta_i \in V_i \ , \ R \in \mathcal{S} .$$

The following lemma can be easily verified.

LEMMA 4.3.13
The following statements hold:

(i) $\mathfrak{g}_0 = \{0\}$ *and each* $\mathfrak{g}_i$, $i \geq 1$, *is an ideal of* $\mathfrak{g}$.

(ii) $\mathfrak{g} = \mathfrak{g}_1 \oplus \ldots \oplus \mathfrak{g}_k$ *and* $[\mathfrak{g}_i, \mathfrak{g}_j] = \{0\}$ *if* $i \neq j$.

(iii) $\mathfrak{g}_i \cdot V_i = V_i$ *for all* $i \geq 1$.

(iv) $\mathfrak{g}_i \cdot V_j = \{0\}$ *if* $i \neq j$.

(v) $\mathfrak{g}_i$ *acts irreducibly on* V_i *for all* $i \geq 1$.

Let now Φ_i be the connected Lie subgroup of Φ_p^* with Lie algebra $\mathfrak{g}_i$. Then

$$\Phi_p^* = \Phi_0 \times \ldots \times \Phi_k ,$$

with Φ_i acting trivially on V_j for all $i \neq j$ and irreducibly on V_i for all $i \geq 1$. Since a connected Lie group of orthogonal transformations on a vector space acting irreducibly is compact, each Φ_i is compact. Thus Φ_p^* is compact.

We can say more. For each $i \geq 1$ we can choose $R_i \in \mathcal{S}$ so that R_i is not identically zero on V_i. Then $[V_i, R_i, \Phi_i]$ is an irreducible holonomy system, R_i has nonvanishing scalar curvature (cf. Remark 4.3.11) and we can apply the following

LEMMA 4.3.14
Let G be a connected Lie subgroup of $SO(V)$ acting irreducibly on a Euclidean vector space V, and let R be an algebraic curvature tensor on V such that $R_{xy} \in \mathfrak{g}$ for all $x, y \in V$. If $k(R) \neq 0$, then G is compact, $S = [V, R, G]$ is an irreducible holonomy system, and G acts on V as an s-representation.

PROOF The compactness of G follows from a general result stating that a connected Lie subgroup of $SO(n)$ acting irreducibly on a vector space is compact (see, for instance, [117], vol 1, Appendix 5). Since G is compact, there exists a Haar measure on G and we can define

$$\hat{R} = \int_G g \cdot R .$$

The scalar curvature of $\hat{R}$ coincides with the one of R, since averaging does not change the scalar curvature (recall that $k(\hat{R}) = k(g \cdot \hat{R})$ for all $g \in G$). Moreover, $\hat{R} \neq 0$ because $k(\hat{R}) \neq 0$. Clearly, $g \cdot \hat{R} = \hat{R}$ for all $g \in G$, so $[V, \hat{R}, G]$ is a symmetric holonomy system. According to previous remarks, G acts on V as an s-representation. ∎

It follows from Lemma 4.3.14 that Φ_i acts on V_i as an s-representation for all $i \geq 1$. This concludes the proof of the Normal Holonomy Theorem 4.2.1.

We now give some geometric applications of the Normal Holonomy Theorem.

4.4 Some geometric applications of the Normal Holonomy Theorem

a) Parallel normal isoparametric sections

Let M be a submanifold of a standard space form $\bar{M}^n(\kappa)$.

DEFINITION 4.4.1 *A parallel normal vector field ξ of M is called a parallel normal isoparametric section if A_ξ has constant eigenvalues.*

In the sequel, we mainly deal with submanifolds of $\mathbb{R}^n$, a situation that includes submanifolds of spheres by regarding the latter as submanifolds of $\mathbb{R}^n$. Most of the results extend to submanifolds of the real hyperbolic space, the latter seen as Riemannian hypersurface of the Lorentzian space (cf. [179]). Therefore, we will concentrate on submanifolds of $\mathbb{R}^n$.

The constancy of the eigenvalues of the shape operator A_ξ is a tensorial property, for it is equivalent to the higher order mean curvatures in direction ξ being constant (cf. Remark4.1.2).

Recall also that, if M is full in $\mathbb{R}^n$, by Theorem 2.5.1 on the reduction of codimension, the only parallel umbilical isoparametric sections are constant multiples of the position vector field with respect to a suitable origin. This situation is, of course, trivial. Thus, we will be interested in nonumbilical isoparametric sections and we will actually see that the existence of one such section has strong consequences on the geometry of the submanifold. A first example in this direction is the following useful lemma.

Let ξ be a parallel isoparametric section, $\lambda_1, \ldots, \lambda_g$ the different (constant) eigenvalues of A_ξ and $TM = E_1 \oplus \ldots \oplus E_g$ the decomposition into the smooth eigendistributions of the shape operator A_ξ.

LEMMA 4.4.2
Let M be a submanifold of $\mathbb{R}^n$ and ξ a parallel normal isoparametric section.

(a) Each eigendistribution E_i of A_ξ is autoparallel and hence integrable with totally geodesic leaves.

(b) Each eigendistribution E_i is invariant under all shape operators of M.

(c) *If we denote by $S_i(q)$ the leaf of E_i through $q \in M$, then $S_i(q)$ is contained in the affine subspace $q + E_i(q) \oplus \nu_q M$ of $\mathbb{R}^n$, and $A^i_\xi = A_\xi | E_i(q)$ for all $\xi \in \nu_q M$, where A^i is the shape operator of the leaf $S_i(q)$, regarded as a submanifold of the affine subspace.*

PROOF (a): The statement is trivial if $g = 1$. Assume that $g \geq 2$. Let X, Y be sections in E_i and Z a section in E_j with $i \neq j$. Using the Codazzi equation we get

$$(\lambda_i - \lambda_j)\langle \nabla_X Y, Z\rangle = \langle (\nabla_X A_\xi)Y, Z\rangle = \langle (\nabla_Z A_\xi)X, Y\rangle = 0 \ .$$

Since this holds for all $i \neq j$ this implies that $\nabla_X Y$ is a section in E_i, which means that E_i is an autoparallel subbundle of TM.

(b): Since $\nabla^\perp \xi = 0$, we have $R^\perp(X,Y)\xi = 0$, and the Ricci equation implies that $[A_\xi, A_\eta] = 0$ for all normal vector fields η. Thus, all shape operators commute with A_ξ and consequently preserve the eigendistributions.

(c) The statement is trivial if $g = 1$. Assume that $g \geq 2$. Let $c : [0,1] \to S_i(q)$ be a smooth curve with $c(0) = q$, v a normal vector of $S_i(q) \subset M$ at q, and V the parallel normal vector field along c with respect to the normal connection of $S_i(q) \subset M$. Since $\alpha(c'(t), V(t)) = 0$ and $S_i(q)$ is totally geodesic in M, the covariant derivative of V with respect to the Levi Civita derivative on $\mathbb{R}^n$ vanishes. This implies that $S_i(q)$ is contained in the affine space $q + E_i(q) \oplus \nu_q M$. The statement about the shape operator A^i follows from the fact that $S_i(q)$ is totally geodesic in M. □

b) Parallel and focal manifolds

Let M be a submanifold of $\mathbb{R}^n$ and assume the existence of a nontrivial parallel normal vector field ξ (which we will view as a smooth map $M \to \mathbb{R}^n$). We define a smooth map

$$t_\xi : M \to \mathbb{R}^n \ , \ x \mapsto x + \xi(x) \ .$$

The differential of t_ξ at x has the same rank as the matrix $\mathrm{id} - A_{\xi(x)}$. We denote by $\nu(x)$ the dimension of the kernel of $\mathrm{id} - A_{\xi(x)}$. Then $x + \xi(x)$ is a focal point of M in direction ξ if $\nu(x) > 0$. In this case, $\nu(x)$ is the multiplicity of the focal point. If $\nu(x) \equiv \nu$ is constant, this happens, for instance, if ξ is an isoparametric section, then the image

$$M_\xi = \{x + \xi(x) \mid x \in M\}$$

of t_ξ is an immersed submanifold of $\mathbb{R}^n$ of dimension $\dim M - \nu$ (cf. [44], [191]). If $\nu(x) \equiv \nu > 0$, then M_ξ is called a *focal (or parallel focal) manifold of M in direction ξ*. If $\nu(x) \equiv 0$, or equivalently, if t_ξ is a regular map, then M_ξ is called a *parallel manifold to M in direction ξ*.

In this chapter, we deal with local properties and hence we can assume that M_ξ is embedded. Global properties will be studied in Section 5.5 (see in particular Exercises 5.6.6, 5.6.7 and 5.6.8) and Chapter 6.

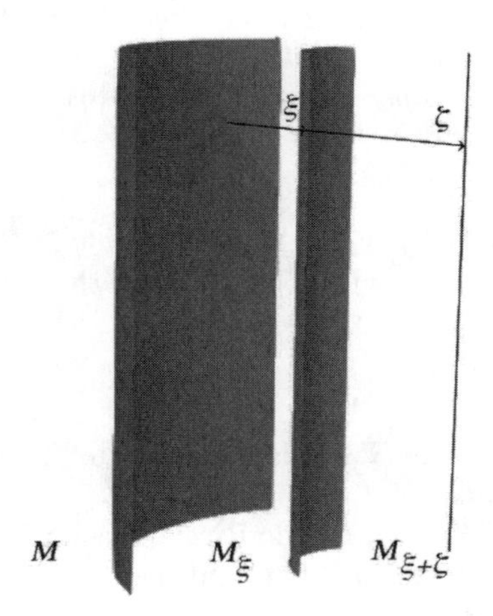

FIGURE 4.2: A piece of a cylinder M with its parallel displacement in direction of vector fields ξ and $\xi + \zeta$ pointing inward. M_ξ a parallel manifold, while $M_{\xi+\zeta}$ is focal.

REMARK 4.4.3 If M is a submanifold of S^n, we might view M as a submanifold of $\mathbb{R}^{n+1}$ and then define parallel and focal manifolds of M via this approach. Note that every parallel normal vector field of $M \subset S^n$ is also a parallel normal vector field of $M \subset \mathbb{R}^{n+1}$. One can easily verify that parallel and focal manifolds of submanifolds of S^n are also contained in a sphere (of different radius in general). $\square$

REMARK 4.4.4 We can also replace the Euclidean space by a Lorentzian space, and view the hyperbolic space H^n sitting inside $\mathbb{R}^{n,1}$ according to our

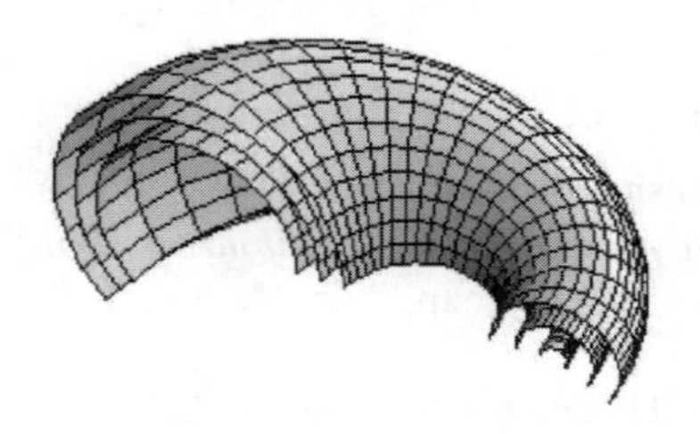

FIGURE 4.3: A piece of a cyclides of Dupin and two of its parallel surfaces.

standard model. Also, in this situation, every parallel normal vector field of $M \subset H^n$ is a parallel normal vector field of $M \subset \mathbb{R}^{n,1}$. Viewing M as a submanifold of $\mathbb{R}^{n,1}$ has the advantage that we can still work in a vector space. Thus $x + \xi(x)$ is a focal point of $M \subset \mathbb{R}^{n,1}$ if $\ker(\mathrm{id} - A_{\xi(x)}) \neq \{0\}$. □

Let M_ξ be an embedded parallel or focal manifold of M. Then the smooth map

$$\pi : M \to M_\xi \ , \ x \mapsto t_\xi(x) = x + \xi(x)$$

is a submersion (and a diffeomorphism if M_ξ is a parallel manifold). We denote by $\mathcal{H}$ and $\mathcal{V}$ the horizontal and vertical distributions that are induced by π. Note that $\mathcal{H}_q$ is isomorphic and parallel to $T_{\pi(q)}M_\xi$ and $\nu_q M \subset \nu_{\pi(q)}M_\xi$. Moreover, $\mathcal{V}_q = \ker \pi_{*q} = \ker(\mathrm{id} - A_\xi)_q = T_q(\pi^{-1}(\pi(q))$.

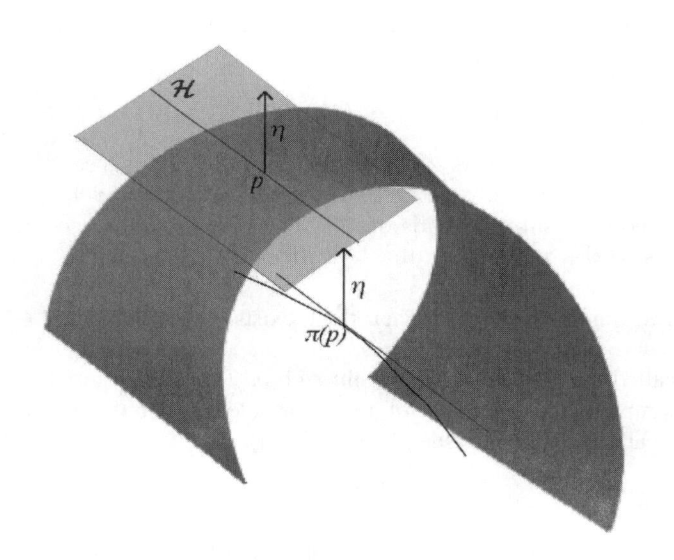

FIGURE 4.4: Submanifold, focal manifold and horizontal direction.

Example 4.2 "Focalization of an eigendistribution".
Let λ_i be a nonzero eigenvalue of a parallel isoparametric section ξ. Then $\xi_i = \xi/\lambda_i$ is a parallel isoparametric section with $A_{\xi_i}|E_i = \mathrm{id}_{E_i}$ and we have $\ker(\mathrm{id} - A_{\xi_i}) = E_i$. Denote by $\pi_i : M \to M_{\xi_i}$ the focal map from M onto M_{ξ_i}. Then the tangent space of M_{ξ_i} at $\pi_i(q)$ is given by

$$E_1(q) \oplus \ldots \oplus \hat{E}_i(q) \oplus \ldots \oplus E_g(q) \ ,$$

where $\hat{E}_i(q)$ means that $E_i(q)$ is omitted. In other words, the vertical and horizontal distribution with respect to $\pi_i : M \to M_{\xi_i}$ is given by $\mathcal{V} = E_i$ and $\mathcal{H} = E_i \oplus \ldots \oplus \hat{E}_i \oplus \ldots \oplus E_g$, respectively. We will say that M_{ξ_i} is the focal manifold that *"focalizes"* E_i (or λ_i). □

The following lemma describes a fundamental property of a focal map $\pi : M \to M_\xi$.

LEMMA 4.4.5
The fibres of the projection π from M onto the parallel focal manifold M_ξ are totally geodesic submanifolds of M, and the shape operator of M leaves the orthogonal decomposition $TM = \mathcal{V} \oplus \mathcal{H}$ invariant.

PROOF The statement is trivial if M_ξ is a point. Assume that $\dim M_\xi \geq 1$. Let X, Y be sections in $\mathcal{V}$ and Z an eigenvector of A_ξ with $A_\xi Z = \lambda Z$, $\lambda \neq 1$. Using the Codazzi equation we get

$$(1-\lambda)\langle \nabla_X Y, Z\rangle = \langle (\nabla_X A_\xi)Y, Z\rangle = \langle (\nabla_Z A_\xi)X, Y\rangle = 0 \ .$$

Since $\mathcal{H}$ is spanned by such eigenvectors Z, this implies that $\nabla_X Y$ is a section in $\mathcal{V}$, which means that $\mathcal{V}$ is an autoparallel subbundle of TM, and hence its leaves are totally geodesic submanifolds of M. The proof for the second statement is analogous to the one for Lemma 4.4.2 (b). □

Let c be a curve in M_ξ and $q = c(t_o)$. Then there exists for each point $p \in \pi^{-1}(\{q\})$ exactly one curve $\tilde{c}$ in M with $\tilde{c}(t_o) = p$, $\pi \circ \tilde{c} = c$ and $\tilde{c}'(t) \in \mathcal{H}_{\tilde{c}(t)}$ for all t. This curve $\tilde{c}$ is called the *horizontal lift* of c through p. The next result is of great importance for comparing the geometry of a submanifold M with the one of the parallel (possibly focal) manifold M_ξ (see [96, page 170]).

LEMMA 4.4.6
Let c be a curve in M_ξ and $q = c(t_o)$. Let $\tilde{c}$ be the horizontal lift of c through $p \in \pi^{-1}(\{q\})$. For each $\zeta \in \nu_p M \subset \nu_q M_\xi$ the parallel transports of ζ along c and $\tilde{c}$ with respect to the normal connections of M_ξ and M, respectively, coincide.

PROOF The parallel transport $\zeta(t)$ of ζ along $\tilde{c}(t)$ with respect to the normal connection of M is orthogonal to M_ξ since $\nu M \subset \nu M_\xi$. Since the horizontal distribution is invariant under the shape operator A of M by Lemma 4.4.5, we get

$$\zeta'(t) = -A_{\zeta(t)}\dot{\tilde{c}}(t) \in \mathcal{H}_{\tilde{c}(t)} \cong T_{c(t)} M_\xi \ .$$

Thus, $\zeta(t)$ is a parallel normal vector field to M_ξ along $c(t)$. Conversely, it easy to see that if $\zeta(t)$ is parallel along $c(t)$ with respect to the normal connection of M_ξ, then it is parallel along $\tilde{c}(t)$ with respect to the normal connection of M. □

We now compare the shape operators A and $\hat{A}$ of M and M_ξ respectively. We have the following important relation between the shape operators in the common normal directions to M and M_ξ.

LEMMA 4.4.7 "Tube formula"
For all $\eta \in \nu_p M \subset \nu_{\pi(p)} M_\xi$ we have

$$\hat{A}_\eta = (A_{\eta|\mathcal{H}}) \circ [(\mathrm{id} - A_{\xi(p)})_{|\mathcal{H}}]^{-1} .$$

PROOF Let $\dot{c}(0) \in T_{\pi(p)} M_\xi \cong \mathcal{H}_p$ be tangent to a curve c in M_ξ with $c(0) = \pi(p)$, and let $\tilde{c}$ be the horizontal lift of c with $\tilde{c}(0) = p$. Then

$$\tilde{c}(t) = c(t) - \xi(t).$$

Set $\eta(t) := \eta(c(t)) = \eta(\tilde{c}(t))$. We have

$$\begin{aligned} A_{\eta(t)} \dot{\tilde{c}}(t) &= -\dot{\eta}(t) = \hat{A}_{\eta(t)} \dot{c}(t) = \hat{A}_{\eta(t)} (\dot{\tilde{c}}(t) + \dot{\xi}(t)) = \\ &= \hat{A}_{\eta(t)} (\dot{\tilde{c}}(t) - A_\xi \dot{\tilde{c}}(t)). \end{aligned}$$

It is now sufficient to set $t = 0$. □

REMARK 4.4.8 One can also derive formulas for shape operators in the vertical normal directions in the focal manifold. Let v be a vertical normal vector to M_ξ, i.e., $v \in \mathcal{V} = \ker(\mathrm{id} - A_\xi)$, and let γ a horizontal curve on M with tangent vector $e \in \mathcal{H}$ at 0. Let $\bar{\gamma}(t) = \gamma(t) + \xi(\gamma(t))$ be the projection of γ on M_ξ; its tangent vector at 0 is $\bar{e} = e - A_\xi e$. Let $v(t)$ denote the vector obtained by parallel transport of v on $\bar{\gamma}$ with respect to the normal connection of M_ξ. Then $v(t)$ is also a vector field on M orthogonal to νM by Lemma 4.4.6. We denote by $\frac{D^\alpha}{dt}$, $\frac{\bar{D}^\alpha}{dt}$ and $\frac{(D^\alpha)^\perp}{dt}$ the covariant derivatives along a curve α induced by ∇, $\bar{\nabla}$ and $\nabla^\perp$ respectively.

On M we have $\dot{v}(0) = \frac{\bar{D}^\gamma}{dt}_{|t=0} v(t) = \frac{D^\gamma}{dt}_{|t=0} v(t) + \alpha(e, v) = \frac{D^\gamma}{dt}_{|t=0} v(t)$, since, by Lemma 4.4.5, the shape operator A of M leaves the splitting $TM = \mathcal{V} \oplus \mathcal{H}$ invariant.

On M_ξ, $\dot{v}(0) = \frac{\bar{D}^{\bar{\gamma}}}{dt}_{|t=0} v(t) = -\hat{A}_v \bar{e}$. Thus

$$\hat{A}_v(e - A_\xi e) = -\left(\frac{D^\gamma}{dt}_{|t=0} v(t) \right)^{\mathcal{H}}, \tag{4.6}$$

where $^{\mathcal{H}}$ denotes the projection on the horizontal subspace.
Furthermore, $A_\xi\left(\frac{D^\gamma}{dt}_{|t=0} v(t)\right) = (\nabla^\perp_e A)_\xi v + \frac{D^\gamma}{dt}_{|t=0}(A_\xi v(t)) = (\nabla^\perp_e A)_\xi v + \frac{D^\gamma}{dt}_{|t=0} v(t)$, so

$$\begin{aligned}\left(\frac{D^\gamma}{dt}_{|t=0} v(t)\right)^{\mathcal{H}} &= \left(A_\xi\left(\frac{D^\gamma}{dt}_{|t=0} v(t)\right)\right)^{\mathcal{H}} - ((\nabla^\perp_e A)_\xi v)^{\mathcal{H}} \\ &= A_\xi\left(\left(\frac{D^\gamma}{dt}_{|t=0} v(t)\right)^{\mathcal{H}}\right) - ((\nabla^\perp_e A)_\xi v)^{\mathcal{H}}.\end{aligned}$$

Thus, $\hat{A}_v(e - A_\xi e) = \hat{A}_v(\mathrm{id} - A_\xi)(e) = -A_\xi((\frac{D^\gamma}{dt}_{|t=0} v(t))^{\mathcal{H}}) + ((\nabla^\perp_e A)_\xi v)^{\mathcal{H}}$ or, in other words,

$$\hat{A}_v e = -A_\xi\left(\left(\frac{D^\gamma}{dt}_{|t=0} v(t)\right)^{\mathcal{H}}\right) + (\nabla^\perp A)_\xi v)^{\mathcal{H}})(\mathrm{id} - A_\xi)^{-1} e\,. \tag{4.7}$$

Taking into account the "tube formula" 4.4.7, the latter implies that *the shape operator $\hat{A}$ of the focal manifold can be expressed in terms of the shape operator A and its covariant derivative $\nabla^\perp A$.* □

REMARK 4.4.9 In Chapter 5, we will consider, for a special case, focal manifolds of submanifolds of spheres (not regarded as submanifolds of Euclidean space). We will examine instances of a "tube formula" in this situation. See Exercise 5.6.3 for the case of isoparametric submanifolds in spheres (and also page 152 for the case of isoparametric hypersurfaces in spheres). □

In the sequel, we will often consider the $\nabla^\perp$-parallel transport along a curve $\gamma : [0,1] \to M$ as a transformation map between the affine normal spaces $\gamma(0)+\nu_{\gamma(0)}M$ and $\gamma(1)+\nu_{\gamma(1)}M$. If M_ξ is a focal manifold of M, we will often identify $p - \xi(p)$ in M ($p \in M_\xi$) with vectors $-\xi(p)$, which are normal to both M at $p - \xi(p)$ and to M_ξ at p (cf. Figure 4.5).

One can relate the geometry of the fibre of the projection $\pi : M \to M_\xi$ with the $\nabla^\perp$-parallel transport in M.

LEMMA 4.4.10
Let M be a submanifold of $\mathbb{R}^n$ and ξ be a parallel normal vector field to M. Assume $\ker(\mathrm{id} - A_\xi)$ has constant dimension, let M_ξ be the parallel (or focal) manifold to M and $\pi : M \to M_\xi$ the corresponding submersion. We have the following.

1. *If $\tilde{\gamma} : [0,1] \to M$ is a horizontal curve in M (i.e. $\tilde{\gamma}'(t)$ is perpendicular to $\ker(\mathrm{id} - A_{\xi(t)})$) then $\tilde{\gamma}(1) = \tau^\perp_\gamma(\tilde{\gamma}(0))$ where $\gamma = \pi \circ \tilde{\gamma}$ and $\tau^\perp_\gamma : \gamma(0) + \nu_{\gamma(0)}(M_\xi) \to \gamma(1) + \nu_{\gamma(1)}(M_\xi)$ is the $\nabla^\perp$-parallel transport.*

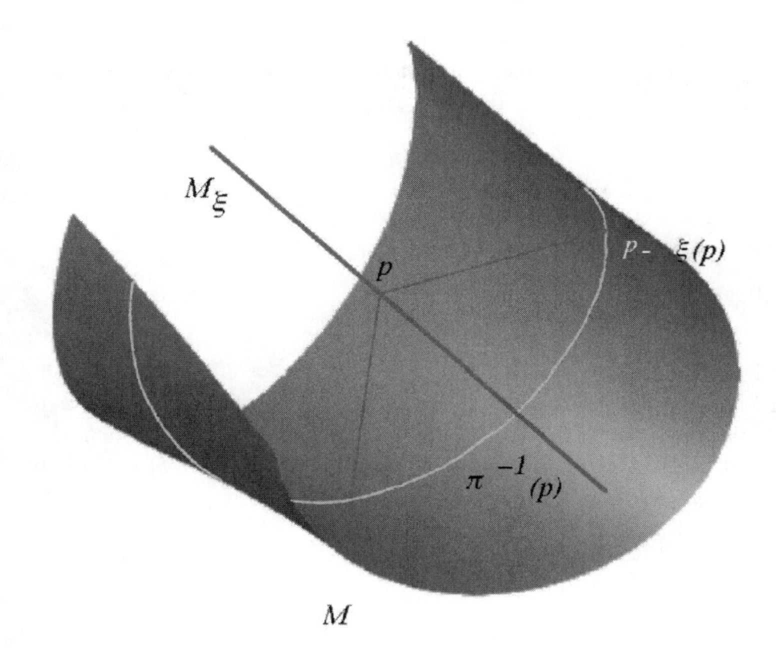

FIGURE 4.5: Focal manifold M_ξ and fibre $\pi^{-1}(p)$ of $\pi : M \to M_\xi$.

2. *The fibre $\pi^{-1}(\bar{p})$ is contained in the affine normal space $\bar{p} + \nu_{\bar{p}}(M_\xi)$ for any $\bar{p} \in M_\xi$.*

3. *Let $p \in M$ and let $\gamma : [0, 1] \to M_\xi$ be a piecewise differentiable curve with $\gamma(0) = \pi(p) = \bar{p}$, $\gamma(1) = \pi(q) = \bar{q}$ and let $\tau_\gamma^\perp$ be the $\nabla^\perp$-parallel transport along γ. Then $q = \tau_\gamma^\perp(p) \in \pi^{-1}(\bar{q})$ and $\tau_\gamma^\perp(\pi^{-1}(\bar{p}))$ coincides with $\pi^{-1}(\bar{q})$ near q.*

PROOF First note that $\tilde{\gamma}(t) = \gamma(t) - \xi(t)$. We identify now $\tilde{\gamma}(t)$ with the vector field $-\xi(t)$. Since $-\xi(t)$ is $\nabla^\perp$-parallel in M along $\tilde{\gamma}$, it follows from Lemma 4.4.6 that $-\xi(t)$ is also $\nabla^\perp$-parallel in M_ξ along γ. Thus, with the above identification $\tau_\gamma^\perp(\tilde{\gamma}(0)) = \tilde{\gamma}(1)$. Parts 2 and 3 follow from 1. ☐

Let M be a Riemannian submanifold of Euclidean (or Lorentzian) space with flat normal bundle. For $q \in M$, let $q + \nu_q(M)$ be the affine normal space. The *focal set* $F_M(q)$ of M at q is given by

$$F_M(q) = \{q + \xi_q \mid \xi_q \in \nu_q(M) \text{ and } (\mathrm{id} - A_{\xi_q}) \text{ is singular}\}$$

Since the normal bundle is flat, the shape operators at q form a commuting family. Let $\lambda_1(\xi_q), \cdots, \lambda_r(\xi_q)$ be the common eigenvalues (which are linear functionals defined on $\nu_q(M)$). Then

$$F_M(q) = \bigcup_{i=1}^{r} \ell_i(q),$$

where

$$\ell_i(q) = \{q + \xi_q \mid \xi_q \in \nu_q(M),\ \lambda_i(\xi_q) = 1\}$$

are called focal hyperplanes. We have the following.

PROPOSITION 4.4.11

Let M be a Riemannian submanifold of Euclidean (or Lorentzian) space with flat normal bundle. Let ξ be a parallel normal vector field to M such that $\ker(\mathrm{id} - A_\xi) \neq 0$ and consider the parallel manifold M_ξ to M. Then, for all $q \in M$

(i) $q + \nu_q(M) = \bar{q} + \nu_{\bar{q}}(M_\xi)$, where $\bar{q} = q + \xi(q)$.

(ii) $F_M(q) = F_{M_\xi}(\bar{q})$.

PROOF Part *(i)* is easy, since $T_{\bar{q}}M_\xi = \mathrm{Im}\,(\mathrm{id} - A_{\xi(q)}) = T_qM$, so $\nu_qM = \nu_{\bar{q}}M_\xi$ and $\bar{q} \in q + \nu_qM$. For *(ii)* we use the tube formula (Lemma 4.4.7) for the shape operators A of M and $\hat{A}$ of M_ξ: let $x \in q + \nu_q(M) = \bar{q} + \nu_{\bar{q}}(M_\xi)$, then $\hat{A}_{x-\bar{q}} = A_{x-\bar{q}}(\mathrm{id} - A_{\xi(q)})^{-1}$. Since $x - \bar{q} = (x - q) - \xi(q)$ we have that

$$\begin{aligned}\hat{A}_{x-\bar{q}} &= (A_{x-q} - A_{\xi(q)})(\mathrm{id} - A_{\xi(q)})^{-1} = \\ &= [(A_{x-q} - \mathrm{id}) + (\mathrm{id} - A_{\xi(q)})](\mathrm{id} - A_{\xi(q)})^{-1} = \\ &= (A_{x-q} - \mathrm{id})(\mathrm{id} - A_{\xi(q)})^{-1} + \mathrm{id}\end{aligned}$$

or, equivalently, $\mathrm{id} - \hat{A}_{x-\bar{q}} = (\mathrm{id} - A_{x-q})(\mathrm{id} - A_{\xi(q)})^{-1}$, which implies that $\mathrm{id} - \hat{A}_{x-\bar{q}}$ is invertible if and only if $\mathrm{id} - A_{x-q}$ is invertible. □

c) The holonomy tubes

Let M be a submanifold of Euclidean space. We consider a tube that can be constructed using the normal holonomy of M. If $\eta_p \in \nu_p(M)$ the *holonomy tube* $(M)_{\eta_p}$ *at* η_p is the image in the exponential map of the subset of the normal bundle, $\mathcal{H}ol_{\eta_p}M$, obtained by parallel translating η_p with respect to $\nabla^\perp$ along any piecewise differentiable curve in M. More explicitly, it is defined by (cf. [96])

$$(M)_{\eta_p} = \{\gamma(1) + \tau_\gamma^\perp \eta_p \mid \gamma : [0,1] \to M \text{ is piecewise differentiable }, \gamma(0) = p\}\ .$$

$\mathcal{H}ol_{\eta_p}M$ is actually a subbundle of νM (but not a vector bundle), called *holonomy subbundle through* η_p. Its fibre at p is the orbit of η_p under the action of the normal holonomy group at p. $\mathcal{H}ol_{\eta_p}M$ is always a 1-1 immersed submanifold of νM and, if the normal holonomy group is compact, in particular if M is simply connected, it is an embedded submanifold. Usually we will need the holonomy tube for local results, so we will assume M to be simply connected.

We define the *focal distance* as the supremum of the positive real numbers ε such that 1 is not an eigenvalue of A_ξ if $||\xi|| < \varepsilon$.

Suppose 1 is not an eigenvalue of $A_{\tau_\gamma^\perp \eta_p}$, for any $\nabla^\perp$-parallel transport $\tau_\gamma^\perp \eta_p$ of η_p along any piecewise differentiable curve γ, or, in particular, $||\eta_p||$ is less than the focal distance. Then the holonomy tube $(M)_{\eta_p}$ is an immersed submanifold of $\mathbb{R}^n$. In this case, there is an obvious submersion $\pi_{\eta_p} : (M)_{\eta_p} \to M$ whose fibres are orbits of the (restricted) normal holonomy group.

Let $\eta := p + \eta_p \in (M)_{\eta_p}$. Then $T_\eta\left((M)_{\eta_p}\right) = T_pM \oplus T_{\eta_p}(\Phi \cdot \eta_p)$. Thus $\nu_\eta\left((M)_{\eta_p}\right)$ can be identified with the normal space in ν_pM to the normal holonomy orbit $\Phi \cdot \eta_p$. In Section 5.4 we will prove that *the normal holonomy at η of the holonomy tube $(M)_{\eta_p}$ is the image in the slice representation of the isotropy subgroup Φ_η on $\nu_\eta\left((M)_{\eta_p}\right) \cong \nu_{\eta_p}(\Phi \cdot \eta_p)$.*

An important local property of the holonomy tube is that it has flat normal bundle if η_p lies on a principal orbit of the restricted normal holonomy group. (This is actually a special case of the above description of the normal holonomy of the holonomy tube.)

PROPOSITION 4.4.12

Suppose $\eta_p \in \nu_pM$ lies on a principal orbit of the restricted normal holonomy group and that $||\eta_p||$ is less than the focal distance of M. Then $(M)_{\eta_p}$ has flat normal bundle.

PROOF Since the result is local, we can assume that $(M)_{\eta_p}$ is embedded. Let $\hat{p} := p + \eta_p$ and $\zeta \in \nu_{\hat{p}}(M)_{\eta_p} \subset \nu_pM$ and consider its parallel transport $\tau_\gamma^\perp \zeta$ (with respect to the normal connection) along any horizontal curve γ. This produces a well defined normal vector $\tilde{\zeta}$ to $(M)_{\eta_p}$. Indeed, by Lemma 4.4.6, $\tilde{\zeta}$ is also $\nabla^\perp$-parallel along the projection $\overline{\gamma}$ of γ in M. Thus, if δ is another horizontal curve with $\gamma(0) = \delta(0) = \hat{p}$, $\gamma(1) = \delta(1)$, then $\tau_\gamma^\perp \zeta(1)$ and $\tau_\gamma^\perp \delta(1)$ differ by an element given by the action of the normal holonomy group Φ_p^* of M, actually by the action of an element fixing η_p, so that it belongs to the isotropy subgroup at η_p of Φ_p^*. Since η_p lies on a principal orbit of Φ_p^*, the isotropy subgroup at η_p of Φ_p^* acts trivially on the normal space, hence on ζ.

Now, $\tilde{\zeta}$ is parallel in the horizontal direction by construction. It is also parallel in the vertical direction. Indeed, a fibre is a principal orbit of Φ_p^*, which acts as an s-representation (in particular, polar). Thus, the normal space to the fibre is flat (cf. Section 3.2) and ζ is a Φ_p^*-equivariant vector field, which is parallel by Corollary 3.2.5. ☐

By means of holonomy tubes, one reduces the study of the geometry of a given submanifold M of Euclidean space to the study of a holonomy tube with flat normal bundle. The first question arising is why not consider the usual spherical tube S_ε (i.e., the normal exponential of the normal vectors of a fixed small length ε), which is a hypersurface and so has flat normal bundle. There are many geometric reasons for considering holonomy tubes instead of spherical tubes (some of them depending on the particular problem). But the most convincing and basic fact is that the holonomy tube of a product of submanifolds is the product of the corresponding holonomy tubes. This is not true for spherical tubes.

Moreover, one can also see that a holonomy tube $(M)_{\eta_p}$ is irreducible if M is irreducible. Indeed, suppose $(M)_{\eta_p}$ was reducible but M irreducible. This would mean that the projection of the holonomy tube on M killed a factor. But then one would not be able to recover this factor by constructing the holonomy tube.

REMARK 4.4.13 For a holonomy tube $(M)_{\eta_p}$, any fibre of the projection $\pi : (M)_{\eta_p} \to M$ is orbit of the (restricted) normal holonomy group. □

d) Combining focalizations with holonomy tubes

We will now somehow combine the constructions of parallel focal manifolds and of holonomy tubes.

Namely, given a parallel normal isoparametric section and a parallel focal manifold M_ξ of M, we pass to a holonomy tube with respect to $-\xi(q)$ (at some q) and then we compare the geometry of M with the tube's.

We do this in the case of a focal manifold that "focalizes" an eigendistribution E_i. Let us fix a nonzero eigenvalue λ_i of the shape operator relative to the parallel normal vector field ξ. Set $\xi_i = \lambda_i^{-1}\xi$ and consider the parallel focal manifold M_{ξ_i}. Fix $q \in M$ and take the holonomy tube $(M_{\xi_i})_{-\xi_i(q)}$. Let us denote by $\mathcal{H}_i$ the horizontal distribution of $\pi_i : M \to M_{\xi_i}$. Then $\mathcal{H}_i = E_1 \oplus \ldots \oplus \hat{E}_i \oplus \ldots \oplus E_g = E_i^\perp$. If A^i is the shape operator of M_{ξ_i}, by the "tube formula" (Lemma 4.4.7)

$$A^i_\zeta = A_\zeta|_{\mathcal{H}_i}[(\mathrm{id} - A_{\xi_i(q)})|_{\mathcal{H}_i}]^{-1}.$$

As one can easily check, 1 is not an eigenvalue of $A^i_{-\xi_i(q)}$, hence

$$f_i = \exp_{\nu M_{\xi_i}|\mathcal{H}ol_{-\xi_i(q)}M_{\xi_i}} : \mathcal{H}ol_{-\xi_i(q)}M_{\xi_i} \to \mathbb{R}^n$$

is an immersion. Let $c : [0,1] \to M_{\xi_i}$ be a piecewise differentiable curve in M_{ξ_i} with $c(0) = \pi_i(q)$ and $\tilde{c} : [0,1] \to M$ its horizontal lift to M with $\tilde{c}(0) = q$. By Lemma 4.4.6, if $\zeta(t)$ is a parallel normal vector field to M_{ξ_i} along c, then $\zeta(t)$ may be regarded as a parallel normal vector field to M along $\tilde{c}$. If $\zeta(0) = -\xi_i(q)$, then $\zeta(t) = -\xi_i(\tilde{c}(t))$ for all $t \in [0,1]$, implying

$$\begin{aligned} f_i((c(1),\zeta(1))) &= c(1) - \xi_i(\tilde{c}(1)) = \pi_i(\tilde{c}(1)) - \xi_i(\tilde{c}(1)) = \\ &= \tilde{c}(1) + \xi_i(\tilde{c}(1)) - \xi_i(\tilde{c}(1)) = \tilde{c}(1), \end{aligned}$$

which, in turn, shows that $f_i\left(\mathcal{H}ol_{-\xi_i(q)}M_{\xi_i}\right) \subseteq M$ and that $\pi f_i = \mathrm{pr}_{|\mathcal{H}ol_{-\xi_i(q)}M_{\xi_i}}$, where pr: $\nu M \to M$ is the projection. It follows immediately that f_i is one to one.

Therefore we identify $\mathcal{H}ol_{-\xi_i(q)}M_{\xi_i}$ with $(M_{\xi_i})_{-\xi_i(q)}$. This shows that *if we pass from M to the focal manifold M_{ξ_i} and then to the holonomy tube with respect to $-\xi_i(q)$, $q \in M_{\xi_i}$, locally $(M_{\xi_i})_{-\xi_i(q)} \subseteq M$.*

Moreover, we can regard $r \in S_i(q)$ as a normal vector to M_{ξ_i} at $\bar{q}$ (i.e., we identify r with the vector $r - \bar{q} = -\xi_i(r)$). Now, by Lemma 4.4.6, parallel transport $\tau^i_\gamma(r)$ in $\nu(M_{\xi_i})$ of r along a short piecewise differentiable curve $\gamma : [0,1] \to M_{\xi_i}$ starting at $\bar{q}$ coincides with $\tau_{\tilde{\gamma}}(-\xi_i(r))$ of $-\xi_i(r)$ in $\nu(M)$ along the horizontal lift $\tilde{\gamma}$ of γ. But $\tau_{\tilde{\gamma}}(-\xi_i(r)) = -\xi_i(\tilde{\gamma}(1))$ because $-\xi_i$ is a parallel normal field to M. So, in our identifications, $\tau^i_\gamma(r) = \tau_{\tilde{\gamma}}(-\xi_i(r)) = -\xi_i(\tilde{\gamma}(1))$ (where $\tilde{\gamma}(1)$ is regarded as a normal vector to M_{ξ_i} at $\gamma(1)$). If γ is a closed null homotopic curve, then $\tilde{\gamma}(1)$ belongs to $S_i(q)$. Since parallel transport along short loops always contains a neighbourhood of the identity of the local normal holonomy group, (see Proposition 4.1.11, cf. [82, Appendix]) we get that the normal holonomy orbit $\Phi^{i*}_{\bar{q}} \cdot r$ is contained, near r, in $S_i(q)$.

This discussion together with Lemma 4.4.10 yields

PROPOSITION 4.4.14

Let M be a submanifold of $\mathbb{R}^n$, λ_i a nonzero eigenvalue of the shape operator relative to the parallel normal isoparametric vector field ξ. Set $\xi_i = \lambda_i^{-1}\xi$ and consider (locally) the focal manifold M_{ξ_i}, the submersion $\pi_i : M \to M_{\xi_i}$ and the holonomy tube $(M_{\xi_i})_{-\xi_i(q)}$. Then

1. *$f_i = \exp_{\nu M_{\xi_i}|\mathcal{H}ol_{-\xi_i(q)}M_{\xi_i}} : \mathcal{H}ol_{-\xi_i(q)}M_{\xi_i} \to \mathbb{R}^n$ is a one to one immersion,*
2. *$\pi_i(q) + \Phi^{i*}_{\pi_i(q)}(-\xi_i(q))$ is locally contained in $\pi_i^{-1}(\pi_i(q))$, where $\Phi^{i*}_{\pi_i(q)}$ denotes the restricted normal holonomy group of M_{ξ_i} at $\pi_i(q)$. In particular, if $r \in S_i(q)$, the normal holonomy orbit $\Phi^{i*}_{\bar{q}} \cdot r$ is contained, near r, in $S_i(q)$,*
3. *$T_q f_i[(M_{\xi_i})_{-\xi_i(q)}] = (E_i(q))^\perp \oplus T_{-\xi_i(q)}\Phi^{i*}_{\pi_i(q)}(-\xi_i(q))$.*

Let $q \in M$, then $q - \pi_i(q) = -\xi_i(q)$ belongs to $\nu_{\pi_i(q)}(M_{\xi_i})$ and (as remarked above) the holonomy tube $(M_{\xi_i})_{-\xi_i(q)}$ is locally contained in M.

Moreover $\xi_i|_{(M_{\xi_i})_{-\xi_i(q)}}$ is also a parallel normal field in the holonomy tube (because it is the vector field defined by the centres of the holonomy tube's fibres). In particular, any tangent space to the holonomy tube is invariant under the shape operator A_ξ of M. Let A, $\tilde{A}$ be the shape operators of M and $(M_{\xi_i})_{-\xi_i(q)}$ respectively, and $c(t)$ be a curve on $(M_{\xi_i})_{-\xi_i(q)}$ with $c(0) = q$. Since ξ_i is a parallel normal field to both submanifolds

$$A_{\xi_i(q)}c'(0) = -\frac{d}{dt}_{|0}\xi_i(c(t)) = \tilde{A}_{\xi_i(q)}c'(0).$$

Hence, $A_{\xi_i}|_{T_q((M_{\xi_i})_{-\xi_i(q)})} = \tilde{A}_{\xi_i}$. This shows that ξ_i restricted to the holonomy tube is also an isoparametric section of $(M_{\xi_i})_{-\xi_i(q)}$ (whose shape operator has no zero eigenvalue).

In the sequel (Section 5.3), we wish apply these techniques to the study of submanifolds with constant principal curvatures.

To begin with, we state a proposition giving a sufficient condition for a submanifold to have constant principal curvatures, namely that the fibres of all focalizations on the submanifolds M_{ξ_i} (which "focalize" the eigendistributions E_i) are homogeneous under the normal holonomy group. We will later see that this condition is also necessary, a consequence of the Homogeneous Slice Theorem [96].

PROPOSITION 4.4.15

Let M be a submanifold of $\mathbb{R}^n$. Let ξ be a parallel isoparametric normal section on M with nonzero eigenvalues $\lambda_1, ..., \lambda_g$ and $\xi_i = \lambda_i^{-1}\xi$. Assume furthermore that, for any i, $S_i(q)$ locally coincides with the orbit $\Phi^{i*} \cdot (-\xi_i(q))$ of the restricted normal holonomy group Φ^{i*} of M_{ξ_i} at $\pi_i(q)$. Then M is a submanifold with constant principal curvatures.

PROOF Let $E_1, ..., E_g$ be the eigendistributions of A_ξ relative to $\lambda_1, ..., \lambda_g$. Recall that (cf. Lemma 4.4.2) the eigendistributions E_i are invariant by all shape operators. We fix the index i and show that, for any piecewise differentiable curve $\gamma(t)$, $A_{\eta(t)|E_i}$ has constant eigenvalues, $\eta(t)$ any parallel normal vector field along γ. Since the property of having constant principal curvatures is tensorial (cf. Remark 4.1.2), we can assume that γ is either vertical or horizontal with respect to the submersions $\pi_i : M \to M_{\xi_i}$.

a) γ is vertical, i.e. $\gamma : [0,1] \to S_i(q)$, for some $q \in M$. Since $S_i(q)$ is totally geodesic in M (Lemma 4.4.5) and E_i is invariant under all shape operators A (Lemma 4.4.2), we have

$$\nabla_X^{\perp S_i(q)}\eta = A_\eta^{S_i(q)}X + \bar{\nabla}_X\eta = A_\eta X + \bar{\nabla}_X\eta = \nabla_X^{\perp}\eta = 0$$

so η is also parallel in the normal bundle of $S_i(q)$ and $A_\eta^{S_i(q)} = A_{\eta|E_i}$. Since $S_i(q)$ is, by assumption, homogeneous under the restricted normal holonomy group, which acts on it as an s-representation (by the Normal Holonomy Theorem 4.2.1), so $S_i(q)$ has constant principal curvatures (by Proposition 4.1.6). Thus $A_{\eta(t)}^{S_i(q)} = A_{\eta(t)}$ has constant eigenvalues.

b) γ is horizontal with respect to $\pi_i : M \to M_{\xi_i}$. Note that by Lemma 4.4.6 $\eta(t)$ is also parallel in νM_{ξ_i} along $\pi_i\gamma$. Let $\tau_{\pi_i\gamma}^{\perp} : \nu_{\pi_i\gamma(0)}M_{\xi_i} \to \nu_{\pi_i\gamma(1)}M_{\xi_i}$ be the $\nabla^\perp$-parallel transport along $\pi_i\gamma$. By Lemma 4.4.10 and Proposition 4.4.14, $\tau_{\pi_i\gamma}^{\perp}(S_i(\gamma(0))) = S_i(\gamma(1))$ and $\tau_{\pi_i\gamma}^{\perp}\gamma(0) = \gamma(1)$. Since $\tau_{\pi_i\gamma}^{\perp}$ is an isometry of affine spaces $\pi_i\gamma(0) + \nu_{\pi_i\gamma(0)}M_{\xi_i}$ and $\pi_i\gamma(1) + \nu_{\pi_i\gamma(1)}M_{\xi_i}$ which maps $S_i(\gamma(0))$ to $S_i(\gamma(1))$, we get

$$\tau_{\pi_i\gamma}^{\perp} A_{\eta(0)}^{S_i(\gamma(0))} (\tau_{\pi_i\gamma}^{\perp})^{-1} = A_{\eta(1)}^{S_i(\gamma(1))},$$

so $A_{\eta(0)}^{S_i(\gamma(0))} = A_{\eta(0)|E_i(\gamma(0))}$ and $A_{\eta(1)}^{S_i(\gamma(1))} = A_{\eta(1)|E_i(\gamma(1))}$ have the same eigenvalues. ∎

REMARK 4.4.16 As a consequence of Proposition 4.4.15, we see that, if all holonomy tubes $(M_{\xi_i})_{-\xi_i(q)}$ locally coincide with M, then M is a submanifold with constant principal curvatures. $\square$

e) Partial tubes about submanifolds of space forms

We now set a common framework for many constructions carried out in this chapter. This also takes into account the classical construction of (spherical) tubes $M_r = \{\exp(r\xi) \mid \xi \in \nu^1 M\}$ of radius r around a submanifold M of a space form.

In the last section, we constructed the parallel manifold M_ξ for any parallel normal vector field ξ, and the holonomy tube $(M)_{\eta_p}$, around any $\eta_p \in \nu_p M$.

All these examples belong to the general class of *partial tubes*, introduced by S. Carter and A. West in [40] for submanifolds of Euclidean space, but can be generalized to submanifolds of space forms by regarding them, as above, as either submanifolds of Euclidean or Lorentzian spaces.

Let M be an m-dimensional submanifold of $\bar{M}^n(\kappa)$ and set $k = \operatorname{codim} M$ as a submanifold of Euclidean space or of Lorentz space. We denote by νM the normal bundle of M as a submanifold of Euclidean space or of Lorentz space. Let Σ be the set of critical normals, i. e. of $\xi_p \in \nu_p M$ such that $p + \xi_p$ is a focal point. A partial tube about M is a subbundle B of νM with typical fibre S, where:

(i) S is a submanifold of $\mathbb{R}^k$

(ii) B does not intersect the set of critical normals Σ

(iii) B is invariant by parallel transport along any curve, with respect to the normal connection

If E is the endpoint map, its restriction $E_{|B}$ is an immersion in Euclidean or Lorentzian space, which follows just from (i) and (ii). The geometric meaning of (iii) is the following: Fix a point $p \in M$ and identify S with the fibre B_p of B at p, or equivalently identify $\nu_p M$ with $\mathbb{R}^k$ via a linear isometry φ_p such that $\varphi_p(S) = B_p \subseteq \nu_p M$. Let $x \in S$, $z = (p, \varphi_p(x)) \in B$ and denote by $\nu_x S$ the normal space to S at x in $\mathbb{R}^k$. Then, as shown in [40, Theorem 1.2], (iii) is equivalent to the property that for any p, x, φ_p, as above the normal space $\nu_z B$ to B at z, coincides with $\varphi_p(\nu_x S)$.

Note that any path γ from p to q determines, by composing parallel transport along γ with φ_p, a linear isometry $\varphi_q : \mathbb{R}^k \to \nu_q M$. One can define the *push-out region* Ω as the subset of $\mathbb{R}^k$ given by those x such that for any φ_q as above $\varphi_q x$ does not belong to set of critical normals Σ_q at q.

Then any submanifold S of $\mathbb{R}^k$ contained in the push-out region Ω invariant under the normal holonomy group defines a unique partial tube with typical fibre S and with $B = \{\varphi_{\gamma(s)} x \mid x \in S, \gamma \text{ a path}\}$.

Examples of partial tubes.

1: r-tubes. Let $r \in \mathbb{R}_+$ and $\nu^1 M$ the unit normal sphere bundle over M, that is, the sphere bundle over M consisting of all unit normal vectors of M. The r-tube or tube with radius r around M is defined by

$$M_r := \{\exp(r\xi) \mid \xi \in \nu^1 M\} = \{p + r\xi(p) \mid p \in M, \xi \in \nu^1 M\} .$$

If r is smaller than the distance between M and its focal set, then M_r identifies with the set of normals of length r. Therefore, M_r is a partial tube about M with typical fibre the hypersphere in $\mathbb{R}^k$ ($k = \operatorname{codim} M$) having centre the origin and radius r.

2: Parallel manifolds. Let M be a submanifold of $\bar{M}^n(\kappa)$ and ξ a parallel normal vector field that does not belong to the set of critical normals Σ. The image of ξ in νM can be thought of as a partial tube about M with typical fibre a single point, and identifies with the parallel manifold M_ξ.

3: Focal manifolds. Let M_ξ be a parallel focal manifold with respect to a parallel normal isoparametric section ξ on a submanifold M of $\mathbb{R}^n$. The fibre $\pi^{-1}(\bar{p})$ is contained in the affine normal space $\bar{p} + \nu_{\bar{p}}(M_\xi)$ and, by Lemma 4.4.10, is invariant by $\nabla^\perp$-parallel transport in M_ξ. By 3 in Lemma 4.4.10 we have that, locally, M can be seen as a partial tube about M_ξ. This actually generalizes the situation of an isoparametric submanifold and one of its focal manifolds.

4: Holonomy tubes. Let $(M)_{\eta_p}$ be a holonomy tube around a submanifold M. Then, if η_p is less than the focal distance, $(M)_{\eta_p}$ is the partial tube with typical fibre the orbit of η_p under the action of the normal holonomy group.

Let B be a partial tube about M and let $\tilde{\gamma}$ be a curve in B joining $\tilde{p} = (p, \varphi_p x)$ and $\tilde{q} = (q, \varphi_q x)$. We can project $\tilde{\gamma}$ down to M, getting a curve γ from p and q. Writing $\tilde{\gamma}(t) = \big(\gamma(t), \varphi_{\gamma(t)}\sigma(t)\big)$, we determine a curve $\sigma(t)$ in the fibre S. Again, parallel transport along σ composed with a fixed isometry $\theta_x : \mathbb{R}^d \to \nu_x S$ determines an isometry $\theta_y : \mathbb{R}^d \to \nu_y S$, where $d = \operatorname{codim} S$ in $\mathbb{R}^k$.

The following result ([40], Lemma 4.2) relates parallel transport along $\tilde{\gamma}$ in the normal bundle to a partial tube to $\nabla^\perp$-parallel transport in M along γ and the parallel transport in the normal bundle of the typical fibre along σ

PROPOSITION 4.4.17

Parallel transport in the normal bundle to the partial tube B along $\tilde{\gamma}$ is given by

$$\varphi_p \theta_x w \mapsto \varphi_q \theta_y w.$$

The proof can be found in [40], page 158, and uses arguments similar to Lemma 4.4.6, together with the description of the normal space to the partial tube, given above. We propose it as an Exercise (Exercise 4.6.3).

Note that this proposition generalizes Lemma 4.4.6. Indeed, if M_ξ is a parallel focal manifold of M with respect to a parallel normal isoparametric section ξ, like in Example 3, one can write a point of the partial tube M as $(p, \xi(p))$, and the typical fibre identifies with the set of vectors $\xi(q)$ such that $q + \xi(q) = p + \xi(p)$. Since ξ is parallel, we have $\varphi_q \xi(p) = \xi(q)$, so that the curve σ identifies with the constant path $t \mapsto \xi(p)$.

4.5 Further remarks

a) Realizations of s-representations as normal holonomy groups

E. Heintze and C. Olmos computed the normal holonomy of orbits of s-representations obtaining that all s-representations arise as normal holonomy representations with 11 exceptions [95]. Up to now, no example has been found of a submanifold realizing one of these exceptions as normal holonomy representation. The simplest exception (it has rank one) is the isotropy representation of the Cayley projective space represented by $F_4/Spin(9)$. K. Tezlaff [218] gave a negative answer to the question of whether this representation is the normal holonomy representation of one of the focal manifolds of the inhomogeneous isoparametric hypersurfaces in spheres of Ferus, Karcher and Münzner [87], which would have been good candidates.

A still open conjecture is that if M is a full irreducible homogeneous submanifold of the sphere that is not the orbit of an s-representation, then the normal holonomy group acts transitively on the unit sphere of the normal space [175].

b) Normal holonomy and irreducibility

The Normal Holonomy Theorem 4.2.1 provides, in particular, an orthogonal decomposition of the normal space at p into invariant subspaces.

The existence of an invariant subspace for the normal holonomy however, does not imply in general that the submanifold splits locally (both extrinsically and intrinsically). For example, for a submanifold of Euclidean space contained in a sphere, the line determined by the position vector p is an invariant subspace under normal holonomy (it always belongs to the flat part of νM), but such a submanifold does not necessarily split.

For submanifolds of space forms, one can get only weaker versions of de Rham's decomposition Theorem (see, for instance, Exercise 4.6.1). But, if M is a complex submanifold of $\mathbb{C}^n$, there is a version of de Rham decomposition theorem available: if Φ splits, M locally splits as a product of submanifolds [67].

c) A bound on the number of factors of the normal holonomy representations

Let M be a submanifold of Euclidean space (or more generally of a space of constant curvature). Let $p \in M$ and let

$$\nu_p M = V_0 \oplus V_1 \oplus \cdots \oplus V_\ell$$

be the decomposition of $\nu_p M$ given by the Normal Holonomy Theorem (applied to the local normal holonomy group).

The following result gives a bound for the number ℓ of irreducible factors of the normal holonomy representation.

THEOREM 4.5.1
Let M be a submanifold of Euclidean space (or, more generally, of a space of constant curvature). Then, in an open and dense subset of M, the local holonomy group has, at most, $\frac{m(m-1)}{2}$ irreducible factors ($m = \dim M$).

PROOF First note that in an open and dense subset Ω the number of factors of the normal holonomy representation is locally constant. From the Normal Holonomy Theorem it follows that, for any $p \in \Omega$, there exists a neighbourhood U and an orthogonal decomposition of the normal holonomy bundle of U into parallel subbundles

$$\nu U = \nu_0 \oplus \nu_1 \oplus \cdots \oplus \nu_\ell$$

such that, for any $q \in U$, $\nu_0(q)$ is the fixed (point) set of Φ_q^{loc}, and Φ_q^{loc} acts irreducibly on $\nu_i(q)$ $(i > 0)$. Note that there always is a point $q \in U$ such that there exist $x_1, y_1, \ldots, x_\ell, y_\ell \in T_qM$ with the property that $\left[R_q^\perp(x_i, y_i)\right]^i$, $(i > 0)$, are different from zero, $[\ ]^i$ denoting the projection on $\nu_i(q)$.

Now, observe that $\mathrm{im}(\mathcal{R}_q^\perp) = \mathrm{im}(R_q^\perp)$. Since $\mathcal{R}_q^\perp(\xi, \eta)\nu_i(q) \subseteq \nu_i(q)$, for all $\xi, \eta \in \nu_qM$, the same argument of the Normal Holonomy Theorem proof (cf. Lemma 4.3.13) gives

$$\mathcal{R}_q^\perp(\xi_i, \xi_j) = 0\,, \qquad \text{if } \xi_i \in \nu_i(q),\ \xi_j \in \nu_j(q)\,,\ i \neq j\,.$$

Then, $\mathrm{im}(\mathcal{R}_q^\perp) = \oplus_{i=1}^{\ell}\left[\mathrm{im}\mathcal{R}_q^\perp\right]^i$, $\mathrm{im}(R_q^\perp) = \oplus_{i=1}^{\ell}\left[\mathrm{im}R^\perp\right]^i$ and consequently, $\left[\mathrm{im}R_q^\perp(x_i, y_i)\right]^i \in \mathrm{im}(R_q^\perp)$.

Thus, $\left[R_q^\perp(x_1, y_1)\right]^1, \ldots, \left[R_q^\perp(x_\ell, y_\ell)\right]^\ell$ are linearly independent, so that $\ell \leq \dim(\mathrm{im}(R_q^\perp))$ and $2\ell \leq m(m-1)$, for $2\dim(\mathrm{im}(R_q^\perp)) \leq m(m-1)$. Here we have used the fact that $R_q^\perp$ can be regarded as a linear map from $\Lambda^2(T_qM)$ to $\Lambda^2(\nu_qM)$ and $2\dim\Lambda^2(T_qM) = m(m-1)$. ∎

d) Normal holonomy of surfaces

Using the above bound on the number of normal holonomy representation components and properties of holonomy systems, we prove the following.

THEOREM 4.5.2
Let M be a surface of Euclidean space that has the property that around any point it is not contained in a sphere or in an affine subspace. Then the local normal holonomy group is either trivial or it acts transitively on the unit sphere of the normal space.

PROOF Assume the local normal holonomy is not trivial and let $p \in M$. We will show that Φ_p^{loc} acts transitively on the unit sphere of the normal space $\nu_p M$.

To begin, observe that there are no parallel umbilical normal sections, or else M would be contained either in a sphere or in an affine subspace. Further, the factor V_0 (i.e., the normal holonomy fixed point set) is trivial, for otherwise there should exist a nonumbilical parallel normal field ξ (around arbitrary close points to p). This is impossible, because it would imply that the normal bundle is flat. In fact, ξ would commute with all other shape operators and, by the Ricci equations, all shape operators would commute, since $\dim(M) = 2$, and so $R^\perp = 0$.

The bound given by Theorem 4.5.1 forces the local normal holonomy to act irreducibly.

We now claim that Φ_p^{loc} is transitive on the unit sphere of $\nu_p M$. Suppose that it is not the case. Then there exists a point q arbitrarily close to p such that $\mathcal{R}_q^\perp \neq 0$ and so $[\nu_q M, \mathcal{R}_q^\perp, \Phi_q^{loc}]$ is an irreducible non transitive holonomy system. By Theorem 4.3.7, this holonomy system is symmetric. In particular, the first normal space $\mathcal{N}_q^1$ coincides with $\nu_q M$. This is because, otherwise, there would exist a $\xi \in \nu_q M$ with $A_\xi = 0$ and so $\mathcal{R}_q^\perp(\xi, \ ) = 0$, contradicting irreducibility. $\xi \mapsto A_\xi$ is injective and so we have that $\dim \nu_q M \leq 3$ (observe that the dimension of the space of 2×2 symmetric matrices is 3). Now, it is not difficult to see that an irreducible symmetric space of dimension at most 3 must be of rank one (Exercise, or else use the classification of symmetric spaces, cf. Appendix). This means the normal holonomy is transitive on the unit sphere of the normal space. □

Note that, for a surface contained in a sphere but not contained in a proper affine subspace (or, equivalently in a smaller dimensional sphere), there is an analogue result (Exercise 4.6.16).

e) Computation of normal holonomy

The description of the Lie algebra $L(\Phi_p)$ given by the Ambrose-Singer Theorem is not very explicit since the normal holonomy algebra depends also on parallel transport $\tau_\gamma^\perp$. Thus it is not very useful for explicit computations.

In some cases, like homogeneous submanifolds, one can compute the normal holonomy by taking the covariant derivatives of the normal curvature tensor.

PROPOSITION 4.5.3
Let $M \to \bar{M}(\kappa)$ be a homogeneous submanifold. The Lie algebra $L(\Phi_p)$ of the normal holonomy group is generated by the skew symmetric operators on $\nu_p M$ of the form

$$(\nabla^\perp R^\perp)^k_{V_1 \dots V_k}(X, Y),$$

where $X, Y, V_1, \dots, V_k \in T_p M$, $k = 0, 1 \dots$.

PROOF See [117, vol. I, Theorem 9.2, page 152] for a proof in the general case of a linear connection. ◻

Example 4.3 normal holonomy of $SO(n) \hookrightarrow \mathfrak{gl}(n,\mathbb{R}) \cong \mathbb{R}^{n^2}$. Using the above proposition one can see that

$$L(\Phi_I^*) = \text{span}\,\{R^\perp(A^*, B^*)\}_{A,B\in\mathfrak{so}(n)} = \mathfrak{so}(n)$$

(Exercise, cf. [25]). Note that $SO(n) \hookrightarrow \mathfrak{gl}(n,\mathbb{R}) \cong \mathbb{R}^{n^2}$ is an orbit of an s-representation (relative to the symmetric space $SO(2n)/SO(n) \times SO(n)$). ◻

In Section 6.2 we will give a description of normal holonomy of orbits in terms of projection of Killing vector fields (Theorem 6.2.7). This yields a very practical tool for computing a homogeneous submanifold's normal holonomy.

4.6 Exercises

Exercise 4.6.1 The following exercise gives a sort of extrinsic version of de Rham's decomposition theorem. Let $M \to \bar{M}(\kappa)$ be a submanifold and Φ^c the combined holonomy (cf. page 97). Let $p \in M$ and suppose that both T_pM and ν_pM split as orthogonal direct sums $T_pM = T_1 \oplus T_2$, $\nu_pM = \nu_1 \oplus \nu_2$ whilst Φ^c splits as a product $\bar{\Phi}_1 \times \bar{\Phi}_2$ with $\bar{\Phi}_1$ acting trivially on $T_2 \oplus \nu_2$ and $\bar{\Phi}_2$ acting trivially on $T_1 \oplus \nu_1$. Assume further that $\alpha(T_1, T_1) \subseteq \nu_1$ and $\alpha(T_2, T_2) \subseteq \nu_2$. Prove that M is locally reducible.

Exercise 4.6.2 We use the same notation as on page 104 and define the loop $\gamma_{s,t}$ by the following procedure: Move on the coordinate lines from $f(0,0)$ to $f(s,0)$, then to $f(s,t)$, then backward to $f(0,t)$ and finally back to $f(0,0)$. Prove that

$$\tau^\perp_{\gamma_{s,0}}\xi = \tau^\perp_{\gamma_{0,t}}\xi = \xi$$

and that

$$R^\perp(u,v)\xi = -\frac{\partial^2}{\partial s\partial t}_{|(0,0)}\tau^\perp_{\gamma_{s,t}}\xi\,.$$

Conclude that

$$R^\perp(u,v)\xi = -\frac{1}{2}\frac{d^2}{dt^2}_{|t=0}\tau^\perp_{\gamma_{t,t}}\xi\,.$$

Exercise 4.6.3 Prove Proposition 4.4.17.

Exercise 4.6.4 Compute the restricted normal holonomy group of the Veronese surface (cf. Section 3.3).

Exercise 4.6.5 Let M be a submanifold of the Euclidean space and ξ a parallel normal vector field. Consider locally the parallel foliation $M_{t\xi}$ (t small). Prove that the shape operator A_ξ of M has constant eigenvalues if and only if, for any t, the shape operator A^t_ξ of $M_{t\xi}$ has constant trace.

Exercise 4.6.6 (cf. [217]) A submanifold M of a space of constant curvature is called a Weingarten submanifold if it has flat normal bundle and its principal curvatures satisfy a polynomial relation [217]. Show that if M is Weingarten, then so are any of its parallel (nonfocal) manifolds. Prove conversely that, if M has a parallel (nonfocal) manifold that is Weingarten, then M is Weingarten.

Exercise 4.6.7 Using Proposition 4.5.3, compute the restricted normal holonomy group of the third standard embedding of $F^3 : S^2(\sqrt{6}) \to S^6 \subset \mathbb{R}^7$ defined by

$$(x, y, z) \in S^2(\sqrt{6}) \mapsto (u_1, u_2, u_3, u_4, u_5, u_6, u_7) \in \mathbb{R}^7,$$

where

$$u_1 = \frac{1}{24}x(-x^2 - y^2 + 4z^2) \qquad u_2 = \frac{1}{24}y(-x^2 - y^2 + 4z^2)$$

$$u_3 = \frac{\sqrt{10}}{24}z(x^2 - y^2) \qquad u_4 = \frac{\sqrt{10}}{12}xyz$$

$$u_5 = \frac{\sqrt{15}}{72}x(x^2 - 3y^2) \qquad u_6 = \frac{\sqrt{15}}{72}y(3x^2 - y^2)$$

$$u_7 = \frac{\sqrt{6}}{72}z(-3x^2 - 3y^2 + 2z^2).$$

Exercise 4.6.8 Using Proposition 4.5.3, compute the restricted normal holonomy group of the orbits of $\mathrm{Ad}(SU(n))$ (cf. Section 3.2).

Exercise 4.6.9 Let $M \to \bar{M}^n$ be an isoparametric submanifold of a space form, i.e, a submanifold with flat normal bundle and such that the eigenvalues of the shape operator with respect to parallel normal fields are constant. Prove that there exist g $\nabla^\perp$-parallel normal vector fields $n_1, ..., n_g$ (the *curvature normals*), with $g \geq \mathrm{codim} M =: h$, such that the eigenvalues of the shape operator A_ξ are $\langle \xi, n_1 \rangle$,..., $\langle \xi, n_g \rangle$ and $n_1, ..., n_h$ is a global parallel frame of νM.

Exercise 4.6.10 Let M be a full isoparametric submanifold of a space form, $\lambda_0(\xi), \lambda_1(\xi),\ldots, \lambda_g(\xi)$ the common eigenvalues of A, with $\lambda_0(\xi) \equiv 0$. Denote by E_i, $i = 0, \ldots, g$ the common eigendistributions of the shape operator (called the *curvature distributions of* M). Prove that any E_i is autoparallel. *Hint:* Use the Codazzi equations. If you are stuck, see the proof of Lemma 4.4.2.

Exercise 4.6.11 Let M be a full isoparametric submanifold of a space form. Prove that the curvature normals $n_1, \ldots n_g$ span the first normal bundle.

Exercise 4.6.12 (cf. [205]) Let M be an m-dimensional isoparametric submanifold of a space form and denote by k_1 the constant dimension of its first normal space and by g the number of distinct curvature normals. Prove that

(a) $k_1 \leq g \leq m$;

(b) if $g \leq 2$ then $\nabla^\perp \alpha = 0$;

(c) if $g = 2$ then M locally splits.

Hints: For (a) use Exercise 4.6.11, for (b) and (c) Exercises 4.6.10 and 2.8.9.

Exercise 4.6.13 Let ξ be a parallel normal vector field. Prove that an eigendistribution E of the shape operator A_ξ, relative to an eigenvector λ, is autoparallel if and only if $d\lambda(v) = 0$ for $v \perp E$. *Hint:* see the proof of Lemma 4.4.2. or compute directly using Codazzi equation.

Exercise 4.6.14 A submanifold $M \to \mathbb{R}^n$ is said to have extrinsic homogeneous normal holonomy bundle if for any $p, q \in M$ and any piecewise differentiable curve $c : [0,1] \to M$ with $c(0) = p$, $c(1) = q$ there is an isometry g of $\mathbb{R}^n$ such that $g(M) = M$, $g(p) = q$ and $g_{*p|\nu_p M} : \nu_p M \to \nu_q M$ coincides with the $\nabla^\perp$ parallel transport along c. Prove that an orbit of an s-representation has extrinsic homogeneous normal holonomy bundle.

Exercise 4.6.15 Let $\tilde{\eta}$ be the normal field on the holonomy tube $\pi : (M)_{\eta_p} \to M$ defined by

$$\tilde{\eta}(\xi) = \xi - \pi(\xi), \qquad \xi \in (M)_{\eta_p}.$$

Prove that $\tilde{\eta}$ is parallel in the normal connection $(M)_{\eta_p}$. *Hint:* see the proof of Proposition 4.4.12.

Exercise 4.6.16 Prove the following analogue of Theorem 4.5.2. Let M be a surface of S^n not contained in a proper affine subspace (or, equivalently, in a smaller dimensional sphere). Then the local normal holonomy group is either trivial or it acts transitively on the unit sphere of the normal space.

Exercise 4.6.17 Let $G \subset SO(3)$. Prove that either G is transitive on the unit sphere of $\mathbb{R}^3$ or it has a fixed vector.

Chapter 5

Isoparametric submanifolds and their focal manifolds

This section is devoted to the study of generalizations of the concept of isoparametric hypersurface to higher codimensions.

The beginning of the history of generalizations of isoparametric hypersurfaces to higher codimensions goes back to the 80s, with the (sometimes independent) work of many authors: J. Eells [76], D. Gromoll and K. Grove [92], Q.-M. Wang [238], C.E. Harle [93], W. Strübing [205] and C.-L. Terng [216]. There are different aspects of these generalizations that are actually strictly related: *isoparametric maps*, *isoparametric submanifolds* and *submanifolds with constant principal curvatures* (for more details on the historical development see [220]).

The general definition of an isoparametric map is credited to C.-L. Terng. It is a smooth map $f = (f_1, \dots, f_k) : \bar{M}^n(\kappa) \to \mathbb{R}^k$, such that,
(a) f has a regular value
(b) $\langle \operatorname{grad} f_i, \operatorname{grad} f_j \rangle$ and Δf_i are functions of f, for any i, j
(c) $[\operatorname{grad} f_i, \operatorname{grad} f_j]$ is a linear combination of $\operatorname{grad} f_1, \dots, \operatorname{grad} f_k$, with functions of f as coefficients, for any i, j.

Condition (b) means that $\langle \operatorname{grad} f_i, \operatorname{grad} f_j \rangle$ and Δf_i are constant on the regular level sets, i.e., the inverse images $f^{-1}(\{x\})$ of regular values x, which exist by (a). This condition assures, as in the case of isoparametric hypersurfaces, that the regular level sets are equidistant and have parallel mean curvature. This latter fact implies that the principal curvatures along parallel normal vector fields are constant (Exercise 5.6.1). Condition (c) says that the normal distribution determined by $\operatorname{grad} f_1, \dots, \operatorname{grad} f_k$ is integrable (and *a fortiori* parallel) and the normal bundle $\nu\left(f^{-1}(\{x\})\right)$ to each regular level set is flat (cf. [187, Chapter 6]). Thus, regular level submanifolds of an isoparametric map are isoparametric submanifolds of codimension (or rank) k (Theorem 5.1.2). Conversely, any isoparametric submanifold M of $\mathbb{R}^n$ determines a polynomial isoparametric map on $\mathbb{R}^n$, which has M as a regular level set (cf. [187, Section 6.4] and here Section 5.2).

The notion of isoparametric submanifold is nowadays also regarded as originally given by C.-L. Terng, even though it was first stated by C.E. Harle in [93]. An important example (by Thorbergsson's Theorem 5.4.5, the only one if the codimension is greater or equal to three) is given by the principal orbits of s-representations, as shown in Section 5.4.

In Section 5.2 we will discuss geometric properties of isoparametric submanifolds. Among them is the important fact due to C.-L. Terng [216] (and S. Carter and

A. West [38], in the particular case of codimension three), that one can associate to isoparametric submanifolds a finite reflection group, the *Coxeter group*.

The singular levels of isoparametric maps are actually focal manifolds of the isoparametric submanifolds. Thus, isoparametric maps determine a singular foliation of the ambient space. If M is a fixed isoparametric submanifold of $\mathbb{R}^n$, the leaves are parallel manifolds $M_\xi = \{p + \xi(p) \mid p \in M\}$, where ξ is an arbitrary parallel normal vector field.

If, in the definition of isoparametric submanifolds, one drops the assumption that the normal bundle is flat and requires only that the shape operator A_ξ has constant eigenvalues for any parallel normal vector field $\xi(t)$ along any piecewise differentiable curve, then one defines a submanifold of a space form with constant principal curvatures (cf. Section 4.1). Strübing in [205] studied these submanifolds (even though he called them isoparametric) and noticed that the focal manifolds of an isoparametric hypersurface are submanifolds with constant principal curvatures. This result was generalized by Heintze, Olmos and Thorbergsson in [96] to isoparametric submanifolds. Indeed, in [96] the converse is proved, namely that a submanifold with constant principal curvatures is either isoparametric or a focal manifold of an isoparametric submanifold (Theorem 5.3.3 here). The paper [205] of Strübing is actually of great importance for the methods adopted. He constructed tubes around isoparametric submanifolds and made use of the normal holonomy for the study of the submanifolds with constant principal curvatures. In fact, these are the techniques we extensively make use of.

In Section 5.5, we examine a more general situation than isoparametricity. We suppose (as in Section 4.4), that there exists a (locally defined) parallel nonumbilical normal section that is isoparametric, i.e., the eigenvalues of the shape operator A_ξ in the direction of ξ are constant. Our aim is to study the geometric consequences of this property. What we will show is that it imposes severe restrictions on the geometry of the submanifold. Namely, if a submanifold of the sphere with such property does not locally split, it is a submanifold with constant principal curvatures [61], Theorem 5.5.2. A global version for complete simply connected submanifolds can be also stated ([70], Theorem 5.5.8 in these notes). By this result we see that the definition of isoparametric submanifolds or, more generally, of submanifolds with constant principal curvatures, cannot be weakened much by assuming the existence of just one nontrivial isoparametric section. Roughly speaking, if one such section exists, then many exist.

5.1 Submersions and isoparametric maps

In this section, we show that isoparametric submanifolds of Euclidean space correspond to regular level sets of isoparametric maps. Moreover, we will sketch the proof of the fact that any isoparametric submanifold is a regular level set of an isoparametric polynomial. To make exposition easier, we restrict the discussion to Euclidean spaces, but everything holds for space forms.

In order to simplify proofs and obtain a better understanding of the concepts involved, we will need some generalities about Riemannian submersions (see [180]). For the sake of completeness, we include a brief account of the topic.

Let M and N be Riemannian manifolds. A surjective $\mathcal{C}^\infty$ map $\pi : M^{n+k} \to N^k$ is called a *Riemannian submersion* if $\pi_{*p} : (\ker(\pi_{*p}))^\perp \to T_{\pi(p)}N^k$ is a linear isometry, for all $p \in M$. The distribution $\mathcal{V}$ on M defined by $q \mapsto \ker(\pi_{*q})$ is called *vertical distribution* and its orthogonal complement $\mathcal{H}$ is called *horizontal distribution*. The vertical distribution is always integrable (and, in general, with non totally geodesic leaves), because any leaf of the vertical distribution is a connected component of the fibre (i.e., the preimage by π of an element in N). Let $c : I \to N$ be a (differentiable) curve. A curve $\tilde{c} : I \to M$ is called *horizontal lift* of c if $\pi \circ \tilde{c} = c$ and $\tilde{c}$ is a horizontal curve, i.e. $\dot{\tilde{c}}(t) \in \mathcal{H}_{\tilde{c}(t)}$, for all $t \in I$. If $c : I \to N$, $t_0 \in I$ and $q \in \pi^{-1}(\{c(t_0)\})$, there exists a unique horizontal lift $\tilde{c}$ of c, defined in a neighbourhood of t_0 such that $\tilde{c}(t_0) = q$ (this is a standard fact involving ordinary differential equations). If β is a curve in M then $\| \frac{d}{dt}\beta(t)\| \geq \|\frac{d}{dt}(\pi \circ \beta)(t)\|$ with equality if and only if β is horizontal. Then, $L(\beta) \geq L(\pi \circ \beta)$ with equality again if and only if β is horizontal. From this, one obtains that the horizontal lift of a geodesic is a geodesic. So, if the horizontal distribution $\mathcal{H}$ is integrable, then it is also totally geodesic (see next Remark 5.1.1).

A (differentiable) vector field X of M is called *projectable* if it is π−related to some vector field Y of N, i.e. $\pi_{*p}X(p) = Y(\pi(p))$, for all $p \in M$. It is standard to show that X is projectable if and only if $\pi_{*p}X(p) = \pi_{*p'}X(p')$ when $\pi(p) = \pi(p')$. Given a vector field Y in N, there exists a unique field X in M that is both horizontal and π−related to Y (the so-called horizontal lift of Y). A projectable horizontal vector field on M is called a *basic* vector field. Observe that a vertical field of M is always projectable, since it is π−related to the null field of N. If X_1 and X_2 are fields of M π−related to fields Y_1 and Y_2 of N, respectively, then the bracket $[X_1, X_2]$ is π−related to $[Y_1, Y_2]$ (see [117]). So the bracket of two projectable fields is projectable, and the bracket of a projectable field by a vertical one is vertical. We can always choose locally an orthonormal frame $X_1, \cdots X_n, \xi_1, \cdots, \xi_k$ of M such that the first n fields are vertical and the last k are basic. Now

$$0 = \xi_i\langle X_l, \xi_j\rangle = \langle \nabla_{\xi_i}X_l, \xi_j\rangle + \langle X_l, \nabla_{\xi_i}\xi_j\rangle,$$

and since $\nabla_{\xi_i}X_l - \nabla_{X_l}\xi_i = [\xi_i, X_l]$ is vertical we obtain that

$$\langle \nabla_{X_l}\xi_i, \xi_j\rangle = -\langle X_l, \nabla_{\xi_i}\xi_j\rangle. \tag{5.1}$$

Observe that the left-hand side changes sign if we interchange ξ_i and ξ_j. Since $[\xi_i, \xi_j] = \nabla_{\xi_i}\xi_j - \nabla_{\xi_j}\xi_i$ we obtain that $\langle \nabla_{\xi_i}\xi_j, X_l\rangle = \frac{1}{2}\langle [\xi_i, \xi_j], X_l\rangle$, for all $i, j = 1, \ldots, k, l = 1, \ldots, n$, so

$$\langle \nabla_\xi \eta, X\rangle = \frac{1}{2}\langle [\xi, \eta], X\rangle \tag{5.2}$$

for all ξ, η basic and X vertical.

REMARK 5.1.1 Formulas (5.1) and (5.2) immediately imply the following equivalence (see [187]):
(a) The horizontal distribution is integrable.
(b) The horizontal distribution is totally geodesic.
(c) Basic vector fields are $\nabla^{\perp}$−parallel when regarded as normal fields to any fibre. □

This finishes our brief general account on Riemannian submersions.

Now let U be an open subset of $\mathbb{R}^{n+k}$ and $f : U \to \mathbb{R}^k$ differentiable with a regular value, i.e., there exists $v \in f(U)$ such that f_{*q} is onto for every q belonging to the level set $f^{-1}(\{v\})$ (such a level set is called regular level set). It follows from the definition in the introduction to this chapter that, in this setup, f is an *isoparametric map* if:

(i) The Laplacian $\Delta f = (\Delta f_1, \cdots, \Delta f_k)$ is constant along any level set of f.

(ii) The inner product $\langle \text{grad}\, f_i, \text{grad}\, f_j \rangle$ is constant along any level set of f, for any $i, j = 1, \cdots, k$.

(iii) The bracket $[\text{grad}\, f_i, \text{grad}\, f_j]$ is a linear combination of $\text{grad}\, f_1, \cdots, \text{grad}\, f_k$, and its coefficients are constant along any level set of f.

THEOREM 5.1.2
A submanifold of Euclidean space is isoparametric if and only if it is locally a regular level set of an isoparametric map.

PROOF Let $f : U \subset \mathbb{R}^{n+k} \to \mathbb{R}^k$ be an isoparametric map and let $v \in \mathbb{R}^k$ be a regular value of f. By making U smaller we can assume that $V := f(U)$ is an open subset containing v and that $f : U \to V$ is a submersion, i.e., f_{*p} is onto for all $p \in U$. (We can also assume that there are coordinates on U and V respectively, where the submersion is the standard one, i.e., $(x_1, \cdots, x_{n+k}) \mapsto (x_1, \cdots, x_k)$). We will define a Riemannian metric on V such that f becomes a Riemannian submersion (the metric on U is the Euclidean one). Let $q \in U$ and $\tilde{q} \in M(q) := f^{-1}(\{q\})$. We copy on T_qV the inner product of $(T_{\tilde{q}}M(q))^{\perp}$ via $f_{*\tilde{q}}$. Let us check that this is well defined. Observe that $f_{*\tilde{q}}((\text{grad}\, f_i)_{\tilde{q}}) = (\langle \text{grad}\, f_1, \text{grad}\, f_i \rangle, \cdots, \langle \text{grad}\, f_k, \text{grad}\, f_i \rangle)$ is constant along level sets. So, if $v \in T_qV$, $v = f_{*\tilde{q}}(\sum_{i=1}^{k} a_i (\text{grad}\, f_i)_{\tilde{q}})$, the coefficients a_i do not depend on $\tilde{q} \in f^{-1}(\{q\})$ (recall $\text{grad}\, f_1, \cdots \text{grad}\, f_k$ is a basis of the normal space to level sets). Since $\langle \text{grad}\, f_i, \text{grad}\, f_j \rangle$ is constant on level sets, the Riemannian metric on V is well defined. So f is a Riemannian submersion. From condition (iii) the horizontal distribution is integrable and hence, with totally geodesic leaves by (5.2). The horizontal distribution is given by the normal spaces to the level sets and, by Exercise 5.6.2, the latter form a foliation by parallel manifolds. Notice that $\text{grad}\, f_i$ is a basic field, for all $i = 1, \cdots, k$, so $\nabla^{\perp}$-parallel along any fibre by Remark 5.1.1. Thus, $(\text{grad } f_1, \cdots, \text{grad } f_k)$ provides a $\nabla^\perp$-parallel frame of the normal bundle at any level set $M(q)$.

To prove that $M(q)$ is isoparametric, it suffices to show that the parallel foliation by level sets has parallel mean curvature H (see Exercise 5.6.1). To show that H is parallel, we will prove that $\langle H, \text{grad } f_i \rangle$ is constant for any $i = 1, \cdots, k$. Let $e_1, \cdots e_n, e_{n+1}, \cdots, e_{n+k}$, be a local orthonormal frame on U such that $e_1, \cdots e_n$ are tangent to the leaves and $e_{n+1}, \cdots, e_{n+k}$ normal. Since $\nabla^\perp \text{grad } f_i = 0$

$$\text{tr}(A_{\text{grad } f_i}) = \sum_{j=1}^{n} \langle \bar{\nabla}_{e_j} \text{grad } f_i, e_j \rangle = \sum_{j=1}^{n+k} \langle \bar{\nabla}_{e_j} \text{grad } f_i, e_j \rangle = \Delta f_i$$

is constant by (i).

Let us now sketch the proof of the converse. Let M be an n-dimensional isoparametric submanifold of $\mathbb{R}^{n+k}$, $q \in M$. Let U be an open subset of the ambient space containing q and such that the foliation by parallel manifolds to M is nonsingular. Let $i : \nu_q(M) \to \mathbb{R}^k$ be a linear isomorphism and restrict it to $U \cap \nu_q(M)$. Now extend this map to $\bar{i} : U \to \mathbb{R}^k$ by defining it to be constant on the parallel manifolds. We leave the reader to prove that $\bar{i}$ is an isoparametric map having M as a regular level set. □

5.2 Isoparametric submanifolds and Coxeter groups

In this section we start the study of the isoparametric submanifold geometry and construct the Coxeter group.

a) Curvature distributions and curvature normals

Let M be an isoparametric submanifold of a space form $\bar{M}^n(\kappa)$, $\kappa = -1, 0$ or 1. If $\kappa = 1$, we will regard $\bar{M}^n(1)$ as a unit sphere, and M as a submanifold of Euclidean space. If $\kappa = -1$, M is regarded as a Riemannian submanifold of Lorentz space. Remark that an isoparametric submanifold of a sphere (resp. hyperbolic space) is still isoparametric when regarded as a submanifold of Euclidean (resp. Lorentz) space, since the radial vector is umbilical (cf. Exercise 5.6.4). Throughout this section, $\bar{\nabla}$ and A will denote the directional derivative and the shape operator in Euclidean or in Lorentz space, respectively.

Since the normal bundle of M is globally flat, normal holonomy is trivial and, by the Ricci equations, the shape operators $\{A_\xi\}_{\xi \in \nu_p M}$ are simultaneously diagonalizable. Let $\lambda_0(\xi), \lambda_1(\xi), \ldots, \lambda_g(\xi)$ be the common eigenvalues of A at p, with $\lambda_0(\xi) \equiv 0$. We have common eigendistributions E_i, $i = 0, \ldots, g$ (the *curvature distributions of* M) with

$$A_\xi X_i = \lambda_i(\xi) X_i, \qquad X_i \in E_i,$$

(autoparallel, Lemma 4.4.2) and $\lambda_i : \nu_p M \to \mathbb{R}$ linear functions, or $\lambda_i \in (\nu_p M)^*$. Using the metric $\lambda_i = \langle n_i(p), . \rangle$, determining normal vector fields n_i called *curvature normals*. Since the eigenvalues $\lambda_i(\xi)$ are constant if ξ is parallel, *the curvature normals are parallel in the normal connection.* If the ambient space is Lorentzian, the curvature normals can be defined as well and lie in Lorentz space.

Choose parallel $\xi \in \nu M$ such that $\langle \xi, n_i \rangle$ are all different. (For a fixed $q \in M$, pick $\xi_q \in \nu_q M$ outside the hyperplanes $\langle n_i(q) - n_j(q), . \rangle = 0, i \neq j$, and extend it to a parallel normal field.) The eigendistributions associated to A_ξ are the common eigendistributions $E_0, \cdots, E_g$ of the commuting family of shape operators ($E_0 = \ker A_\xi$ is the possibly trivial nullity distribution).

REMARK 5.2.1 If M were an isoparametric submanifold of hyperbolic space, we could have defined the curvature normals by regarding M as a submanifold of H^n. Denoting them by $\tilde{n}_i$, we have relations $n_i = \tilde{n}_i + p$, where p is the position vector field. In particular, $n_i \neq 0$ for any $i = 1, \cdots, g$. Moreover, the eigendistributions associated to n_i and $\tilde{n}_i$ coincide, since the position vector field is umbilical. □

By Lemma 4.4.2, the eigendistributions E_i are autoparallel and invariant by all shape operators that operate on them as a multiple of the identity. Thus, the leaf $S_i(q) \subset q + E_i(q) + \nu_q(M)$ of E_i through q is a totally umbilical submanifold of the ambient space. In the Euclidean case, $S_i(q)$ is the open part of an affine subspace, if $n_i = 0$, or of a sphere, otherwise. In the Lorentzian case $S_i(q)$ is the open part of a sphere, if $\langle n_i(q), n_i(q) \rangle > 1$, of a horosphere, if $\langle n_i(q), n_i(q) \rangle = 1$ or of a hyperbolic space if $\langle n_i(q), n_i(q) \rangle < 1$.

The leaf of E_i at q will be denoted by $S_i(q)$ and called *curvature leaf at q relative to λ_i*.

The dimensions $m_i := \dim E_i$ are called *multiplicities* of the isoparametric submanifold.

***LEMMA* 5.2.2**

Let M be an isoparametric submanifold of Euclidean or hyperbolic space (regarded as a Riemannian submanifold of Lorentz space). Then M is full if and only if the curvature normals span the normal bundle, i.e., $\operatorname{span}(n_i(x)) = \nu_x M$*, for any $x \in M$. In particular,* $\operatorname{codim} M \leq \dim M$.

PROOF The curvature normals $n_1, \ldots, n_g$ do not span the normal space at any point if and only if there exists a parallel normal field $\xi \neq 0$ on M orthogonal to all curvature normals n_i. By Theorem 2.5.1 on the reduction of the codimension (which also applies to Lorentzian submanifolds) this is equivalent to M not to be full, since $A_\xi = 0$. □

REMARK 5.2.3 In view of Lemma 4.1.3, the above can be restated as follows: *the first normal space to an isoparametric submanifold is spanned by the curvature normals.* □

Actually, using the above Lemma, $S_i(q)$ is contained in the affine space $q+E_i(q)+\mathbb{R}n_i(q)$. This is due to Lemma 4.4.2 and the fact that the first normal space is generated by the n_i. So, in the Euclidean case, we have

LEMMA 5.2.4
Let M be an isoparametric submanifold of $\mathbb{R}^n$.

1. *If the curvature normal n_i does not vanish, then $S_i(q)$ is a an open part of a sphere of the affine subspace $q + E_i(q) + \mathbb{R}\, n_i(q)$.*
2. *If the curvature normal $n_i = 0$ then $S_i(q)$ is an open part of the affine subspace $q + E_i(q)$.*

Let ξ be any parallel normal vector field on an isoparametric submanifold M of Euclidean or Lorentz space. Then ξ is a parallel isoparametric section and (cf. Section 4.4) we can consider the parallel (possibly focal) manifolds M_ξ. Recall that $x + \xi(x)$ is a focal point if $\ker(\mathrm{id}_{T_xM} - A_{\xi(x)}) \neq 0$. We can write down this condition taking into account the curvature normals n_i. Indeed $A_{\xi|E_i} = \langle n_i, \xi\rangle \mathrm{id}_{E_i}$, so the matrix of $\mathrm{id}_{T_xM} - A_{\xi(x)}$ is given by

$$\begin{pmatrix} 1-\langle n_0,\xi\rangle \ldots & 0 \\ \ldots & \\ \cdot & \\ \cdot & \\ \cdot & \\ 0 & \ldots 1-\langle n_g,\xi\rangle \end{pmatrix}.$$

The focal points belong to the hyperplanes of $\nu_x M$

$$\ell_i(x) := \{1 - \langle n_i(x), \xi(x)\rangle = 0\}.$$

Each $\ell_i(x)$ is called *focal hyperplane associated to E_i.*

REMARK 5.2.5 Focal hyperplanes and focal set of an isoparametric submanifold are invariant by parallel transport in the normal bundle, since the curvature normals are $\nabla^\perp$-parallel. □

Thus, an isoparametric submanifold determines a singular foliation of Euclidean or Lorentz space, where each leaf is either isoparametric or a focal manifold of an isoparametric submanifold.

Note that, *for a compact immersed full isoparametric submanifold, all curvature normals are nonzero.* In fact, since M is complete, any leaf of the 0-eigendistribution is an affine subspace of the ambient space, which must be trivial since M is compact.

The *rank of an isoparametric submanifold* of Euclidean space is the maximal number of linear independent curvature normals.

Thus, for a full isoparametric submanifold of Euclidean space, the rank coincides with the codimension. By Lemma 5.2.2, Lemma 4.1.3 and the theorem on the reduction of the codimension (cf. Remark 5.2.3), we can assume that an m-dimensional isoparametric submanifold of rank k is contained in $\mathbb{R}^n$, with $n = m + k$.

b) The Coxeter group

Next, we associate a Coxeter group to a complete isoparametric submanifold M of Euclidean space (cf. [187]). A similar construction can be carried out in the Lorentzian case.

Let $p, q \in M$ and let $\tau_{p,q} : \nu_p M \to \nu_q M$ be the parallel transport. The affine parallel transport $\tilde{\tau}_{p,q} : p + \nu_p M \to q + \nu_q M$ is the unique isometry defined by: $\tilde{\tau}_{p,q}(p) = q$ and $(\tilde{\tau}_{p,q})_{*p} = \tau_{p,q}$.

From the isoparametric condition

$$\tilde{\tau}_{p,q}(F_M(p)) = F_M(q),$$

where $F_M(r)$ denotes the focal set of M at r.

Let σ_i^q denote the reflection in the affine space $q + \nu_q M$ across the hyperplane $\ell_i(q)$ corresponding to a (nonzero) space-like curvature normal.

LEMMA 5.2.6

$\sigma_i^q(q)$ is the antipodal point of q in the sphere $S_i(q)$. Moreover,

$$\sigma_i^q = \tilde{\tau}_{q,\sigma_i^q(q)}.$$

PROOF Using the fact that the curvature normal $n_i(q)$ is in the radial direction of $S_i(q)$, the first part is easily verified. Let us compute $\tau_{q,\sigma_i^q(q)}$. Consider a curve $c(t)$ in $S_i(q)$ from q to $\sigma_i^q(q)$. Recall that $S_i(q)$ is totally geodesic in M and invariant by the shape operator of M (Lemma 4.4.2). Therefore, parallel transport in the normal space to $S_i(q)$ (regarded as a submanifold of the whole ambient space) restricted to vectors in $\nu_q M$ coincides with parallel transport in the normal space to M. Since parallel transport in the sphere maps $n_i(q)$ to $n_i(\sigma_i^q(q)) = -n_i(q)$, we have $\tau_{q,\sigma_i^q(q)}(n_i(q)) = -n_i(q)$. Moreover, if ξ is in the orthogonal complement to $n_i(q)$ in $\nu_q M$ and $\xi(t)$ its parallel transport along $c(t)$ in the normal connection

$$\frac{d}{dt}\xi(t) = A^{S_i(q)}_{\xi(t)}\dot{c}(t) + \nabla^{\perp}_{\dot{c}(t)}\xi(t) = 0,$$

so $\xi(t)$ is constant and $\tau_{q,\sigma_i^q(q)}(\xi) = \xi$.

Thus, $(\tilde{\tau}_{q,\sigma_i^q(q)})_{*q} = \tau_{q,\sigma_i^q(q)} = (\sigma_i^q)_{*q}$. ∎

We now relate the antipodal maps with the focal structure. Define $\psi_i := 2\frac{n_i}{\langle n_i, n_i\rangle}$. So $\varphi_i(q) := q + \psi_i(q) = \sigma_i^q(q)$ is the antipodal map with respect to the curvature sphere $S_i(q)$. The isoparametric condition implies $\tilde{\tau}_{q,\varphi_i(q)}(F_M(q)) = F_M(\varphi_i(q))$. Note that φ_i can be regarded as the projection map sending M to the parallel manifold M_{ψ_i} (which coincides with M). From Proposition 4.4.11

$$F_M(q) = F_{M_{\psi_i}}(q + \psi_i(q)).$$

Since $M = M_{\psi_i}$ and $\varphi_i(q) = q + \psi_i(q)$, we conclude that $\tilde{\tau}_{q,\varphi_i(q)}(F_M(q)) = F_M(q)$.

Therefore, any reflection σ_i^q permutes the focal hyperplanes $\ell_1(q), \cdots, \ell_g(q)$, which generate a finite group of reflections [104]

THEOREM 5.2.7 (Terng)
The reflections σ_i^q through the hyperplanes $\ell_i(q)$ (corresponding to space-like curvature normals) generate a finite reflection group W^q, called Coxeter group of the isoparametric submanifold at q.

Note that, for any $q, q' \in M$, $W^q \cong W^{q'}$, because parallel transport along any curve joining q and q' conjugates W^q and $W^{q'}$. We then write W for W^q.

Observe that the reflections σ_i^q determine permutations of the curvature distributions. To see this, let $c(t)$ be a (piecewise differentiable) curve in M based at q and $\tilde{c}(t) := c(t) + \psi_i(t)$ a curve based at $\varphi_i(q)$. Then $\nu_q M = \nu_{\varphi_i(q)} M$ and by Lemma 4.4.6 a normal vector field ζ is $\nabla^\perp$-parallel along $c(t)$ if and only if it is $\nabla^\perp$-parallel along $\tilde{c}(t)$. Moreover

$$A_{\zeta(\varphi_i(q))} = A_{\zeta(q)}(\mathrm{id} - A_{\psi_i(q)})^{-1} = A_{\zeta(q)}(\varphi_i)_{*q}^{-1}.$$

Thus, if E_j is a curvature distribution of A at q, it is a curvature distribution of A at $\varphi_i(q)$, as well. Hence, there is a permutation p_i of $\{1, ..., g\}$ such that

$$(\varphi_i)_{*q} E_j(q) = E_{p_i(j)}(\varphi_i(q)) \tag{5.3}$$

Observe that $E_j(q)$ equals $E_{p_i(j)}(\varphi_i(q))$ as linear subspace of the ambient space. The curvature normals at $\varphi_i(q)$ are given by

$$n_{p_i(j)}(\varphi_i(q)) = \frac{1}{1 - \langle \psi_i(q), n_j(q)\rangle} n_j(\varphi_i(q)) \qquad j = 1, \ldots, g. \tag{5.4}$$

REMARK 5.2.8 The above relation imposes severe restrictions on the geometry of M. Indeed, as a consequence of (5.3) $m_i = \dim E_i = \dim E_{p_i(j)} = m_{\sigma_i(j)}$. ∎

In general, if G is a reflection group generated by reflections through hyperplanes π_j orthogonal to vectors ν_j, one defines the rank of G as the maximal number of independent vectors in $\{\nu_j\}$. Thus, the Coxeter group W associated to an isoparametric submanifold of Euclidean space has rank equal to the rank of the isoparametric submanifold. If M is a homogeneous isoparametric submanifold, we will see (Section 5.4) that M is a principal orbit of the isotropy representation of a symmetric space. The Coxeter group coincides then with the Weyl group of the corresponding symmetric space.

The complement in $q + \nu_q M$ of the union of the focal hyperplanes $\ell_i(q)$ is not connected. Let U be one of its connected components. Its closure $\overline{U}$ is a simplicial cone and a fundamental domain for the W-action on $q + \nu_q M$ (i.e., each W-orbit meets $\overline{U}$ at exactly one point). $\overline{U}$ is called the *Weyl chamber* for W.

Example 5.1 Isoparametric hypersurfaces in spheres

If M is a compact isoparametric hypersurface of the sphere, it has codimension 2 in Euclidean space. Its associated Coxeter group W has rank 2 (i.e., it is a reflection group of the plane). By the classification of finite Coxeter groups (see, e.g., [104, Chapter 2]) or direct inspection, W is the dihedral group of order $2g$ (group of symmetries of a regular g-gon), where g is the number of different curvature normals. The picture below shows the case $g = 3$.

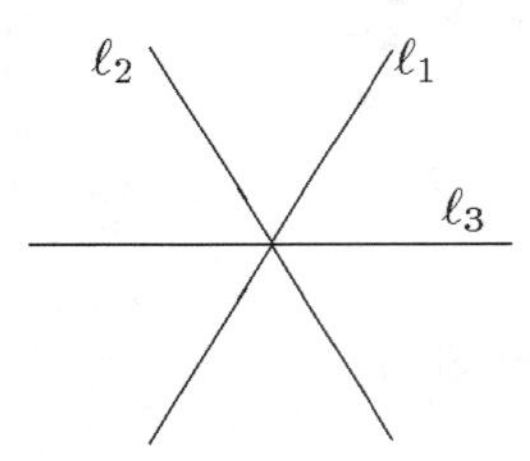

If g is odd, the Coxeter group is transitive on the set of focal lines. Thus, the multiplicities are all equal. If g is even, there are two orbits and two multiplicities. If we put indices so that ℓ_i and ℓ_{i+1} are adjacent, then $m_i = m_{i+2}$ for any $i \mod g$ ([144], cf. also Section 3.8).

By the remarkable results of Münzner [144,145], relying on delicate cohomological arguments, the number g of curvature normals can only be 1, 2, 3, 4 or 6 (in case $g = 1$ M is a sphere, and for $g = 2$ it is a product of spheres). In other terms, the Coxeter group of an isoparametric hypersurface of the sphere is crystallographic (i.e., it stabilizes a lattice in $\mathbb{R}^2$, see [104, Section 2.8]), cf. [187]. □

The fact that W is a finite reflection group forces the intersection $\bigcap_i \ell_i(q)$ to consist of one point, or, equivalently, W to have a fixed point.

This implies that if M is a rank k isoparametric submanifold of $\mathbb{R}^n$, there exists a parallel normal field ζ such that $\langle \zeta, n_i \rangle = 1$ for $i = 1, ..., g$ (indeed $q + v \in \ell_i(q)$

if and only if $\langle v, n_i \rangle = 1$). This observation will be crucial for the next reduction results.

c) Reduction theorems for isoparametric submanifolds of Euclidean space

The Coxeter group allows one to prove reduction results on isoparametric submanifolds. We begin with Euclidean ambient space.

Roughly speaking, what happens is that, given an isoparametric submanifold of Euclidean space, one can always split off its Euclidean factor. What is left is a product of compact isoparametric submanifolds with an irreducible Coxeter group.

As stated, we start with the following strong result of the fact that the Coxeter group has a fixed point.

PROPOSITION 5.2.9
Let $X : M \to \mathbb{R}^n$ be a rank k complete isoparametric submanifold such that all curvature normals do not vanish (e.g., M compact). Then M is contained in a sphere of radius $||\zeta||$ and centre $c_o = X + \zeta$.

PROOF Let ζ be a parallel normal field such that $\langle \zeta, n_i \rangle = 1$ for $i = 1, ..., g$, as above. Then $X + \zeta$ is a constant vector. Indeed, for any vector field v on M, $\bar{\nabla}_v (X + \zeta) = v - A_\zeta v$ vanishes because if $v_i \in E_i$, then $A_\zeta v_i = \langle n_i, \zeta \rangle v_i = v_i$ and $TM = \oplus_i E_i$. Therefore, $||X - c_o|| = ||\zeta||$. □

COROLLARY 5.2.10
Let M be a complete isoparametric submanifold of $\mathbb{R}^n$. Then the following are equivalent

1. *M is compact.*

2. *All curvature normals are non-zero.*

3. *M is contained in a sphere of $\mathbb{R}^n$.*

More in general, if M is not compact, one can split off a Euclidean factor, writing M locally as an extrinsic product of an isoparametric submanifold of the sphere with a Euclidean factor. Namely [187]

THEOREM 5.2.11
If M is a rank k complete isoparametric submanifold of $\mathbb{R}^n$ with nullity distribution E_0, there exists a rank k isoparametric submanifold M_1 of the sphere in $\mathbb{R}^{n-m_0}$ such that M splits as the extrinsic direct product of M_1 and E_0.

PROOF Let ζ be a parallel vector field such that $\langle n_i, \zeta \rangle = 1$ for any curvature normal. Then ζ focalizes every eigendistribution by the nullity. In other words, $\ker(\mathrm{id}-A_\zeta)$ is the eigendistribution corresponding to the nullity's complement. Since both nullity and $\ker(\mathrm{id}-A_\zeta)$ are autoparallel and invariant by all shape operators, Moore's Lemma implies that M splits. □

A local statement of the theorem is straightforward.

There is one more splitting result for isoparametric submanifolds. Suppose that M_1 and M_2 are isoparametric submanifolds of $\mathbb{R}^{m_1+k_1}$ and $\mathbb{R}^{m_2+k_2}$ respectively. Let W_i indicate the Coxeter group of M_i ($i = 1,2$). Then $M_1 \times M_2$ is an isoparametric submanifold with Coxeter group $W_1 \times W_2$ (of $\mathbb{R}^{m_1+m_2+k_1+k_2}$). The converse is also true.

THEOREM 5.2.12

Let M be an isoparametric submanifold of $\mathbb{R}^n$ with Coxeter group W. Then M is reducible if and only if W is reducible.

PROOF We have only to show that if W splits as $W_1 \times W_2$ where W_i is a Coxeter group on $\mathbb{R}^{k_i}$ ($i = 1,2$), then M splits as an extrinsic product of two isoparametric submanifolds M_1 and M_2. The converse is clear.

By Theorem 5.2.11, we can assume that all curvature normals $n_1, ..., n_g$ are nonzero. Suppose that the first p span W_1 and those remaining span W_2. Since W splits, these two sets of curvature normals are perpendicular to each other.

Let ζ be a parallel vector field such that $\langle n_i, \zeta\rangle = 1$ for any curvature normal. Then $\zeta = \zeta_1 + \zeta_2$, where ζ_1 (resp. ζ_2) is perpendicular to all curvature normals relative to W_2 (resp. W_1).

Now, the eigenstributions $\mathcal{D}_1 := \ker(\mathrm{id} - A_{\zeta_1})$ and $\mathcal{D}_2 := \ker(\mathrm{id} - A_{\zeta_2})$ are mutually orthogonal, autoparallel and invariant by all shape operators. Hence, they are both parallel and Moore's Lemma applies. □

d) The Slice Theorem

Essentially by the same proof as Lemma 4.4.2, we have the following

LEMMA 5.2.13

Let M be an isoparametric submanifold of Euclidean space, let ξ be a parallel normal vector field and consider the parallel focal manifold $\pi : M \to M_\xi$. For any $\bar{q} \in M_\xi$, each connected component of $\pi^{-1}(\bar{q})$ is an isoparametric submanifold of $V := \nu_{\bar{q}} M_\xi$.

Recall that, for a fixed isoparametric submanifold M, the parallel (possibly focal) manifolds M_ξ determine a singular foliation of the ambient space $\mathbb{R}^n$. Isoparametric leaves correspond to ξ_p in the interior of a Weyl chamber of W_p.

From the previous lemma we can deduce the important slice theorem of Hsiang, Palais and Terng [103].

THEOREM 5.2.14 (Slice Theorem)
Let M_ξ be a focal manifold of an isoparametric submanifold M of Euclidean space. Fix $\bar{q} \in M_\xi$, and let $V := \nu_{\bar{q}} M_\xi$. Then the parallel manifolds M_η of M intersect V in an isoparametric foliation of V.

We now compare the normal spaces of parallel focal manifolds, getting a generalization for isoparametric submanifolds of Lemma 4.4.11.

LEMMA 5.2.15
Let M be an isoparametric submanifold, ξ a parallel normal vector field and M_ξ be the corresponding parallel (focal) manifold. Let $\pi : M \to M_\xi$ be the projection and $\bar{q} = \pi(q)$ Then

$$\nu_{\bar{q}} M_\xi = \bigcup_{\tilde{q} \in \pi^{-1}(\bar{q})} \nu_{\tilde{q}} M .$$

PROOF From Lemma 5.2.13, we know each connected component of $\pi^{-1}(\bar{q})$ is an isoparametric submanifold of $\nu_{\bar{q}} M_\xi$. Thus, $\nu_{\bar{q}} M_\xi$ is the union of the leaves of the singular foliation determined by $\pi^{-1}(\bar{q})$. Let $y \in \nu_{\bar{q}} M_\xi$. There are $\hat{q} \in \pi^{-1}(\bar{q})$ and a parallel normal vector field η in $\pi^{-1}(\bar{q})$ such that $y = \hat{q} + \eta(\hat{q})$.

Since $\hat{q} \in \nu_{\hat{q}} M$ and $\eta(\hat{q}) \in \nu_{\hat{q}} \pi^{-1}(\bar{q}) = \nu_{\hat{q}} M$, y belongs to $\nu_{\hat{q}} M$. □

e) Applications to isoparametric hypersurfaces of spheres

Let us consider the case of isoparametric hypersurfaces in spheres more closely. We can actually use Coxeter groups to write down an explicit formula for the principal curvatures. We will, moreover, have a formula for the principal curvatures of the focal manifolds, showing they are minimal. This will also yield another proof of Cartan's fundamental formula 3.8.5.

Let $M \hookrightarrow S^n$ be an isoparametric hypersurface of the unit sphere S^n, $p \in M$ and ξ be a unit normal vector to M in S^n oriented so that (p, ξ) is a positive orthonormal frame of $\nu_p M$. Let $\lambda_1, ..., \lambda_g$ be the distinct principal curvatures (in direction of ξ). Set $y = -p$ (then y is pointing toward the centre of the sphere). Then a generic unit normal vector to M in $\mathbb{R}^{n+1}$ can be written as

$$\rho = \cos(t)\, \xi + \sin(t)\, y$$

and the shape operator A_ρ has eigenvalues $\lambda_i \cos(t) + \sin(t)$. Thus, it is easy to see that the curvature normals are

$$n_i = \lambda_i \xi + y \,.$$

If we set $\lambda_i =: \cot(\theta_i)$, with $0 < \theta_1 < ... < \theta_g < \pi$, θ_i is the angle between ξ and n_i and, by rank 2 Coxeter groups properties (see page 148), $\theta_i - \theta_{i-1} = \frac{\pi}{g}$. We have proved the following

THEOREM 5.2.16 (Münzner)
Let M be an isoparametric hypersurface of the sphere with distinct principal curvatures $\lambda_i = \cot(\theta_i)$ with $0 < \theta_1 < ... < \theta_g < \pi$ and multiplicities m_i. Then

$$\theta_k = \theta_1 + \frac{k-1}{g}\pi \qquad k = 1, ..., g$$

and the multiplicities satisfy $m_i = m_{i+2}$, (modulo g indexing).

Next, we consider parallel and focal manifolds in S^n. So, given $M \to S^n$ and a fixed unit normal field ξ to M in S^n, we move along the geodesic from any $p \in M$ in direction $\xi(p)$. We pass to the parallel (perhaps focal) manifolds $M_t := \{\cos(t)\, p + \sin(t)\, \xi(p) \mid p \in M\}$ and have maps $\varphi_t : M \to M_t$, $\varphi_t(p) = \cos(t)\, p + \sin(t)\, \xi(p)$.

The differential of φ_t at p is given by $\cos(t)\,\mathrm{id} - \sin(t)\, A_\xi$. If X is taken in the eigendistribution E_i relative to a principal curvature $\lambda_i = \cot(\theta_i)$,

$$(\varphi_t)_{*p}X = (\cos(t) - \sin(t)\cot(\theta_i))X = \frac{\sin(\theta_i - t)}{\sin(\theta_i)}X \,.$$

Thus, $\varphi_t(p)$ is focal if $t = \theta_i$, $i = 1, ..., g$ and $M_i := M_{\theta_i}$ focalizes E_i.

As in the case of submanifolds of Euclidean space (cf. Lemma 4.4.7), we can write a "tube formula" for the shape operator of M_t: if $\xi(p)$ is normal to M at p, then $\bar{\xi}(\varphi(p)) := -\sin(t)\, p + \cos(t)\, \xi(p)$ is normal to M_t at $\varphi(p)$ and the shape operator $\bar{A}$ of M_t is given by

$$\bar{A}_{\bar{\xi}} = (\sin(t)\,\mathrm{id} + \cos(t) A_\xi)(\cos(t)\,\mathrm{id} - \sin(t) A_\xi)^{-1} \,,$$

where we have to restrict to horizontal spaces, in case of focal manifolds (Exercise 5.6.3).

An easy computation shows that the principal curvatures in direction $\bar{\xi}$ of M_t are given by

$$\cot(\theta_k - t) \quad k = 1, ..., g \,,$$

(cf. also [44], page 246).

In case of a focal manifold M_i the principal curvatures in direction $\bar{\xi}$ at $\bar{p} = \varphi_i(p)$ are given by

$$\cot(\theta_k - \theta_i) = \frac{1 + \lambda_i \lambda_k}{\lambda_i - \lambda_k} \quad k = 1, ..., g,\, k \neq i \,,$$

and an easy generalization of Lemma 5.2.15 shows that normal vectors $\bar{\xi}(\bar{p})$, $p \in \varphi_i^{-1}(\bar{p})$, generate the normal space at $\bar{p}$ of the focal manifold M_i. In particular, we see that *the principal curvatures of the focal manifold are independent of the (unit) normal vector.*

We now apply Theorem 5.2.16 to prove the minimality of focal manifolds.

COROLLARY 5.2.17
Each focal manifold M_i of an isoparametric hypersurface in the sphere is minimal.

PROOF Let $\bar{\xi}$ be a unit normal to M_i. Then

$$\begin{aligned}\operatorname{tr}(A_{\bar{\xi}}) &= \sum_{h\neq i} m_h \cot(\theta_h - \theta_i) = \sum_{k=1}^{g-1} m_k \cot\left(\frac{k\pi}{g}\right) = \\ &= -\sum_{k=1}^{g-1} m_k \cot\left(\pi - \frac{k\pi}{g}\right) = -\sum_{k=1}^{g-1} m_k \cot\left(\frac{(g-k)\pi}{g}\right) = \\ &= -\sum_{j=1}^{g-1} m_{g-j} \cot\left(\frac{j\pi}{g}\right).\end{aligned}$$

If g is even, then $m_{g-j} = m_j$, so the last term above equals $-\operatorname{tr}(A_{\bar{\xi}})$, therefore $\operatorname{tr}(A_{\bar{\xi}}) = 0$. The same holds also if g is odd, since all multiplicities are equal in this case. ∎

Observe that, since $\cot(\theta_j - \theta_i) = \dfrac{1+\lambda_i\lambda_j}{\lambda_i - \lambda_j}$, $\operatorname{tr}(A_{\bar{\xi}}) = \sum_{j\neq i} m_j \dfrac{1+\lambda_i\lambda_j}{\lambda_i - \lambda_j}$, so minimality of the focal manifolds is equivalent to Cartan's fundamental formula. Therefore, Corollary 5.2.17 yields a proof of Cartan's fundamental formula.

REMARK 5.2.18 There is actually a proof of minimality of the focal manifolds that uses geometric considerations only. Here is an outline of this proof. Notice that the mean curvature of a focal manifold M_i is parallel, because M_i has constant principal curvatures. If the mean curvature of a focal manifold M_i were not trivial, since it is a parallel isoparametric section, one could focalize M_i again, which is impossible. ∎

f) Reduction theorem for isoparametric submanifolds of hyperbolic space

To study isoparametric submanifolds of hyperbolic space, we regard H^n as a subspace of Lorentz space L^{n+1}. We also need to consider, more generally, Riemannian submanifolds of L^{n+1}, i.e., submanifolds of Lorentz space whose tangent space is Riemannian. We just mention that for such submanifolds one can define the same basic objects, such as shape operators and normal connection, and the fundamental equations still hold.

We now want to prove a reduction theorem from Bingle Wu [244] for complete isoparametric submanifolds of real hyperbolic space. This states that any isoparametric submanifold of H^n, which is full when regarded in Lorentz space, splits as an extrinsic product of a smaller dimensional hyperbolic space and an isoparametric submanifold of a sphere. Recall that when the submanifold is not full (in Lorentz space) then it is also an isoparametric submanifold of a proper totally umbilical submanifold of hyperbolic space (which is also a space of constant curvature).

The proof follows A. Will, whose ideas [241] simplify the original arguments of [244].

If ξ is a parallel normal vector field, then, as in the Euclidean case, we can consider the parallel manifold $M_\xi = \{\bar{q} = q + \xi(q)\}$. This is always a Riemannian manifold (perhaps of lower dimension) since its tangent space at $\bar{q}$ is given by the Riemannian $T_{\bar{q}}M_\xi = (\mathrm{id} - A_\xi)_{\bar{q}}(T_qM)$.

Let ξ be a parallel normal vector field that distinguishes all eigenvalues of the shape operator (cf. the beginning of this section). By adding a suitable multiple of the position vector field (which is umbilical) we can assume that ξ is timelike and that $\langle \xi, n_i \rangle \neq 0$ for any $i = 1, \cdots, g$.

Fix such a ξ and let

$$\xi_i = \frac{1}{\langle \xi, n_i \rangle}\xi.$$

Then $\langle \xi_i, n_i \rangle = 1$ and $\langle \xi, n_j \rangle \neq 1$ for any $j \neq i$. Consider the focal manifold M_{ξ_i}. As in the Euclidean case, the normal space $\nu_{\bar{q}}M_{\xi_i}$ equals $\nu_qM \oplus E_i(q)$, which contains the leaf $S_i(q)$. More precisely, $S_i(q)$ is contained in the hyperbolic space of radius $\langle q - \bar{q}, q - \bar{q} \rangle = \langle \xi_i, \xi_i \rangle$.

The following lemma is the key to proving that the theories of isoparametric submanifolds of hyperbolic space and Euclidean space are related [241].

LEMMA 5.2.19
Let $A^i : \eta \mapsto A^i_\eta$ be the map from the normal space $\nu_{\bar{q}}M_{\xi_i}$ to the space of symmetric endomorphisms of $T_{\bar{q}}M_{\xi_i}$ determined by the shape operator A^i_η of M_{ξ_i} at $\bar{q} = q + \xi_i(q)$. Then $A^i|_{S_i(q)}$ is constant, and consequently, $A^i_v = 0$ for v tangent to $S_i(q)$.

PROOF For the proof we will need the following

Sublemma 5.2.20
Let S be a noncompact totally umbilical submanifold of the hyperbolic space $H^r(\kappa) \subset L^{r+1}$ (i.e., either a horosphere or a lower dimensional hyperbolic space) and let $l : L^{r+1} \to \mathbb{R}$ be a linear map. Then $l(S)$ either consists of one element only or it is unbounded.

PROOF of the sublemma Let γ be a geodesic in S. It suffices to show that $l(\gamma)$ is either constant or unbounded. Using the explicit description of geodesics given in Exercises 2.8.13 and 2.8.14, it is easy to get the result. □

Endow the space of symmetric automorphisms with the Riemannian metric given by $\langle B, C\rangle = \mathrm{tr}(BC)$. Observe that $\|B\| = \sqrt{\lambda_1^2 + \cdots + \lambda_m^2}$, where $\lambda_1, \cdots, \lambda_m$ are the eigenvalues of B. From the tube formula we deduce that the eigenvalues of A^i_r do not depend on $r \in S_i(q)$, so $\|A^i_r\|$ is constant on $S_i(q)$. Then by the sublemma, A^i is constant on $S_i(q)$. ☐

We now go on to the reduction theorem, for which we need

LEMMA 5.2.21

Let $M \to H^n \subset L^{n+1}$ be a complete, full, irreducible isoparametric submanifold. Then either all curvature normals are space-like or $g = 1$.

PROOF Suppose $g \geq 2$ and that there is a curvature normal, say n_1, with $\langle n_1, n_1\rangle \leq 0$. Let $q \in M$ and let $S_1(q)$ be the leaf through q associated to the eigendistribution E_1 corresponding to n_1. $S_1(q)$ is either a horosphere or a hyperbolic space contained into the normal space $\nu_{\bar{q}} M_{\xi_1}$. We will show that E_1 is parallel, so that M would split by the Lorentzian version of Moore's Lemma 2.7.4.

Since E_1 is autoparallel, it suffices to prove that $\frac{D}{dt}Y(t)$ belongs to $E_1(c(t))$ for any curve $c(t)$ lying in $E_1^\perp$ and any vector field $Y(t)$ along c lying in E_1. $\frac{D}{dt}$ denotes the covariant derivative along the curve $c(t)$ induced by the Levi-Civita connection on M. Such a $Y(t)$ can be regarded as a normal vector field to M_{ξ_1} along $\pi(c(t))$ (where $\pi : M \to M_{\xi_1}$ is the projection). By Lemma 5.2.19, $A^1_{Y(t)}$ vanishes, so

$$\begin{aligned}\tfrac{D}{dt}Y(t) = [\tfrac{d}{dt}Y(t)]_{T_{c(t)}M} = -A^1_{Y(t)}(\pi \circ c)'(t) + [\tfrac{d}{dt}Y(t)]_{E_1(c(t))} = \\ = [\tfrac{d}{dt}Y(t)]_{E_1(c(t))} \subset E_1(c(t)).\end{aligned}$$

Then all curvature normals are space-like. ☐

As a consequence, we have the following theorem from Bingle Wu [244]:

THEOREM 5.2.22 (Bingle Wu)

Let $M \to H^n \subset L^{n+1}$ be a complete full irreducible isoparametric submanifold. Then $M = H^n$.

PROOF Suppose that M has $g \geq 2$ different shape operator eigenvalues. We will derive a contradiction. By Lemma 5.2.21, all curvature normals are space-like. Let W be the Weyl group associated to the isoparametric submanifold, constructed exactly as in the Euclidean case.

The following sublemma is a fundamental tool for studying isoparametric submanifolds of hyperbolic space (see [241]):

Sublemma 5.2.23
Let $V_0 \subset \nu_q M$ ($\nu_q M$ denotes the normal space in Lorentz space) be the fixed point set of the Coxeter group W. Then $V_0^\perp$ is Euclidean and so $\dim V_0 \neq 0$.

PROOF Let $\prec \cdot, \cdot \succ$ be any positive definite scalar product on the vector space $\nu_q M$ and let $(\cdot, \cdot)$ be the scalar product given by

$$(x, y) := \sum_{w \in W} \prec wx, wy \succ .$$

Obviously $(\cdot, \cdot)$ is W-invariant.

Let B be the symmetric endomorphism of $\nu_q M$ relating the inner product $(\cdot, \cdot)$ with the original Lorentzian metric $\langle \cdot, \cdot \rangle$, i.e., $(Bx, y) = \langle x, y \rangle$ for any $x, y \in \nu_q M$. Any Lorentzian symmetry s of W is a Euclidean symmetry with respect to $(\cdot, \cdot)$, since $s^2 = \mathrm{id}$ and the fixed point set of s is a hyperplane.

Decompose orthogonally (relatively to the Euclidean metric) $\nu_q M = V_0 \oplus V_1 \oplus \cdots \oplus V_k$ into irreducible subspaces with respect to the W-action and let W_i be the group generated by reflections with trivial action on V_j, $j \neq i$. Then W splits as a product $W_1 \times \cdots \times W_k$, W_i acts irreducibly on V_i and trivially on V_j, if $j \neq i$. Observe that, if $i \neq j$, V_i is perpendicular to V_j with respect to the Lorentzian metric. In fact, if $j \neq 0$, and $x \in V_i$, $y \in V_j$ we have, for all $w \in W_j$,

$$\langle x, y \rangle = \langle w \cdot x, w \cdot y \rangle = \langle x, w \cdot y \rangle$$

Summing up over the elements of W_j we obtain that

$$|W_j| \langle x, y \rangle = \langle x, h \rangle = 0 ,$$

where $|W_j|$ is the cardinality of W_j and $0 = h = \sum_{w \in W_j} w \cdot y$, since it is a vector in V_j fixed by the W-action. Since B commutes with all elements in W, $B_{|V_i}$ is a constant multiple of the identity, for any i. Therefore, the restriction of the Lorentzian metric to any V_i is either positive definite, negative definite or zero.

Fix an index $i \neq 0$. We have two cases.

If $\dim V_i \geq 2$, the restriction of the Lorentzian metric to V_i has to be a positive multiple of the Euclidean one.

If $\dim V_i = 1$, $V_i = \mathbb{R}v$, then v must be a multiple of a (space-like) curvature normal. So V_i is Riemannian, for any $i = 1, \cdots, k$, and therefore $V_0^\perp = V_1 \oplus \cdots \oplus V_k$ is Euclidean. ∎

We continue with the proof.

The curvature normals, all space-like by Lemma 5.2.21, are perpendicular to V_0 (which is the intersection of all reflection hyperplanes). By Sublemma

5.2.23, the curvature normals span a Euclidean subspace of the normal space. So M cannot be full by Lemma 5.2.2, a contradiction. Therefore $g = 1$ and M is a totally umbilical submanifold. Since M is full and complete, M must be the whole hyperbolic space. □

g) Isoparametric submanifolds and polynomial isoparametric maps

Recall that an isoparametric submanifold M of $\mathbb{R}^n$ is a regular level of an isoparametric map f. Moreover, this map can be taken to be a polynomial map $f : \mathbb{R}^n \to \mathbb{R}^k$, where $k = \operatorname{codim}(M)$; see [187, 216]. Crucial in this result is the role played by the Coxeter group W. Any W-invariant homogeneous polynomial of $\mathbb{R}^k$ extends uniquely to an isoparametric polynomial map defined in $\mathbb{R}^k$. This is attained by defining the extension to be constant on the leaves of the singular foliation of $\mathbb{R}^n$ determined by the isoparametric submanifolds parallel to M. By the well known Chevalley Restriction Theorem (see [216]), the ring of W-invariant polynomials on $\mathbb{R}^k$ has exactly k generators $P_1, \cdots, P_k$. Then $f = (\tilde{P}_1, \cdots, \tilde{P}_k)$ is the desired isoparametric polynomial map, $\tilde{P}_i$ denoting the extension of P_i to $\mathbb{R}^n$.

The following results of C-L Terng [216, 217] depend strongly on the existence of these polynomial isoparametric maps.

(1) *Isoparametric submanifolds have globally flat normal bundle:* if in the isoparametric submanifold definition the normal bundle is required only to be locally flat, it follows that it is globally flat.

(2) *Extendability of isoparametric submanifolds:* a local isoparametric submanifold is an open part of a complete and embedded isoparametric submanifold.

5.3 Geometric properties of submanifolds with constant principal curvatures

Recall that a submanifold M of Euclidean space is a *submanifold with constant principal curvatures* if the eigenvalues of the shape operator $A_{\xi(t)}$ are constant for any parallel normal vector field $\xi(t)$ along any piecewise differentiable curve [96], cf. also Section 4.1, page 98.

a) Geometric characterization of submanifolds with constant principal curvatures

Let M be a submanifold of $\mathbb{R}^n$ and consider, for $\xi_p \in \nu_p M$, the holonomy tube $(M)_{\xi_p}$. We will need the following application of the Normal Holonomy Theorem

LEMMA 5.3.1
Let M be a submanifold of $\mathbb{R}^n$ and let $\xi \in \nu_p M$ belong to a principal orbit of the restricted normal holonomy group Φ_p^*. If ζ, η are in the normal space to the orbit $\Phi_p^* \cdot \xi$, then $[A_\zeta, A_\eta] = 0$.

PROOF We use the Normal Holonomy Theorem 4.2.1. Orthogonally decompose $\nu_p M = V_0 \oplus ... \oplus V_k$ into Φ_p^*-invariant subsets such that V_0 is the fixed set and Φ_p^* acts irreducibly on V_i, for $i = 1, ..., k$. Then $\Phi_p^* = \Phi_0 \times ... \times \Phi_k$, where $\Phi_0 = \{1\}$, Φ_i acts trivially on V_j if $i \neq j$ and Φ_i acts irreducibly on V_i, if $i \geq 1$. Moreover, for $i \geq 1$, either Φ_i acts transitively on the unit sphere of V_i or it acts as an s-representation (and more precisely as the isotropy representation of an irreducible Riemannian symmetric space of rank ≥ 2). Set $\xi := \xi_0 + ... + \xi_k$, where $\xi_i \in V_i$. Then $\Phi_p^* \cdot \xi = \{\xi_0\} \times \Phi_1 \cdot \xi_1 \times ... \times \Phi_k \cdot \xi_k$. Since $\Phi_p^* \cdot \xi$ is principal, each orbit $\Phi_i \cdot \xi_i$ is principal in V_i. Set $W_i :=$ {normal space at ξ_i of $\Phi_i \cdot \xi_i$ in V_i}. Then $\nu_\xi(\Phi_p^* \cdot \xi) = W_0 \times ... \times W_k$.

If $\zeta \in W_i$, $\eta \in W_j$, with $i \neq j$, Then $\mathcal{R}_p^\perp(\zeta, \eta) = 0$, so $[A_\zeta, A_\eta] = 0$.

Suppose $\zeta, \eta \in W_i$. If $\Phi_i \cdot \xi_i$ is a sphere of radius $||\xi_i||$, then ζ and η are linearly dependent and hence $[A_\zeta, A_\eta] = 0$. If Φ is not transitive on the sphere, then Φ_i acts on V_i as the isotropy representation of an irreducible Riemannian symmetric space of rank ≥ 2. Then $[V_i, \mathcal{R}^\perp_{p|V_i \times V_i}, \Phi_i]$ is a symmetric holonomy system, in which case, the normal space to a principal orbit at a point is a Cartan subalgebra. Then $\mathcal{R}_p^\perp(\zeta, \eta) = [\zeta, \eta] = 0$, and $[A_\zeta, A_\eta] = 0$. □

THEOREM 5.3.2
Suppose $\xi_p \in \nu_p M$ lies on a principal orbit of the restricted normal holonomy group and that $||\xi_p||$ is less than the focal distance of M. Then $(M)_{\xi_p}$ is isoparametric if and only if M has constant principal curvatures.

PROOF Recall (Proposition 4.4.12) that $(M)_{\xi_p}$ has flat normal bundle.

a) If $(M)_{\xi_p}$ is isoparametric then M has constant principal curvatures.

Let $\zeta(t)$ be a $\nabla^\perp$-parallel normal vector field in M along a curve $\gamma(t)$.

Since $\nu_p M = \cup_{\hat{p} \in \pi^{-1}(p)} \nu_{\hat{p}}(M)_{\xi_p}$ (Lemma 5.2.15) we can suppose that $\zeta(t) \in \nu_{\hat{\gamma}(t)}(M)_{\xi_p}$, where $\hat{\gamma}(t)$ is a horizontal lift of $\gamma(t)$. Let us denote with A and $\hat{A}$ the shape operators in M and $(M)_{\xi_p}$ respectively. By the "tube formula" 4.4.7,

$$A_{\zeta(t)} = \hat{A}_{\zeta(t)}|_{\mathcal{H}}[(\mathrm{id} - \hat{A}_{-\xi(t)})|_{\mathcal{H}}]^{-1}, \tag{5.5}$$

where $\xi(t)$ is the $\nabla^\perp$-parallel transport of ξ_p along γ or $\hat{\gamma}$ (cf. Lemma 4.4.6). Since $(M)_{\xi_p}$ has flat normal bundle, the $\hat{A}$'s are simultaneously diagonalizable and $A_{\zeta(t)}$ has constant eigenvalues by (5.5).

b) If M has constant principal curvatures then $(M)_{\xi_p}$ is isoparametric.

Let ζ be a parallel normal vector field to $(M)_{\xi_p}$. The restriction of $\hat{A}_\zeta$ to the vertical subspaces is equal to the shape operator of the fibres (because these are totally geodesic in $(M)_{\xi_p}$, cf. Lemma 4.4.5). Since the fibres are orbits of s-representations, they have constant principal curvatures (Proposition 4.1.6). So the restriction of $\hat{A}_\zeta$ to the vertical spaces has constant eigenvalues. Now, just as in Proposition 4.4.15, part *b*), the eigenvalues of $\hat{A}_\zeta$ do not change along a horizontal curve from one fibre to another. Hence $\hat{A}_{\zeta|\mathcal{V}}$ has constant eigenvalues. It remains to show that $\hat{A}_{\zeta|\mathcal{H}}$ has constant eigenvalues. Like the "tube formula" 4.4.7, one obtains

$$\hat{A}_{\zeta|\mathcal{H}} = A_\zeta(\mathrm{id} - A_\xi)^{-1} \tag{5.6}$$

Now, by Lemma 5.3.1, A_ξ and A_ζ commute, so equation (5.6) yields the result. ∎

We now discuss the following complete characterization of submanifolds with constant principal curvatures [96].

THEOREM 5.3.3
A submanifold M of Euclidean space has constant principal curvatures if and only if it is either isoparametric or a focal manifold to an isoparametric submanifold.

PROOF a) If M is a submanifold with constant principal curvatures, then it suffices to take a holonomy tube as in Theorem 5.3.2's proof.
b) If M is isoparametric, it clearly has constant principal curvatures. We show that if M is a focal manifold of an isoparametric submanifold $\hat{M}$, it must have constant principal curvatures. Proceed as in part *a*) of the proof of Theorem 5.3.2. Indeed, let $\eta(t)$ be a $\nabla^\perp$-parallel normal vector field in M along a curve $\gamma(t)$. Since $\nu_p M = \cup_{\hat{p}\in\pi^{-1}(p)} \nu_{\hat{p}} \hat{M}$, we can suppose that $\zeta(t) \in \nu_{\hat{\gamma}(t)} \hat{M}$, where $\hat{\gamma}(t)$ is a horizontal lift of $\gamma(t)$. Let A and $\hat{A}$ be the shape operators in M and $\hat{M}$ respectively. By the "tube formula" 4.4.7 $A_{\zeta(t)} = \hat{A}_{\zeta(t)}|_{\mathcal{H}}[(\mathrm{id} - \hat{A}_{-\xi(t)})|_{\mathcal{H}}]^{-1}$, where $\xi(t)$ is the $\nabla^\perp$-parallel transport of ξ_p along γ or $\hat{\gamma}$ (cf. Lemma 4.4.6). Since $\hat{M}$ has flat normal bundle, the $\hat{A}$'s are simultaneously diagonalizable and by (5.5), $A_{\zeta(t)}$ has constant eigenvalues. ∎

REMARK 5.3.4 The above theorem is also true for submanifolds of a space form (similar proofs). In the case of hyperbolic space, due to Theorem 5.2.22, one will have the following result: *Let M be a complete submanifold of the hyperbolic space $H^n \subset L^{n+1}$ which is full and irreducible in L^{n+1}. If M has constant principal curvatures, then $M = H^n$.* ∎

b) A Lemma on normal holonomy of submanifolds with constant principal curvatures

The following Lemma expresses the fact that, if ζ and η belong to different orbits of the normal holonomy group at q, then A_ζ and A_η have different eigenvalues.

LEMMA 5.3.5 ("Holonomy Lemma")
Let M be a full submanifold of $\mathbb{R}^n$ with constant principal curvatures. For any $q \in M$, the eigenvalues of the shape operator A locally distinguish different orbits of the restricted normal holonomy group Φ_q^*.

PROOF M is a full parallel manifold (possibly focal) of a full isoparametric submanifold M' of $\mathbb{R}^n$, i.e. $M = M'_\eta$ for some parallel vector field η. We denote by $\pi : M' \to M$ the projection of the isoparametric submanifold on the (possibly) focal one.

Let $n_1, ..., n_g$ be the different curvature normals of M'. We can assume without loss of generality that $\langle \eta, n_i \rangle \neq 1$, $i = 1, \ldots, r$, $\langle \eta, n_j \rangle = 1$, $j = r+1, \ldots, g$. By the "tube formula", if $q \in M$ and $q' \in \pi^{-1}(q)$, the eigenvalues of the shape operator A_ψ of M for $\psi \in \nu_{q'}(M') \subset \nu_q(M)$ are given by

$$\lambda_i(\psi) = \langle (1 - \langle \eta, n_i \rangle)^{-1} n_i, \psi \rangle \qquad i = 1, \ldots, r \, .$$

Let $\omega \in \nu_q(M)$ and consider the orbit $\Phi_q^* \cdot \omega$. By Lemma 5.2.15 $\nu_q(M) = \cup_{\tilde{q} \in \pi^{-1}(q)} \nu_{\tilde{q}}(M')$, so we can assume that ω belongs to $\nu_{q'}(M')$ and hence, the normal space at ω to $\Phi_q^* \cdot \omega$ (in $\nu_q(M)$) coincides with $\nu_{q'}(M')$. Note that we have used the fact that the eigenvalues of A are the same along the normal holonomy orbits (this follows immediately from the definition of submanifold with constant principal curvatures).

We want to show that nearby orbits to $\Phi_q^* \cdot \omega$ correspond to different eigenvalues for the shape operators.

We first assume that $\Phi_q^* \cdot \omega$ is a principal orbit. Since $\nu_{q'}(M')$ provides a slice to the normal holonomy action (which is polar) of Φ_q^* on $\nu_q(M)$, any orbit of the normal holonomy group Φ_q^* can be written as $\Phi_q^* \cdot (\omega + \mu)$ where μ belongs to $\nu_{q'}(M')$ and the distance between the two orbits is less or equal to $|\mu|$.

If $\{\lambda_i(\omega), \quad i = 1, ..., r\}$ is equal to $\{\lambda_i(\omega + \mu), \quad i = 1, ..., r\}$, then for $|\mu|$ small

$$\lambda_1(\omega) = \lambda_1(\omega + \mu), ..., \lambda_r(\omega) = \lambda_r(\omega + \mu) \; ,$$

which implies $\langle n_1, \mu \rangle = ... = \langle n_r, \mu \rangle = 0$. But $n_1(q), ..., n_r(q)$ generate $\nu_{q'}(M')$ since M (hence M') is full (Lemma 5.2.2). Thus, $\mu = 0$, i.e., the orbits coincide.

If the orbit $\Phi_q^* \cdot \omega$ is not principal, there are principal orbits that are arbitrarily close to it. Using this fact, it is not difficult to see that the lemma holds in this case as well, ending the proof. ∎

c) The Homogeneous Slice Theorem

As a result of Lemma 5.3.5, we get the following important result taken from [96]:

THEOREM 5.3.6 (Homogeneous Slice Theorem)
The fibres of the projection of a complete isoparametric submanifold onto a full focal manifold are orbits of an s-representation.

PROOF If M' is an irreducible full isoparametric submanifold, $\pi : M' \to M$ is a focal manifold, a fibre F of π is a union of normal holonomy orbits of the focal manifold. The eigenvalues of the shape operator of M on the whole fibre F are constant. Hence, by the Holonomy Lemma, the latter's connected component should consist of only one orbit. The result then follows from the Normal Holonomy Theorem. □

An alternative proof can be found in [96].

In general, if ξ is a parallel normal isoparametric section and one builds the focal manifold M_ξ and then goes back to make up the holonomy tube $(M_\xi)_{-\xi(p)}$, the holonomy tube is contained in M (Proposition 4.4.14). With isoparametric submanifolds we get equality. Namely, as a corollary of the Holonomy Lemma:

COROLLARY 5.3.7
Let M_ξ be a full focal manifold of a complete isoparametric submanifold M. Then $M = (M_\xi)_{-\xi(p)}$. Thus, a full focal manifold determines the isoparametric foliation.

5.4 Homogeneous isoparametric submanifolds

a) Homogeneous isoparametric submanifolds and orbits of s-representations

We saw in Section 3.2 that the principal orbits of an s-representations are isoparametric.

Conversely, if an orbit of an orthogonal representation of a compact Lie group is isoparametric, then the representation is polar and the orbit is principal. More precisely (cf. [186], Theorem 6.5),

THEOREM 5.4.1
Let M be a full compact isoparametric submanifold of $\mathbb{R}^n$. Suppose that $G = \{g \in I(\mathbb{R}^n) \mid g(M) \subseteq M\}$ acts transitively on M. Let G^o be the connected component at the identity of G. Then the representation $\rho : G^o \to O(n)$ is polar and M is a principal G^o-orbit.

PROOF Let $p, q \in M$ and $g : [0,1] \to G^o$ be a curve such that $g(0) = \mathrm{id}$ and $g(1)p = q$. Let τ_t be the $\nabla^\perp$-parallel transport along $\gamma(s) := g(s)p$ from 0 to t. Then $h(t) := \tau_t^{-1} \circ g(t)_{*p}$ permutes the curvature normals at p. Since the curvature normals span $\nu_p M$ and $h(0)$ is the identity transformation of $\nu_p M$, we conclude that $h(t) = \mathrm{id}$ for any t. Parallel transport is then given by the group. Therefore, M is a principal orbit and the action is locally polar hence polar (because the ambient space is Euclidean). □

We can reformulate the above result in the following way:

THEOREM 5.4.2 (Palais and Terng) [187]
Let M be an orbit of an orthogonal representation ρ of a compact Lie group. Then M is isoparametric if and only if ρ is polar and M is a principal orbit.

As a consequence, we get an important property of polar actions.

COROLLARY 5.4.3
Let $G \to O(V)$ be a polar action. Then a maximal dimensional orbit is principal.

PROOF Let $G \cdot p$ be a maximal dimensional orbit and take a principal orbit $G \cdot \xi$. Thus, $G \cdot p$ belongs to the isoparametric foliation determined by $G \cdot \xi$ and is parallel to $G \cdot \xi$. For dimensional reasons, $G \cdot p$ cannot be focal, so it is isoparametric and thus principal. □

In other words, the corollary tells us that polar actions have no exceptional orbit (cf. Section 3.1).

Moreover, using Dadok's Theorem we get:

THEOREM 5.4.4
Every extrinsic homogeneous isoparametric submanifold is a principal orbit of an s-representation.

Under high rank assumptions, any full isoparametric submanifold is homogeneous, and thus, by the previous results, an orbit of an s-representation. More precisely, one has the following theorem, which was first proved by Thorbergsson [219] (another proof [174] of which will be given in Section 7.3)

THEOREM 5.4.5 (Thorbergsson)
Any full irreducible isoparametric submanifold of $\mathbb{R}^n$ of rank at least three is an orbit of an s-representation.

In Section 6.1, we will introduce the idea of rank for homogeneous submanifolds. We will see (Theorem 6.1.7) that this is related to orbits of s-representations, namely, a full irreducible homogeneous submanifold of the sphere of rank greater than one is an orbit of an s-representation [175].

b) Transvections

The *transvection group* of a Riemannian manifold N is the group $\mathrm{Tr}(N)$ of isometries of N that preserve the holonomy subbundle at any point. In other words, $\mathrm{Tr}(N)$ is the group of all isometries φ such that, for any $p \in M$, there exists a piecewise differentiable curve γ joining p and $\varphi(p)$ for which the differential $\varphi_{*p} : T_pM \to T_{\varphi(p)}M$ coincides with the parallel displacement along γ.

Now, a symmetric space can be characterized by the fact that the transvection group acts transitively on the holonomy bundle. This is to say that, for any $p, q \in M$ and any piecewise differentiable curve γ from p to q, there exists an isometry g such that $g(p) = q$ and $g_{*p} : T_pM \to T_qM$ coincides with the parallel transport along γ.

We will see that there is a similar property for orbits of s-representations, this time involving the transvections of the normal connection.

Let M be a submanifold of $\mathbb{R}^n$ and let $g \in I(\mathbb{R}^n)$ be such that $g(M) = M$.

DEFINITION 5.4.6 *g is a transvection of M with respect to the normal connection $\nabla^\perp$ if for any $p \in M$ there exists a piecewise differentiable curve $c : [0,1] \to M$ with $c(0) = p, c(1) = g(p)$ such that*

$$g_{*p|\nu_pM} = \tau_c^\perp$$

where $\tau_c^\perp$ is the parallel transport along c with respect to $\nabla^\perp$.

The set of all transvections of M (with respect to $\nabla^\perp$) is a subgroup of $I(\mathbb{R}^n)$ and will be denoted by $\mathrm{Tr}(M, \nabla^\perp)$.

In the definition of a transvection, one can replace the word "any" with "some". In fact, if q is any other point, let γ be a curve joining q with p. Then $g_{*q|\nu_qM}$ coincides with the parallel transport along the curve $\gamma * c * (g \circ \tilde{\gamma})$, obtained by glueing γ with c and then with the opposite curve $g \circ \tilde{\gamma}$ of $g \circ \gamma$. In fact, since g is an isometry, it maps parallel normal fields along curves into parallel normal fields along curves. So

$$\tau_{g\circ\gamma}^\perp \circ g_{*q} = g_{*p} \circ \tau_\gamma^\perp = \tau_c^\perp \circ \tau_\gamma^\perp \,, \tag{5.7}$$

wherefore

$$g_{*q} = \tau_{g\circ\tilde{\gamma}}^\perp \circ \tau_c^\perp \circ \tau_\gamma^\perp \,,$$

with g_* always restricted to the normal space.

REMARK 5.4.7 For the reader who is familiar with the notion of connections in principal bundles: let $\mathcal{B}(\nu M) \to M$ be the principal bundle of orthonormal frames of the normal bundle (the structure group is $O(k)$,

$k = \operatorname{codim}(M)$). If g is an isometry of the ambient space such that $g(M) = M$, then g induces a map $g_* : \mathcal{B}(\nu M) \to \mathcal{B}(\nu M)$ in a natural way. Namely, $g_*(u_1, \cdots, u_k) = (g_{*p}u_1, \cdots, g_{*p}u_k)$, where $u_1, \cdots, u_k$ is an orthonormal basis of $\nu_p M$. One has that g_* maps holonomy subbundles into holonomy subbundles of $\mathcal{B}(\nu M)$, since g_* preserves both the horizontal and the vertical distribution. Then g is a transvection if and only if it leaves invariant any holonomy subbundle of $\mathcal{B}(\nu M)$. By right $O(k)$-invariance of the connection, g_* is a transvection if and only if g_* leaves invariant some holonomy subbundle (which corresponds to replacing the "any" with "some" in the definition of a transvection). In this context, it is now easy to prove that the transvections form a Lie subgroup of the extrinsic isometries of M. □

REMARK 5.4.8 Let $g \in I(\mathbb{R}^n)$ with $g(M) = M$. If, for some $p \in M$, $g(p) = p$ and $g_{*|\nu_p M} = \mathrm{id}$, then g is a transvection of M. In fact, $g_{*|\nu_p M}$ coincides with the $\nabla^\perp$-parallel transport along the curve $c(t) \equiv p$. Thus, the isotropy of a principal orbit at any given point is always contained in the group of transvections of the orbit. □

c) Homogeneous submanifolds with constant principal curvatures

The main class of examples of homogeneous submanifolds with constant principal curvatures is given by orbits of s-representations. Indeed, recall that, by Proposition 4.1.6, an orbit of an s-representation has constant principal curvatures.

Now, orbits of s-representations can be characterized by the fact that $\mathrm{Tr}(M, \nabla^\perp)$ acts transitively on the normal holonomy bundle. More explicitly, for any $p, q \in M$ and any curve γ on M joining p and q, there exists an isometry g of Euclidean space, leaving the submanifold M invariant, mapping p to q and such that

$$g_{*p|\nu_p M} : \nu_p M \to \nu_q M$$

coincides with the $\nabla^\perp-$ parallel transport along γ [178]. In this case, one also says that M *has extrinsic homogeneous normal holonomy bundle.*

THEOREM 5.4.9
Let M be a full connected compact submanifold of $\mathbb{R}^n$. The following are equivalent:

1. *M has extrinsic homogeneous normal holonomy bundle.*
2. *M is an orbit of an s-representation.*

PROOF We first assume that 1 holds, i.e., M has extrinsic homogeneous normal holonomy bundle. Then it is a submanifold with constant principal curvatures and thus, by Theorem 5.3.3, it is either isoparametric or a focal manifold of an isoparametric submanifold.

a) If M is isoparametric, since it also homogeneous (under the group $I(M, \mathbb{R}^n)$ of isometries of $\mathbb{R}^n$ leaving M invariant), by Theorem 5.4.2 it is a principal orbit of a polar representation. Moreover, by Dadok's Theorem 3.2.15, M is an orbit of an s-representation.
b) If M is the focal manifold of an isoparametric submanifold P, then P is a holonomy tube $(M)_{\xi(p)}$ obtained $\nabla^\perp$-parallel displacing $\xi(p) \in \nu_p M$ along piecewise differentiable curves in M. Let $I(M, \mathbb{R}^n)$ be the group of isometries of $\mathbb{R}^n$ leaving M invariant and $I(P, M) := \{g \in I(M, \mathbb{R}^n) \mid g(P) \subseteq P\}$. We show that P is a orbit of $I(P, M)$. Indeed, let $\hat{p} = p + \xi(p), \hat{q} = q + \tau_\gamma^\perp \xi(p) \in P$, where γ is some curve in M. We want to prove that there is some $g \in I(P, M)$ such that $g(\hat{p}) = \hat{q}$. Given γ, since M has extrinsic homogeneous normal holonomy bundle, there is a $g \in I(M)$ such that $g(p) = q$ and the restriction of g_* to the normal spaces is the parallel transport $\tau_\gamma^\perp$. Hence

$$g(\hat{p}) = g(p + \xi(p)) = g(p) + g_*(\xi(p)) = q + \tau_\gamma^\perp(\xi(p)) = \hat{q}.$$

We can now proceed in a fashion similar to *a)*, getting that P is a principal orbit of an s-representation. Hence M is an orbit of an s-representation as well.

For the converse statement, see Exercise 4.6.14. □

REMARK 5.4.10 In [178] the submanifolds with extrinsic homogeneous normal holonomy bundle are called homogeneous submanifolds with constant principal curvatures. We avoid this terminology because the classes of submanifolds with extrinsic homogeneous normal holonomy bundle (or equivalently, of orbits of s-representations) and of submanifolds that are at the same time homogeneous and have constant principal curvatures do not agree. Examples of homogeneous submanifolds with constant principal curvatures that are not orbits of s-representations can be found looking at nonhomogeneous isoparametric hypersurfaces in spheres. Indeed, it can happen that, although an isoparametric hypersurface M is not homogeneous, one of its focal manifolds M_ξ is homogeneous. M_ξ cannot be an orbit of an s-representation, otherwise this would be the case for M as well. An example of this type is provided in [87, page 497, Satz 6.4]. □

d) Normal holonomy of holonomy tubes

Let M be a submanifold of $\mathbb{R}^n$, which we assume to be simply connected.

We use the results obtained on the geometry of submanifolds with constant principal curvatures to describe the normal holonomy of the holonomy tube of M. The reason for the connection with the geometry of submanifolds with constant principal curvatures, or, more precisely, of orbits of s-representations, is that if $\pi : (M)_{\eta_p} \to M$ is a holonomy tube, then its fibres are orbits of the normal holonomy group, which, by the Normal Holonomy Theorem, are orbits of s-representations.

Recall that, if $\eta_p \in \nu_p(M)$, the holonomy tube $(M)_{\eta_p}$ at η_p is the image by the exponential map of the holonomy subbundle $\mathcal{H}ol_{\eta_p}M$. The latter is the subset of the normal bundle obtained by parallel translating η_p with respect to $\nabla^\perp$, along any piecewise differentiable curve in M; cf. Section 4.4. There is an obvious submersion $\pi : (M)_{\eta_p} \to M$, whose fibres are orbits of the normal holonomy group of M.

Let $\eta := p + \eta_p \in (M)_{\eta_p}$. Then

$$T_\eta\left((M)_{\eta_p}\right) = T_pM \oplus T_{\eta_p}(\Phi \cdot \eta_p).$$

Thus, $\nu_\eta\left((M)_{\eta_p}\right)$ can be identified with the normal space in ν_pM to the normal holonomy orbit $\Phi \cdot \eta_p$.

Let $\tilde{\eta}$ be the normal field on the holonomy tube defined by

$$\tilde{\eta}(\xi) = \xi - \pi(\xi), \qquad \xi \in (M)_{\eta_p}.$$

Then $\tilde{\eta}$ is parallel in the normal connection $(M)_{\eta_p}$ (the proof is similar to the one of Proposition 4.4.12, Exercise 4.6.15). Note that the mapping $\pi : (M)_{\eta_p} \to M$ can be viewed as endpoint map $\xi \to \xi - \tilde{\eta}$, so that

$$\ker(\pi_*) = \ker(\mathrm{id} + A_{\tilde{\eta}}),$$

where A is the shape operator of the holonomy tube.

The mapping $\pi : (M)_{\eta_p} \to M$ is a submersion, whose horizontal distribution $\mathcal{H} = \ker(\pi_*)^\perp$ can be identified with the tangent space at p to M, while the vertical space is given by $\mathcal{V} = \ker(\mathrm{id} + A_{\tilde{\eta}})$. Observe that, since $\tilde{\eta}$ is parallel, $A_{\tilde{\eta}}$ commutes with all shape operators of $(M)_{\eta_p}$. So, all shape operators of $(M)_{\eta_p}$ leave $\ker(\mathrm{id} + A_{\tilde{\eta}})$, and hence $\mathcal{H}$, invariant.

Using once again the Ricci identity

$$R^\perp(X,Y) = 0, \qquad \text{if } X \in \mathcal{H},\ Y \in \ker(\mathrm{id} + A_{\tilde{\eta}}).$$

We are thus in the general situation where we have a submersion $N \to M$ on which every piecewise $\mathcal{C}^1$ curve c has a unique horizontal lifting $\tilde{c}_s$ (for a fixed basepoint $c(s)$) and a connection ∇ (in our case, the normal connection on the holonomy tube) on a vector bundle over N whose curvature R^∇ satisfies $R^\nabla(X,Y) = 0$ if X is horizontal and Y vertical. In this setting, we have the following

LEMMA 5.4.11

Let $\pi : N \to M$ be a submersion with unique horizontal lifting property. Let E be a vector bundle over N with a connection ∇ whose curvature satisfies $R^\nabla(X,Y) = 0$ if X is horizontal and Y vertical. Let $c : [0,1] \to N$ be a piecewise differentiable curve. Then there exists a vertical curve γ (lying in $\pi^{-1}(\pi(c(1)))$) in N such that

$$\tau_c = \tau_\gamma \circ \tau_{\tilde{c}_0},$$

where τ denotes the parallel transport in the connection on E and $\tilde{c}_0$ is the horizontal lift of c with $\tilde{c}_0(0) = c(0)$.

PROOF Let $f : [0,1] \times [0,1] \to N$ be defined by $f(s,t) = \tilde{c}_s(t)$, where $\tilde{c}_s(t)$ is the horizontal lifting of c such that $\tilde{c}_s(s) = c(s)$. Then $\gamma := f(\,\cdot\,, 1)$ is vertical. Moreover, $\tau_c = \tau_\gamma \circ \tau_{\tilde{c}_0}$ because $\dfrac{D}{\partial t}\dfrac{D}{\partial s} - \dfrac{D}{\partial s}\dfrac{D}{\partial t} = R^\nabla(X,Y) = 0$. □

Using this we get

THEOREM 5.4.12

Let M be a simply connected submanifold of $\mathbb{R}^n$. The normal holonomy at $\eta = p + \eta_p$ of the holonomy tube $(M)_{\eta_p}$ is the image under the slice representation of the isotropy subgroup Φ_η on $\nu_\eta\left((M)_{\eta_p}\right) \cong \nu_{\eta_p}(\Phi \cdot \eta_p)$.

PROOF The submersion $\pi : (M)_{\eta_p} \to M$ can be viewed as a map arising from focalization with respect to the parallel normal field $\tilde{\eta}$. Thus, by Lemma 4.4.6, the $\nabla^\perp$-parallel transport on $\nu_\eta\left((M)_{\eta_p}\right) \cong \nu_{\eta_p}(\Phi \cdot \eta_p)$ along a curve on M coincides with the one on its horizontal lift to the holonomy tube.

Let $c : [0,1] \to (M)_{\eta_p}$ be a piecewise differentiable closed curve with $c(0) = c(1) = \xi$. We know that $\tau_c^\perp = \tau_\gamma^\perp \circ \tau_{\tilde{c}_0}^\perp$, where γ is a vertical curve and $\tilde{c}_0$ horizontal. By Theorem 5.4.9, since the fibre is orbit of an s-representation, there exists $g \in \Phi$ such that

$$\tau_\gamma^\perp = g_{|\nu_{\gamma(0)}(\Phi \cdot \eta_p)}.$$

Moreover, since $\nabla^\perp$-parallel transport along a curve on M coincides with $\nabla^\perp$-parallel transport along its horizontal lift on the holonomy tube,

$$\tau_{\tilde{c}_0}^\perp = \tau^\perp_{h|\nu_\eta\left((M)_{\eta_p}\right)} = \tilde{g}_{|\nu_{\eta_p}(\Phi \cdot \eta_p)},$$

where $h = \pi \circ \tilde{c}_0 = \pi \circ c$ and $\tilde{g} = \tau_h^\perp \in \Phi$. Thus, $\tau_c^\perp = g \circ \tilde{g}$ and $g \circ \tilde{g} \in \Phi_{\eta_p}$, because η_p extends to a parallel normal vector field to the holonomy tube, so $(g \circ \tilde{g}) \cdot \eta_p = \eta_p$. Hence, the normal holonomy at η of the holonomy tube $(M)_{\eta_p}$ is contained the image in the slice representation of the isotropy subgroup Φ_η.

To prove the other inclusion, let $g \in \Phi_\eta$ and $r : [0,1] \to M$ be a curve such that $g = \tau_r^\perp$. Let $\tilde{r}$ be the horizontal lifting of r on the holonomy tube such that $\tilde{r}(0) = \eta_p$. Since $g \cdot \eta_p = \eta_p$ and $\tilde{r}(0) = \tilde{r}(1)$, we have

$$\tau_{\tilde{r}}^\perp = \tau^\perp_{r|\nu_{\eta_p}(\Phi \cdot \eta_p)} = g_{|\nu_{\eta_p}(\Phi \cdot \eta_p)}.$$

Thus, the image in the slice representation of the isotropy subgroup Φ_η is contained in the normal holonomy at η of the holonomy tube $(M)_{\eta_p}$. □

5.5 Isoparametric rank

The existence of a (nontrivial) isoparametric section on a submanifold M of a space form has consequences on its geometry. Instances of this occurred in Section 4.4.

DEFINITION 5.5.1 *Let M be a submanifold of a space form $\bar{M}(\kappa)$ and $q \in M$. The maximal number of linearly independent parallel isoparametric normal sections (defined in a neighbourhood of q) is called local isoparametric rank of M at q and is denoted by $\text{iso-rank}^{loc}(M)_q$. The local isoparametric rank is defined by*

$$\text{iso-rank}^{loc}(M) = \min\{\text{iso-rank}^{loc}(M)_q \mid q \in M\}$$

In the case of global normal isoparametric sections, we speak of global isoparametric rank or simply isoparametric rank. The global isoparametric rank is denoted by $\text{iso-rank}(M)$.

In the global case, it is sometimes important to stress the immersion rather than the submanifold alone. Thus, if $f : M \to \bar{M}(\kappa)$ is an immersed submanifold the (global) isoparametric rank is denoted by $\text{iso-rank}_f(M)$.

Note that, for isoparametric submanifolds, the isoparametric rank coincides with the usual notion of rank. For a principal orbit of an s-representation, the rank coincides with that of the corresponding symmetric space.

a) Local higher isoparametric rank rigidity

We recall from Chapter 2 that a submanifold is *locally reducible at q* if there is some neighbourhood U of q such that U is an extrinsic product. We call a submanifold *locally irreducible* if it is not locally reducible at any point.

The following higher rank rigidity result for submanifolds of the Euclidean sphere S^n holds [61].

THEOREM 5.5.2
Let $M \to S^n$ be a locally irreducible full submanifold with isoparametric rank greater than or equal to one. Then M has constant principal curvatures.

Thorbergsson's Theorem (Theorem 5.4.5) implies that any irreducible isoparametric submanifold of the sphere with codimension at least two is an orbit of an s-representation. Thus,

COROLLARY 5.5.3

Let $M \to S^n$ be a locally irreducible full submanifold with isoparametric rank greater than or equal to one, which is not an isoparametric hypersurface of the sphere. Then M is an orbit of an s-representation.

The key point to prove Theorem 5.5.2 is a generalization, given in [179], of the Homogeneous Slice Theorem of [96]. We state only the Euclidean version, but the same geometric proof applies to the Lorentzian case with timelike parallel normal sections. Furthermore, it can be adapted to proper Fredholm submanifolds of Hilbert space.

THEOREM 5.5.4

Let M be a submanifold of Euclidean space and assume that M is not reducible at any point. Let ξ be an isoparametric parallel normal field to M that is not umbilical. Assume that 1 is an eigenvalue of the shape operator A_ξ. Then, if $q \in M$, the holonomy tube $(M_\xi)_{-\xi(q)}$ around the parallel focal manifold $M_\xi \subset \mathbb{R}^n$ coincides locally with M.

Before dwelling into the proof we briefly explain main ideas. If we had two maximal dimensional holonomy tubes around M_ξ, through points of M, they would be parallel manifolds of the ambient space. If $\eta \neq 0$ is the parallel normal field which relates the second tube to the first by parallelism, an easy computation involving the "tube formula" (relating shape operators of parallel manifolds), proves that the shape operator of η is zero in the horizontal directions. Any two points in the holonomy tube can be joined by horizontal curves, so η is constant in the ambient space. This implies that the normal distribution to the orbits of the polar action (on the fibres) of the normal holonomy group of M_ξ is parallel in the ambient space. Thus, M splits and there can be only one maximal dimensional holonomy tube.

We need the following consequence of the "tube formula":

LEMMA 5.5.5

Let M be a submanifold of Euclidean space and let ξ, η be parallel normal fields. Assume that 1 is not an eigenvalue of the shape operator A_η and $A_{\xi(q)} = \tilde{A}_{\xi(q)-\eta(q)}$ for some $q \in M$, where $\tilde{A}$ is the shape operator of the parallel manifold M_η (identifying T_qM with $T_{q+\eta(q)}M_\eta$). Then $A_{\eta(q)}X = 0$ for all $X \in (\ker(\mathrm{id} - A_{\xi(q)}))^\perp$.

PROOF From the tube formula we have $\tilde{A}_{\xi(q)} = A_{\xi(q)}(\mathrm{id} - A_{\eta(q)})^{-1}$ and $\tilde{A}_{\eta(q)} = A_{\eta(q)}(\mathrm{id} - A_{\eta(q)})^{-1}$. Then $\tilde{A}_{\xi(q)-\eta(q)} = (A_{\xi(q)} - A_{\eta(q)})(\mathrm{id} - A_{\eta(q)})^{-1} = A_{\xi(q)}$ (last equality holds by assumption). Then $A_{\xi(q)} - A_{\eta(q)} = A_{\xi(q)}(\mathrm{id} - A_{\eta(q)})$, so $A_{\eta(q)} = A_{\xi(q)}A_{\eta(q)}$. Thus $(\mathrm{id} - A_{\xi(q)})A_{\eta(q)} = 0$ and the lemma follows, for both shape operators commute, by the Ricci equation. ∎

PROOF of Theorem 5.5.4 Let $E = \ker(\mathrm{id} - A_\xi)$ be the autoparallel eigendistribution on M associated with eigenvalue 1 of the shape operator A_ξ. Using arguments like those in Section 4.4 (cf. also [61, 174]) it is not difficult to see that given a leaf $S(x)$ of E, any other close leaf $S(x')$ can be obtained by parallel transporting $S(x)$ in the normal space to M_ξ, along any short curve in M_ξ joining $x + \xi(x)$ with $x + \xi(x')$. In particular, $S(x)$ is locally invariant under the restricted normal holonomy group of M_ξ at $x+\xi(x)$ (see the results on local normal holonomy, Section 4.1, part f; cf. also [82, Appendix]). (Since our discussion will be local, no distinction will be made between restricted and non restricted normal holonomy groups.)

Let $q \in M$ and consider the (local) orbit $\Phi_{\bar{q}} \cdot q$ through q of the normal holonomy group $\Phi_{\bar{q}}$ of M_ξ at $\bar{q} = q + \xi(q)$, where $\Phi_{\bar{q}}$ is regarded as acting on the affine normal space $\bar{q} + \nu_{\bar{q}}(M_\xi)$.

We have two possibilities: either $\Phi_{\bar{q}} \cdot q$ is of local maximal dimension in $S(q)$ or it is not.

We start with the former case. We can assume, by considering a neighbourhood of q if necessary, that all local orbits of $\Phi_{\bar{q}}$ in $S(q)$ have the same dimension. Since the leaves $S(x)$ move parallelly in the normal space of M_ξ, we suppose the local orbit $\Phi_{\bar{r}} \cdot r$ in $S(r)$ is of maximal dimension for any $r \in M$. In this way, we have a distribution $\tilde{\nu}$ in M given by the normal space $\tilde{\nu}(r)$ in $S(r)$ to the orbit $\Phi_{\bar{r}} \cdot r$. Observe that $\tilde{\nu}^\perp(r)$ is the tangent space to the holonomy tube $(M_\xi)_{-\xi(r)} \subset M$ at r. We will show that the distribution $\tilde{\nu}$ satisfies the condition of Moore's Lemma 2.7.1, hence $\tilde{\nu} = 0$ since M is not locally reducible ($\tilde{\nu} \neq TM$, otherwise ξ would be umbilical). $\tilde{\nu}$ is an autoparallel distribution by the Normal Holonomy Theorem and Proposition 3.2.9. By the proof of the latter proposition, if $q_2 \in \mathcal{L}(q_1)$, where $\mathcal{L}(q_1)$ denotes the leaf of $\tilde{\nu}$ through q_1, then $\eta_{q_1} = q_2 - q_1$ belongs to the normal space of $\Phi_{\bar{q}_1} \cdot q_1$ at q_1 (regarded as a submanifold of the affine normal space $\bar{q}_1 + \nu_{\bar{q}_1}(M_\xi)$). If q_2 is close to q_1 then η_{q_1} is fixed by the isotropy subgroup of $\Phi_{\bar{q}_1}$ at q_1, since $\Phi_{\bar{q}_1} \cdot q_1$ and $\Phi_{\bar{q}_1} \cdot q_2$ have the same dimension (note that $\bar{q}_1 = \bar{q}_2$, since $\mathcal{L}(q_1) \subset S(q_1)$). By Proposition 3.2.4 (see Remark 3.2.9) η_{q_1} extends to a $\Phi_{\bar{q}_1}$–invariant, $\nabla^\perp$-parallel normal field to the orbit $\Phi_{\bar{q}_1} . q_1$ in $\bar{q}_1 + \nu_{\bar{q}_1}(M_\xi)$. It is now standard to show that η_{q_1} extends to a parallel normal field η to the holonomy tube $(M_\xi)_{-\xi(q_1)}$. Thus $(M_\xi)_{-\xi(q_2)}$ coincides with the parallel manifold $((M_\xi)_{-\xi(q_1)})_\eta$ to $(M_\xi)_{-\xi(q_1)}$.

Observe that the distribution $\tilde{\nu}$ consists of fixed points $\tilde{\nu}(r)$ of the isotropy subgroup of $\Phi_{\bar{r}}$ at r. Considering horizontal and vertical curves and applying similar arguments as in Section 5.4 (part d, page 165), one obtains that $\tilde{\nu}$ defines by restriction a $\nabla^\perp$-parallel, flat subbundle of the normal bundle of any holonomy tube $(M_\xi)_{-\xi(r)}$ (regarded as a submanifold of the ambient space). Moreover $x + \xi(x)$ belongs to the leaf $\mathcal{L}(x)$ for any $x \in (M_\xi)_{-\xi(r)}$.

Let A, A^1, A^2 and $\bar{A}$ be shape operators of M, $(M_\xi)_{-\xi(q_1)}$, $(M_\xi)_{-\xi(q_2)}$ and M_ξ respectively. If $\mathcal{H}$ is the distribution given by the horizontal spaces of the holonomy tubes $(M_\xi)_{-\xi(r)} \subset M$ $(r \in M)$, then $\mathcal{H}(q_1) = T_{\bar{q}_1}(M_\xi) =$

$\mathcal{H}(q_2)$, regarded as subspaces of the ambient space. Note that $\mathcal{H} = E^\perp$. The distribution $\mathcal{H}$ is invariant under the shape operator of M and also of the holonomy tubes. The restriction of ξ to any holonomy tube $(M_\xi)_{-\xi(r)} \subset M$, $r \in M$, is also a parallel normal vector field to this submanifold of the ambient space. This is a consequence of the invariance of $\mathcal{H}$ under shape operators (of M and the holonomy tube) and the fact that $A_{|\mathcal{H}^\perp} = \mathrm{id}$. Moreover, $A^1_{\xi(q_1)}$ and $A^2_{\xi(q_2)}$ have the same eigenvalues as A_ξ (which are constant), with the possible exception of 1 if $\tilde{\nu} = \ker(\mathrm{id} - A_\xi)$. This implies $A^1_{\xi(q_1)} = A^2_{\xi(q_2)}$, if η is near 0. In fact, both shape operators are simultaneously diagonalizable by the "tube formula". Notice $\xi(q_i) = \bar{q}_i - q_i$, $i = 1, 2$, so $\xi(q_2) = \xi(q_1) - \eta(q_1)$. By Lemma 5.5.5, we have $A^1_{\eta|\mathcal{H}} = 0$, since $\mathcal{H} = E^\perp = (\ker(\mathrm{id} - A_\xi))^\perp = (\ker(\mathrm{id} - A^1_\xi))^\perp$ (the distributions are restricted to $(M_\xi)_{-\xi(q_1)}$). η is thus constant along horizontal curves (with respect to $\mathcal{H}$). But any two points in $(M_\xi)_{-\xi(q_1)}$ can be joined by a horizontal curve (by the construction of the holonomy tube), so η is constant along $(M_\xi)_{-\xi(q_1)}$ in the ambient space.

Therefore, the leaves of the autoparallel distribution $\tilde{\nu}$ are parallel, in the ambient space, along any holonomy tube $(M_\xi)_{-\xi(r)}$. Since $\tilde{\nu}^\perp$ is the distribution tangent to the holonomy tubes, $\alpha(\tilde{\nu}, \tilde{\nu}^\perp) = 0$, where α is the second fundamental form of M. Furthermore, $\tilde{\nu}$ is a parallel distribution of M, for it is autoparallel and its leaves are parallel (in the ambient space) along curves in M lying in $\tilde{\nu}^\perp$. Moore's Lemma implies $\tilde{\nu} = 0$, since M is not locally reducible.

We must still analyze when the orbit through q of the normal holonomy group $\Phi_{\bar{q}}$ of M_ξ at $\bar{q} = q + \xi(q)$ in $S(q)$ is not of local maximal dimension. This orbit is always contained in $S(q)$ near q (see the first part of the proof). Moreover, there are elements q' of $S(q)$ arbitrary close to q and such that the normal holonomy orbit in $S(q)$ is of maximal dimension. Furthermore, $\Phi_{\bar{q}'}.q'$ is a complete Riemannian submanifold of the ambient space that locally coincides with $S(q')$, by what we have proved.

A standard argument now shows that near q the orbit $\Phi_{\bar{q}'} \cdot q'$ locally coincides with $S(q)$ and the theorem follows. □

PROOF of Theorem 5.5.2 Since M is contained in a sphere S^{n-1}, without loss of generality, we can assume that all eigenvalues λ_i of the shape operator A_ξ are different from zero (otherwise we can add a suitable constant multiple of the position vector field to ξ).

Let $\xi_i := \lambda_i^{-1}\xi$. We can consider the focal manifold $M_{\xi_i} = \{q + \xi_i(q) \mid q \in M\}$ and, for any $q \in M$, the holonomy tube, $(M_\xi)_{-\xi(q)}$. By Theorem 5.5.4, $(M_\xi)_{-\xi(q)}$ coincides locally with M. It is enough to recall Remark 4.4.16, according to which, if all holonomy tubes $(M_{\xi_i})_{-\xi_i(q)}$ locally coincide with M, then M is a submanifold with constant principal curvatures. □

A different proof of Theorem 5.5.2, via the Holonomy Lemma, can be found in

[61].

b) Global higher isoparametric rank rigidity

The global version of Theorem 5.5.2 is not trivial, since a simply connected irreducible Riemannian manifold can be locally reducible at any point. The same pathology probably exists in the context of submanifolds as well. The key fact for this global version is the following result:

LEMMA 5.5.6
Let M be a complete Riemannian manifold and let G act local polarly on M. Let $\mathcal{O}$ be the open and dense subset of M for which the G-orbit has maximal dimension. Assume that the distribution ν on $\mathcal{O}$ defined by the normal spaces to the orbits is (not only autoparallel but also) parallel. Then $\mathcal{O} = M$ (i.e., all orbits are maximal dimensional and ν defines a parallel distribution on M).

PROOF Let $q \in M$ and $v \in \nu_q(G \cdot q)$ be a principal vector for the slice representation of G_q. Then there exits $\varepsilon > 0$ such that $G \cdot \exp_q(tv)$ is a principal orbit of G for all $t \in (0, \varepsilon)$ (see Exercise 3.10.2). Let X be a Killing field on M induced by G_q with associated flow ϕ_s. Observe that $\phi_s(q) = q$ and $(\phi_s)_* v$ is a principal vector for the slice representation, for all $s \in \mathbb{R}$. Let us consider $f : \mathbb{R} \times [0, \varepsilon) \to M$, $f(s,t) = \phi_s(\gamma_v(t)) = \gamma_s(t)$, where $\gamma_s := \gamma_{(\phi_s)_* v}$. Let $\nu_{s,t} = \nu_{\gamma_s(t)}$ if $t \in (0, \varepsilon)$ and let $\nu_{s,0} = (\tau_t^s)^{-1} \nu_{s,t}$ $(t \in (0, \varepsilon))$, where τ_t^s denotes the parallel transport from 0 to t along the geodesic γ_s. Note that $\nu_{s,0}$ is well defined since $\nu_{s,t}$, $t \in (0, \varepsilon)$, is parallel along f and in particular along the geodesic γ_v. It is clear that $\nu_{s,t}$ varies smoothly along f. Moreover, being parallel along the curve $s \mapsto f(s,t)$ for $t \neq 0$, $\nu_{s,0}$ must be parallel along the curve $s \mapsto f(s, 0) \equiv q$. This means that $\nu_{s,0} = \nu_{0,0}$ for all $s \in \mathbb{R}$. But $\nu_{s,0}$ is a slice of the isotropy action that contains $(\phi_s)_*(v)$ (see Exercise 3.10.4). Then $(\phi_s)_*(v) \in \nu_{0,0}$ for all $s \in \mathbb{R}$. Then $X \cdot v = \frac{d}{ds}_{|0} (\phi_s)_* v \in \nu_{0,0}$. But $X \cdot v \perp \nu_{0,0}$, since $\nu_{0,0}$ is a slice for the isotropy action. Thus, $X \cdot v = 0$. Since X is an arbitrary Killing field induced by G_q, we conclude that the slice representation at q is discrete. Then $G \cdot q$ has maximal dimension for any $q \in M$. □

For the global version of Theorem 5.5.2 we need the global version of 5.5.4. First of all, we make some observations.

Let $f : M \to \mathbb{R}^n$ be a simply connected complete submanifold and let ξ be a non-umbilical isoparametric parallel normal section of M. Assume 1 is an eigenvalue of the shape operator A_ξ. Endow M with the "bundle-like" metric g defined in Exercise 5.6.6. Consider the quotient space $M_{/\sim}$, where $x \sim y$ if x and y are both in the same leaf of $\ker(\mathrm{id} - A_\xi)$. Then $M_{/\sim}$ is endowed with a natural differentiable Hausdorff manifold structure such that the projection $\pi : M \to M_{/\sim}$ is a C^∞ submersion. Moreover, if $f_\xi : M_{/\sim} \to \mathbb{R}^n$ is defined by $f_\xi(\pi(q)) = f(q) + \xi(q)$, then f_ξ is an immersion and $M_{/\sim}$ with the induced metric is a complete Riemannian manifold (Exercise 5.6.7). With this procedure, starting from a complete submanifold M, we construct a parallel focal manifold $f_\xi : M_{/\sim} \to \mathbb{R}^r$, which is also complete with the induced metric.

As in Theorem 5.5.4, the (global) restricted normal holonomy group $\Phi^*_{\pi(q)}$ of $M_{/\sim}$ acts on the fibre $\pi^{-1}(\pi(q))$. Let $c : [0,1] \to M_{/\sim}$ be piecewise differentiable with $c(0) = c(1) = \pi(q)$ and let $\tilde{c}$ be its horizontal lift to M with $\tilde{c}(0) = q$. As in the local case, $\tilde{c}(t) - c(t)$ can be regarded as a parallel normal field to $f_\xi : M_{/\sim} \to \mathbb{R}^n$. As in the proof of Theorem 5.5.4, the normal space to maximal dimensional orbits of $\Phi^*_{\pi(q)}$ form a parallel distribution on $\pi^{-1}(\pi(q)))$. Applying Lemma 5.5.6, we obtain that this parallel distribution is never singular. Using the same ideas of Theorem 5.5.4 in the category of immersions, the global version of Moore's Lemma yields:

THEOREM 5.5.7 [70]
Let $f : M \to \mathbb{R}^n$ be a full simply connected complete submanifold and ξ a nonumbilical isoparametric parallel normal section of M. Then, for any $q \in M$, the holonomy tube $(M_\xi)_{-\xi(q)}$ around the parallel focal manifold $f_\xi : M_{/\sim} \to \mathbb{R}^n$ coincides with $f : M \to \mathbb{R}^n$.

As in the local case, there is a corollary, the global version of Theorem 5.5.2

THEOREM 5.5.8 [70]
Let $f : M \to S^n$ be a full and irreducible isometric immersion, where M is a simply connected complete Riemannian manifold with (global) isoparametric rank $\text{iso-rank}_f(M) \geq 1$. Then M is a submanifold with constant principal curvatures.

As an immediate corollary, we have that if M is not an isoparametric hypersurface, it is an orbit of an s-representation.

c) Higher isoparametric rank rigidity for submanifolds of Euclidean and hyperbolic spaces.

We now examine the case of submanifolds of Euclidean and hyperbolic spaces.

As a consequence of a Lorentzian version of Theorem 5.5.4, C. Olmos and A. Will proved that irreducible and full submanifolds of hyperbolic space must have isoparametric rank zero [179].

THEOREM 5.5.9
Let M be a full submanifold of the hyperbolic space that is locally irreducible at any point. Then any isoparametric parallel normal field vanishes.

Using the same methods, regarding a submanifold M of Euclidean space as a submanifold of a horosphere $Q \subset H^n \subset L^{n+1}$, C. Olmos and A. Will proved the following [179]:

THEOREM 5.5.10
Let M be a locally irreducible full submanifold of Euclidean space. If M admits an isoparametric parallel normal field $\xi \neq 0$, then M is contained in a sphere.

As a consequence, if ξ is not a multiple of the radial vector, by Theorem 5.5.2, M has constant principal curvatures.

In other words, Theorem 5.5.2 is true for submanifolds of Euclidean space. If $\text{iso-rank}^{loc}(M) \geq 1$ then M is contained in a sphere and if $\text{iso-rank}^{loc}(M) \geq 2$ then M has constant principal curvatures.

Moreover, the global version (Theorem 5.5.8) of Theorem 5.5.2 is true in the more general context of submanifolds of Euclidean space [70]. Namely, let $f : M \to \mathbb{R}^n$ be a full and irreducible isometric immersion, where M is a simply connected complete Riemannian manifold. Then $f(M)$ is contained in a sphere if $\text{iso-rank}_f(M) \geq 1$ and $f(M)$ is a submanifold with constant principal curvatures if $\text{iso-rank}_f(M) \geq 2$.

By a result of A.J. Di Scala (Theorem 3.4.1; [68]) any minimal homogeneous submanifold of Euclidean space is totally geodesic. The same is true for the hyperbolic space by the results explained in Section 3.5 ([69]). So, we have the following (cf. [176]).

COROLLARY 5.5.11
Let M be a homogeneous irreducible full submanifold (of positive codimension) of hyperbolic or Euclidean space which is not contained in a sphere. Then the mean curvature vector of M is not parallel.

5.6 Exercises

Exercise 5.6.1 Prove that, if all leaves of a parallel foliation have parallel mean curvature, then each leaf has constant principal curvatures.

Exercise 5.6.2 Let $\mathcal{D}$ be an integrable distribution on a open subset of $\mathbb{R}^n$ and assume that the distribution $\mathcal{D}^\perp$ is integrable with totally geodesic leaves (or, equivalently, it is autoparallel). Then any two (nearby) leaves of $\mathcal{D}$ are parallel. Equivalently, if M denotes one leaf, the other is the parallel manifold M_ξ with respect to some parallel vector field ξ. In particular, any leaf has flat normal bundle.

Exercise 5.6.3 Let $M \to S^n$ be an isoparametric submanifold and fix a unit parallel normal field ξ to M in S^n (more generally, let ξ be a parallel normal section). Moving along the geodesic from any $p \in M$ in direction $\xi(p)$, consider the parallel (possibly focal) manifolds $M_t := \{\varphi_t(p) := \cos(t)p + \sin(t)\xi(p) \mid p \in M$. Prove the following "tube formula": if $\xi(p)$ is normal to M at p, then $\bar{\xi}(\varphi(p)) := -\sin(t)p + \cos(t)\xi(p)$ is normal to M_t at $\varphi(p)$ and the shape operator $\bar{A}$ of M_t is given by

$$\bar{A}_{\bar{\xi}} = (\sin(t)\mathrm{id} + \cos(t)A_\xi)(\cos(t)\mathrm{id} - \sin(t)A_\xi)^{-1},$$

where we have to restrict to horizontal spaces, in case of a focal manifold (Exercise 5.6.3).

Exercise 5.6.4 Let M be a submanifold of a space form $\bar{M}(\kappa)$ contained in a totally umbilical submanifold N of $\bar{M}(\kappa)$. Prove that M is isoparametric in N if and only if it is isoparametric in $\bar{M}(\kappa)$.

Exercise 5.6.5 Prove that for a compact immersed full isoparametric submanifold all curvature normals are nonzero.

Exercise 5.6.6 (cf. [70]) Let $f : M^n \to \mathbb{R}^n$ be an isometric immersion, where $(M, \langle\, , \,\rangle)$ is a complete Riemannian manifold. Let ξ be an isoparametric parallel normal field to M and $\lambda \neq 0$ an eigenvalue of the shape operator A_ξ. Consider the autoparallel distribution $\ker(\lambda\mathrm{id} - A_\xi)$ on M and define a following Riemannian metric g on M by requiring:

(i) $\ker\,(\mathrm{id} - A_{\lambda^{-1}\xi})$ and $(\ker\,(\mathrm{id} - A_{\lambda^{-1}\xi}))^\perp$ are also perpendicular with respect to g.

(ii) $g(X, Y) = \langle X, Y\rangle$ if X, Y lie in $\ker\,(\mathrm{id} - A_{\lambda^{-1}\xi})$.

(iii) $g(X, Y) = \langle(\mathrm{id} - A_{\lambda^{-1}\xi})X, (\mathrm{id} - A_{\lambda^{-1}\xi})Y\rangle$, if X, Y lie in $(\ker\,(\mathrm{id} - A_{\lambda^{-1}\xi})^\perp$.

Prove that:

(a) (M, g) is a complete Riemannian manifold.

(b) Any leaf of $\ker\,(\mathrm{id} - A_{\lambda^{-1}\xi})$ is an embedded closed submanifold of M

(c) Any two leaves S_1, S_2 of $\ker\,(\lambda\mathrm{id} - A_\xi)$ are equidistant with respect to g (i.e. the distance $d_g(x, S_2)$ does not depend on $x \in S_1$).

Hints: (b): consider the parallel map $f_{\lambda^{-1}\xi} : M \to \mathbb{R}^n$, $f_{\lambda^{-1}\xi}(p) = f(p) + \lambda^{-1}\xi(p)$. Then $d f_{\lambda^{-1}\xi} = \mathrm{id} - \lambda^{-1}A_\xi$ has constant rank and $(f_{\lambda^{-1}\xi})^{-1}(v)$ is an embedded closed submanifold of M, whose connected component is an integral manifold of $\ker(\lambda\mathrm{id} - A_\xi)$.
(c): let $S(x)$, $S(y)$ be any two different leaves of $\ker(\lambda\mathrm{id} - A_\xi)$ and let $\gamma : [0, r] \to M$ be a unit speed geodesic (with respect to g) with $\gamma(0) := p \in S(x)$, $\gamma(r) \in S(y)$ and such that $r = L(\gamma) = d_g(S(x), S(y))$. Let $p' \in S(x)$ and $c : [0, 1] \to S(x)$ with $c(0) = p$, $c(1) = p'$ (observe that γ must be a horizontal geodesic). Prove that:

(1) There exist $\varepsilon > 0$ such that, for all $s \in [0, 1]$, $f_{\lambda^{-1}\xi}(B_\varepsilon(c(s)))$ is an embedded submanifold of $\mathbb{R}^N$ and so $f_{\lambda^{-1}\xi} : (M, g) \to f_{\lambda^{-1}\xi}(B_\varepsilon(c(s)))$ is a Riemannian submersion, where $f_{\lambda^{-1}\xi}(B_\varepsilon(c(s)))$ carries the Riemannian metric induced by $\mathbb{R}^N$.

(2) For s small, let γ_s denote the horizontal lift of the geodesic $f_{\lambda^{-1}\xi} \circ \gamma_{|[0,\varepsilon]}$ of $f_{\lambda^{-1}\xi}(B_\varepsilon(c(0)))$ with the initial condition $\gamma_s(0) = c(s)$. Use this idea to construct gluing horizontal lifts, geodesics $\gamma_s : [0, r] \to M$ starting at $c(s)$ and such that $\gamma_s(t) \in S(\gamma(t))$ for all $s \in [0, 1]$. Then

$$\begin{aligned} d_g(p, S(y)) &= d_g(p, \gamma(r)) = d_g(p', \gamma_s(r)) \geq \\ &\geq d_g(p', S(y)) \geq d_g(S(x), S(y)) = \\ &= d_g(p, \gamma(r)). \end{aligned}$$

Therefore, $d_g(p, S(y)) = d_g(p', S(y))$.

The metric g on M is an example of *bundle-like metric* in the sense of Reinhart [193] (cf. Exercise 5.6.8).

Exercise 5.6.7 (cf. [70]) Let $f : M^n \to \mathbb{R}^n$ be an isometric immersion, where $(M, \langle\, ,\, \rangle)$ is a complete Riemannian manifold. Let ξ be an isoparametric parallel normal field to M and $\lambda \neq 0$ an eigenvalue of the shape operator A_ξ. Let $M_{/\sim}$ be the quotient space, where $x \sim y$ if x and y are both in the same leaf of $\ker(\mathrm{id} - A_{\lambda^{-1}\xi})$. Prove that:

(i) $M_{/\sim}$ is endowed with a natural differentiable Hausdorff manifold structure such that the projection $\pi : M \to M_{/\sim}$ is a C^∞ submersion.

(ii) Let $f_\xi : M_{/\sim} \to \mathbb{R}^n$ be defined by $f_\xi(\pi(q)) = f(q) + \xi(q)$. Then f_ξ is an immersion. Moreover, $M_{/\sim}$ with the induced metric is a complete Riemannian manifold.

Exercise 5.6.8 (cf. [193]) Using the ideas of Exercises 5.6.6 and 5.6.7, prove the following: let $f : M \to N$ be a map of constant rank, from a complete Riemannian manifold M into a Riemannian manifold N. Assume that locally f is a Riemannian submersion from M into its image. Then there exists a Riemannian submersion $\phi : M \to \hat{M}$ and an immersion $i : \hat{M} \to N$ such that $f = i \circ \phi$.

Chapter 6

Rank rigidity of submanifolds and normal holonomy of orbits

In the previous chapter, we have seen that orbits of s-representations agree, up to codimension two, with isoparametric submanifolds and their focal manifolds (or, equivalently, submanifolds with constant principal curvatures). It is therefore natural to look for geometric invariants distinguishing orbits of s-representations from different orbits (or submanifolds with constant principal curvatures from other submanifolds). In Chapters 4 and 5, we observed that admitting a (nontrivial) parallel isoparametric normal field has strong consequences on the geometry of a submanifold.

In this chapter, we weaken this condition, requiring only that the submanifold admits "enough" parallel normal fields, in other words, that the normal holonomy group has a nontrivial point-wise fixed subspace whose dimension is called rank of the immersion. In the case of a homogeneous submanifold M of Euclidean space, it was proved in [175] that, if the rank is bigger than or equal to two, then M is an orbit of an s-representation. In the original proof, a crucial fact was that the curvature normals (defined as in the isoparametric case, considering only directions in the flat part of the normal bundle, as we will explain) of a homogeneous submanifold have constant length. In [70] it is actually shown that this property, together with the same higher rank assumption, yields a generalization (stated here as Theorem 6.1.7) of the above higher rank rigidity result. Unlike theorems on higher isoparametric rank rigidity (Theorem 5.5.2 and 5.5.8), Theorem 6.1.7 is global and fails without the completeness assumption. As a consequence, one can derive a global characterization of an isoparametric submanifold: a complete immersed and irreducible submanifold $f : M \to \mathbb{R}^n$, $\dim M \geq 2$ with flat normal bundle is isoparametric if and only if the distances to focal hyperplanes are constant on M.

In the last part of this chapter, we will apply these higher rank rigidity results to investigate normal holonomy (and, more generally, $\nabla^\perp$-parallel transport) of a homogeneous submanifold. In a more general setting of homogeneous (pseudo)-Riemannian vector bundles, the holonomy algebra can be described in terms of projection of Killing vector fields on the homogeneous bundle (see [60], for more details). In the case of Riemannian manifolds, this yields Kostant's method for computing the Lie algebra of the holonomy group of a homogeneous Riemannian manifold. Here, we explain how to compute normal holonomy of homogeneous submanifolds by projecting the Killing vector fields determined by the action on the normal spaces (Theorem 6.2.7).

6.1 Submanifolds with curvature normals of constant length and rank of homogeneous submanifolds

In this section, we will be concerned with immersed submanifolds admitting parallel normal sections. In other words, we assume that the normal holonomy group has a nontrivial point-wise fixed subspace, whose dimension is called rank of the immersion. The aim is to prove a global higher rank rigidity theorem, which is false in the local setup. In particular, this leads to a definition of isoparametricity that coincides with the one used for complete submanifolds only. This condition can be formulated in terms of the induced metric by the usual Gauss map, or equivalently, in terms of the so-called third fundamental form.

a) Rank of submanifolds

Let $f : M \to \bar{M}^n(\kappa)$ be an immersed submanifold of a space form (with the induced metric) and consider the following subspaces of the normal space at $p \in M$

$$(\nu_p M)_0 := \{\xi \in \nu_p M \mid \Phi_p^* \cdot \xi = \xi\}$$
$$(\nu_p M)_s := [(\nu_p M)_0]^\perp$$

where Φ_p^* is the restricted normal holonomy group at p. Note that $(\nu M)_0$ is the maximal flat $\nabla^\perp$-parallel subbundle of νM, where $((\nu M)_0)_p = (\nu_p M)_0$.

DEFINITION 6.1.1 *The dimension over M of the bundle $(\nu M)_0$ is called rank of the submanifold and it is denoted by $\mathrm{rank}_f(M)$. When there is no possible confusion, we will write $\mathrm{rank}(M)$ instead of $\mathrm{rank}_f(M)$.*

If M is simply connected, then $(\nu M)_0$ must be globally flat and so $\mathrm{rank}_f(M)$ is the maximal number of linearly independent parallel normal vector fields to M. Since we are working in the category of immersions, we always assume that M is simply connected. Otherwise, we consider the immersed submanifold $f \circ \pi : \tilde{M} \to \bar{M}^n$, where $\pi : \tilde{M} \to M$ is the universal cover of M. In this case, we have $((\nu M)_0)_{pr(p)} = ((\nu \tilde{M})_0)_p$.

If M has flat normal bundle then $\mathrm{rank}(M)$ is just the codimension $\mathrm{codim}(M)$. Thus, if M is a full isoparametric submanifold of Euclidean space, the above notion of rank coincides with the usual notion of rank for isoparametric submanifolds. If, in addition, M is homogeneous then, by [63], it is a principal orbit of an s-representation. Then $\mathrm{rank}(M)$ coincides with the rank of the corresponding symmetric space (cf. Section 5.4).

The general philosophy is always that, for submanifolds of Euclidean space, normal holonomy plays a similar role to Riemannian holonomy, and orbits of s-representations are the extrinsic analogue of symmetric spaces.

By replacing the normal holonomy group with the local normal holonomy group, one obtains the notion of the local rank $\mathrm{rank}_f^{loc}(M)_q$ at a given point $q \in M$. In other words, $\mathrm{rank}_f^{loc}(M)_q$ is the maximal number of linearly independent normal fields defined in a neighbourhood of q. The local rank is defined by

$$\mathrm{rank}_f^{loc}(M) = \min\{\mathrm{rank}_f^{loc}(M)_q \mid q \in M\}.$$

If M is simply connected, then $\mathrm{rank}_f^{loc}(M) = \mathrm{rank}_f(M)$ if and only if any locally defined parallel normal field extends globally.

The above equality holds in the following two important cases: when M is an analytic submanifold or M has flat normal bundle.

Example 6.1
If M is an isoparametric hypersurface of the sphere S^n and M_i $(i = 1, \ldots, g)$ are their focal submanifolds

$$(\nu M_i)_0 = \{0\}$$

for $i = 1, \ldots, g$ (Exercise 6.3.1). □

As we will see, this notion of rank turns out to be particularly useful when $M = G \cdot p$ is a homogeneous submanifold of Euclidean space contained in a sphere and hence, $\mathrm{rank} M \geq 1$ (in particular this is the case if M is a compact homogeneous submanifold of $\mathbb{R}^n$).

On the other hand, in the case of submanifolds of the real hyperbolic space, A. Will [240] found a family of homogeneous, irreducible and full submanifolds with flat normal bundle and codimension at least 2 (non isoparametric by the classification of Section 5.2, see also [244]). In this case, the rank does not interfere with the geometry of the homogeneous submanifold. We will therefore concentrate on (homogeneous) submanifolds of Euclidean space (and the sphere) where the existence of a nontrivial parallel normal field has strong influence on the geometry. This is a particular case of so-called *submanifolds with curvature normals of constant length* that we will discuss right now.

b) Submanifolds with curvature normals of constant length

Let $f : M \to \mathbb{R}^n$ be an immersed submanifold and assume that $\mathrm{rank}_f(M) \geq 1$. Let ξ be a section of $(\nu M)_0$.
Since $(\nu M)_0$ is $\nabla^\perp$-parallel and flat, then $R^\perp(X,Y)\xi = 0$ for any tangent vectors X and Y and so, by the Ricci equation, A_ξ commutes with all shape operators and in particular with those relative to sections of $(\nu M)_0$. In this way, for any $p \in M$, the set $\{A_\xi : \xi \in (\nu_p M)_0\}$ is a commuting family of symmetric endomorphisms. Simultaneous diagonalization induces a decomposition of the tangent space at p

$$T_p M = E_1(p) \oplus \cdots \oplus E_{g(p)}(p)$$

into distinct common eigenspaces. Associated to this decomposition are well defined normal vectors $\eta_1(p), \cdots, \eta_{g(p)}(p)$, called *curvature normals*, such that

$$A_{\xi\,|E_i(p)} = \langle \eta_i(p), \xi \rangle \mathrm{id}_{E_i(p)}$$

for any $\xi \in (\nu_p M)_0$, $i = 1, \cdots, g(p)$. The dimension of $E_i(p)$ is called *multiplicity* of the curvature normal $\eta_i(p)$. This corresponds to the multiplicity of the eigenvalue $\langle \eta_i(p), \xi \rangle$ of the shape operator A_ξ, for generic $\xi \in (\nu_p M)_0$ (i.e., ξ is not in the union of the hyperplanes given by $\langle \eta_j(p) - \eta_r(p), \cdot \rangle = 0$, $j, r \in \{1, \cdots, g(p)\}$). Sometimes it is convenient to regard curvature normals at p as an n-tuple $(\eta_1(p), \cdots, \eta_m(p))$, where each curvature normal is counted with multiplicity (with $m = \dim(M)$).

The curvature normals have the following continuity property whose proof is standard:

PROPOSITION 6.1.2 (Continuity property of curvature normals)
Let $(p_k)_{k \in \mathbb{N}}$ be a sequence in M converging to p and let $(\eta_1(p_k), \cdots, \eta_m(p_k))$ be the curvature normals at p_k (chosen in any order and counted with multiplicity). Then there exists a subsequence $(p_{k_j})_{j \in \mathbb{N}}$ such that $(\eta_1(p_{k_j}), \cdots, \eta_m(p_{k_j}))$ converges to the curvature normals $(\eta_1(p), \cdots, \eta_m(p))$ at p (order is not necessarily preserved).

One can also show that there is an open and dense subset $\Omega \subset M$ where the number of eigendistributions is constant (or equivalently, the number of distinct curvature normals is locally constant). In Ω, the eigenspaces locally define $\mathcal{C}^\infty$ distributions and their associated curvature normals define $\mathcal{C}^\infty$ locally defined normal sections. It is standard to show, using the Codazzi equation, that any eigendistribution in Ω is integrable (in general, the leaves are not totally umbilical unless $(\nu M)_0 = \nu M$), cf. Lemma 4.4.2). If $\dim(E_i) \geq 2$, then $\nabla^\perp_X \eta_i = 0$ if X lies in E_i, as can be shown using the Codazzi equation of the shape operator. $\nabla^\perp$-parallelism of η_i in the directions orthogonal to E_i is equivalent to autoparallelism of E_i. Once more, the main ingredient is the Codazzi equation. Namely:

LEMMA 6.1.3
Let $f : M \to \mathbb{R}^n$ be an immersed submanifold with $\operatorname{rank}_f(M) \geq 1$. Let U be an open subset of M where the common eigenspaces define $\mathcal{C}^\infty$ distributions $E_1, \cdots, E_g$ with associated curvature normal fields $\eta_1, \cdots, \eta_g$. Then, for any $i = 1, \cdots, g$,

(a) E_i is autoparallel at $q \in U$ if and only if $\nabla^\perp_w \eta_i = 0$ for all $w \in E_i^\perp(q)$

(b) If $\dim(E_i) \geq 2$ then E_i is autoparallel at $q \in U$ if and only if η_i is parallel at q.

PROOF Let $q \in U$, X, Y, Z_j be arbitrary vector fields of M such that X, Y lie both in E_i and Z_j lies in E_j for $j \neq i$ arbitrary. Let ξ be a parallel section of $(\nu U)_0$ generic at q, i.e.,

$$\lambda_i(\xi(q)) \neq \lambda_j(\xi(q)), \quad \text{for any } j \neq i$$

where $\lambda_r = \langle \eta_r , \cdot \rangle$. Then

$$\begin{aligned}\langle (\nabla_X A)_\xi Y, Z_j\rangle_q &= X(q)(\lambda_i(\xi))\langle Y, Z_j\rangle_q + \lambda_i(\xi)(q)\langle \nabla_X Y, Z_j\rangle_q - \\ &\quad - \langle A_\xi(\nabla_X Y), Z_j\rangle_q \\ &= \lambda_i(\xi)(q)\langle \nabla_X Y, Z_j\rangle_q - \langle \nabla_X Y, A_\xi(Z_j)\rangle_q \\ &= (\lambda_i(\xi)(q) - \lambda_j(\xi)(q))\langle \nabla_X Y, Z_j\rangle_q .\end{aligned}$$

Thus, for arbitrary X, Y lying in E_i, Z lying in $E_i^\perp$ $\langle (\nabla_X A)_\xi Y, Z\rangle_q = 0$ if and only if $(\nabla_X Y)_q \in E_i(q)$, which is equivalent to E_i autoparallel at q.

Now, by the Codazzi equation

$$\begin{aligned}\langle (\nabla_X A)_\xi Y, Z\rangle_q &= \langle (\nabla_Z A)_\xi X, Y\rangle_q \\ &= Z(q)(\lambda_i(\xi))\langle X, Y\rangle_q + \lambda_i(\xi)(q)\langle \nabla_Z X, Y\rangle_q - \langle A_\xi(\nabla_Z X), Y\rangle_q \\ &= Z(q)(\lambda_i(\xi))\langle X, Y\rangle_q .\end{aligned}$$

Therefore, E_i is autoparallel at q if and only if

$$0 = Z(q)(\lambda_i(\xi)) = \langle \nabla^\perp_Z \eta_i, \xi\rangle_q,$$

for any parallel section ξ of $(\nu U)_0$ generic at q. Since generic vectors at q form an open and dense subset of $(\nu_q(M))_0$, we have part (a).

To prove (b) it is enough to observe that, if $\dim E_i \geq 2$ we can choose X and Y lying in E_i such that $||X||_q = ||Y||_q = 1$ and $\langle X, Y\rangle_q = 0$. Then

$$\begin{aligned}\langle (\nabla_X A)_\xi Y, Y\rangle_q &= X(q)(\lambda_i(\xi))\langle Y, Y\rangle_q = \\ \langle (\nabla_Y A)_\xi X, Y\rangle_q &= Y(q)(\lambda_i(\xi))\langle X, Y\rangle_q = 0 ,\end{aligned}$$

and $0 = X(q)(\lambda_i(\xi)) = \langle \nabla^\perp_X n_i, \xi\rangle_q$. Since X is arbitrary with $||X|| = 1$ we obtain $\nabla^\perp_{E_i} \eta_i = 0$. □

REMARK 6.1.4 Observe that $R^\perp(X,Y)\xi = 0$ for all sections $\xi \in \nu_0(M)$. By the Ricci equation, $\langle R^\perp(X,Y)\xi, \eta\rangle = \langle [A_\xi, A_\eta]X, Y\rangle$, so A_ξ commutes with all shape operators. Any eigendistribution E_i is then invariant under all shape operators, or equivalently, $\alpha(E_i, E_j) = 0$, if $i \neq j$, α being the second fundamental form. □

Let $\mathtt{L}_p$ be the subset of $\mathbb{R}$ of the lengths of the curvature normals at p

$$\mathtt{L}_p = \{||\eta_1(p)||, \cdots, ||\eta_{g(p)}(p)||\} .$$

DEFINITION 6.1.5 *An immersed submanifold $f : M \to \mathbb{R}^n$ is said to have curvature normals of constant length if the set of lengths $\mathtt{L}_p$ does not depend on $p \in M$.*

It is interesting to note that an (extrinsic) homogeneous submanifold M with rank at least one has curvature normals of constant length. In fact, if h is an extrinsic isometry mapping p into q then $h_{*p}(\nu_p M) = \nu_q M$ and clearly, the linear isometry h_{*p} maps curvature normals at p into curvature normals at q.

Let $\mathtt{L}_p^2 = \{||\eta_1(p)||^2, \cdots, ||\eta_{g(p)}(p)||^2\}$. Then M has curvature normal of constant length if and only if $\mathtt{L}_p^2$ does not depend on p. An interesting fact is that the set $\mathtt{L}_p^2$ is related to the eigenvalues of the so-called *adapted third fundamental form*, which we will introduce next.

Let $k = \mathrm{rank}_f(M)$ and consider the map G^0 from M into the Grassmannian $G(k,n)$ of k-planes in $\mathbb{R}^n$: $p \mapsto (\nu_p M)_0$.

The map G^0 will be called *adapted Gauss map*. If νM is flat, then the adapted Gauss map coincides with the usual Gauss map G. Let g^0 be the possibly degenerate metric induced on M by the adapted Gauss map and let B^0 be the symmetric tensor on M relating g^0 with the Riemannian metric $\langle \cdot, \cdot \rangle$

$$g^0(X,Y) = \langle B^0 X, Y \rangle .$$

We call B^0 the *adapted third fundamental form*. If νM is flat, then $B^0 = B$, the so-called *third fundamental form*, classically defined by means of the usual Gauss map (see [168]). If $e_1, \cdots, e_k$ is an orthonormal basis of $(\nu_p M)_0$ then (see Exercise 6.3.4)

$$B_p^0 = \sum_{j=1}^{k} A_{e_j}^2 . \tag{6.1}$$

Since $A_{e_j \,|E_i(p)} = \langle \eta_i(p), e_j \rangle \mathrm{id}_{E_i(p)}$, formula (6.1) yields

$$B^0_{p\,|E_i(p)} = \sum_{j=1}^{k} \langle \eta_i(p), e_j \rangle^2 \mathrm{id}_{E_i(p)} = ||\eta_i(p)||^2 \mathrm{id}_{E_i(p)}$$

for any $i = 1, \cdots, g(p)$. So we have the following result.

PROPOSITION 6.1.6
An immersed submanifold $f : M \to \mathbb{R}^n$ has curvature normals of constant length if and only if its adapted third fundamental form has constant eigenvalues.

c) Higher rank rigidity

We now state a global result due to A.J. Di Scala and C. Olmos, for which the assumption of completeness is fundamental [70].

THEOREM 6.1.7 [70]
Let M, $\dim M \geq 2$, be a simply connected and complete Riemannian manifold, let $f : M \to \mathbb{R}^n$ be a full and irreducible isometric immersion with

$\text{rank}_f(M) \geq 1$ *and such that the curvature normals have constant length. Assume, furthermore, that the number of curvature normals is constant on* M *or that* $\text{rank}_f(M) = \text{rank}_f^{loc}(M)$*. Then* $f(M)$ *is contained in a sphere.*

Moreover, if $\text{rank}_f(M) \geq 2$*, then* M *is a submanifold with constant principal curvatures (and hence* $f(M)$ *is either an isoparametric hypersurface of the sphere or an orbit of an s-representation).*

COROLLARY 6.1.8 [175, 176]
Let M *(*$\dim M \geq 2$*) be an extrinsically homogeneous irreducible and full submanifold of Euclidean space with rank at least one. Then* M *is contained in a sphere. Moreover, if rank*$(M) \geq 2$*, then* M *is an orbit of an s-representation.*

By Theorem 3.4.2 ([68]), there exist no minimal homogeneous submanifolds of Euclidean space besides the totally geodesic ones, hence the following corollary.

COROLLARY 6.1.9 [68, 175, 176]
Let M *(*$\dim M \geq 2$*) be an extrinsically homogeneous irreducible and full submanifold of Euclidean space with parallel mean curvature. Then* M *is either a minimal submanifold of a sphere or an orbit of an s-representation.*

The above corollary cannot be strengthened, since any representation of a compact Lie group has a minimal orbit in the sphere (e.g., a principal orbit with maximal volume, see [102]).

We will explain the main steps used in the proof of Theorem 6.1.7. All the details can be found in [70].

The aim is to demonstrate that the curvature normals are parallel in the normal connection and hence, by Theorem 5.5.8 and its extension to submanifolds of Euclidean spaces (see part c) of Section 5.5, page 174), the theorem follows. The strategy of the proof is to show that if there is a nonparallel curvature normal then the submanifold must split off a curve, contradicting irreducibility.

Simplifying hypothesis: we will make some extra assumptions that spare the technical details. These assumptions are automatically fulfilled if M satisfies the hypothesis of Corollary 6.1.8, i.e., a homogeneous submanifold with rank at least 1 (simply connectedness is not important, since one can pass to the universal cover). Here are the assumptions:

(ex_1) *The number of curvature normals is constant on* M *and*

$$\text{rank}_f(M) = \text{rank}_f^{loc}(M).$$

(ex_2) *If a curvature normal is parallel in an open and non-empty subset of* M *then it is globally parallel.*

Observe that, if the number of curvature normals is constant on M, then the curvature normals and their associated eigendistributions are globally defined and smooth (see Exercise 6.3.5).

Also note that the assumption (ex_1) can be replaced by the assumption that the local rank $\mathrm{rank}_f^{loc}(M)_q$ is constant on M (see Exercise 6.3.3).

Suppose that the simplifying hypothesis of Theorem 6.1.7, ex_1, ex_2 hold. Let $E_1, \cdots, E_g$ be the (globally defined) eigendistributions with associated curvature normals $\eta_1, \cdots, \eta_g$. We can assume that $||\eta_i|| \geq ||\eta_j||$ if $i < j$. If all curvature normals are parallel, then any of them provides an isoparametric global normal section and, by Theorem 5.5.8, we are finished. Let us then assume that the curvature normals are not all parallel. Without loss of generality, we can assume that the first $k-1$ are the only parallel ones. In other words, η_k is a nonparallel curvature normal of maximal length.

For $i, j \in \{1, \cdots, g\}$ let $h_{ij} : M \to \mathbb{R}$ be defined by

$$h_{ij} = \langle \eta_j, \eta_j \rangle - \langle \eta_i, \eta_j \rangle = \langle \eta_j - \eta_i, \eta_j \rangle .$$

By the Cauchy-Schwarz inequality

$$h_{ij} > 0 \qquad \text{if} \quad i > j .$$

Let J be an arbitrary subset of $\{1, \cdots, g\} - \{k\}$ ($J = \emptyset$ is allowed!) and let

$$\Omega_J = (\{p \in M \mid h_{jk}(p) = 0 \Leftrightarrow j \in J\})^o ,$$

superscript denoting the interior. Observe that $\Omega_J = \emptyset$ if J is not contained in $\{1, \cdots, k-1\}$. Notice also that $\Omega_\emptyset = (\{p \in M : h_{jk}(p) \neq 0 \; \forall j = 1, \cdots, k-1, k+1, \cdots, g\})^o$. In particular, $\Omega_\emptyset = M$, if $k = 1$. It is a standard fact that

$$\Omega = \bigcup_{J \subseteq \{1, \cdots, k-1\}} \Omega_J$$

is an open and dense subset of M.

We will show that the eigendistribution E_k associated to the curvature normal η_k is autoparallel. It suffices to show that the restriction $E_k|_{\Omega_J}$ is autoparallel, for any $J \subseteq \{1, \cdots, k-1\}$. For this we will follow the outline of Section 2 of [175]. Let $J \subseteq \{1, \cdots, k-1\}$. Without loss of generality, we can assume that $J = \{1, \cdots, s\}$, $s < k$. Now

(i) $\langle \eta_i, \eta_j \rangle$ is constant if $i, j < k$ (in particular, if $i, j \leq s$), since η_i, η_j are parallel.

(ii) $\langle \eta_i, \eta_i \rangle > \langle \eta_i, \eta_k \rangle$ for any $i \leq s$, since $||\eta_i|| \geq ||\eta_k||$.

(iii) $\langle \eta_k, \eta_k \rangle > \langle \eta_l, \eta_k \rangle$ for any $l > k$, since $||\eta_k|| \geq ||\eta_l||$.

(iv) For $i \leq s$ $\langle \eta_i, \eta_k \rangle$ is constant in Ω_J, since $\langle \eta_i, \eta_k \rangle = ||\eta_k||^2$ and the curvature normals have constant length.

Let $p \in \Omega_J$ and ξ^p be the parallel normal section $\xi^p(p) = \eta_k(p)$. The shape operator A_{ξ^p} does not distinguish, near p, the eigendistribution E_k, unless $J = \emptyset$. This is because of the definition of Ω_J which implies $\lambda_k(\xi^p(p)) - \lambda_i(\xi^p(p)) = h_{ik}(p) = 0$, for $i = 1, \cdots, s$. Also note that, in Ω_J, $\lambda_k(\xi^p(p)) - \lambda_l(\xi^p(p)) = h_{lk}(p) \neq 0$, for all $k \neq l > s$. It is clear by (i), (ii) (iii) and (iv) that there exists a parallel normal section $\bar{\xi}$ that is a linear combination of $\eta_1, \cdots, \eta_s$, and such that $\bar{\xi} + \xi^p$ distinguishes at p, hence near p, the eigendistribution E_k (i.e. $\lambda_j(\bar{\xi}(p) + \xi^p(p)) \neq \lambda_k(\bar{\xi}(p) + \xi^p(p))$ if $j \neq k$). From (iv), $\langle \bar{\xi}, \eta_k \rangle = c$ is a constant, so $\langle \bar{\xi} + \xi^p, \eta_k \rangle = c + \langle \xi^p, \eta_k \rangle$. Using the Cauchy-Schwarz inequality, since $||\xi^p|| = ||\eta_k||$ is constant, the function $\langle \xi^p, \eta_k \rangle$ has a maximum at p and $\langle \bar{\xi} + \xi^p, \eta_k \rangle$ achieves its maximum at p, too. Hence its differential is zero at p. The symmetric tensor

$$T^p = A_{\bar{\xi}+\xi^p} - \langle \bar{\xi} + \xi^p, \eta_k \rangle \mathrm{id}$$

then satisfies the Codazzi equation (only) at p, since $A_{\bar{\xi}+\xi^p}$ and id satisfy the Codazzi equation. Namely, $\nabla_X(T^p)(Y) = \nabla_Y(T^p)(X)$, where ∇ is the Levi-Civita connection of M, and $\nabla_X(T^p)(Y) = \nabla_X T^p(Y) - T^p(\nabla_X Y)$. Equivalently, the tensor $\langle \nabla_X(T^p)(Y), Z \rangle$ is symmetric in all its three entries. Since $E_k = \ker(T^p)$ near p, E_k is autoparallel at p. In fact, if X, Y are tangent fields lying in E_k and Z is arbitrary, then

$$\langle \nabla_X(T^p)(Y), Z \rangle_p = \langle \nabla_X T^p(Y) - T^p(\nabla_X Y), Z \rangle_p = -\langle T^p(\nabla_X Y), Z \rangle_p .$$

The Codazzi equation gives

$$-\langle T^p(\nabla_X Y), Z \rangle_p = \langle \nabla_Z(T^p)(Y), X \rangle_p = \langle \nabla_Z T^p(Y) - T^p(\nabla_Z Y), X \rangle_p =$$

$$-\langle T^p(\nabla_Z Y), X \rangle_p = -\langle \nabla_Z Y, T^p(X) \rangle_p = 0 .$$

Then $-\langle T^p(\nabla_X Y), Z \rangle_p = 0$ for Z arbitrary and $(\nabla_X Y)_p$ lies in $(\ker(T^p))_p = E_k(p)$. Since p is arbitrary we conclude that E_k is autoparallel in Ω_J. But J is arbitrary, so we conclude that E_k is an autoparallel distribution in Ω, hence in M. Applying Lemma 6.1.3 $\dim(E_k) = 1$, since η_k is not parallel. Moreover, the distribution $E_k^\perp$ is integrable, as shown by the next lemma.

In order to reinforce these ideas, it would be convenient for the reader to reproduce the above arguments in the important case $k = 1$ (proving that E_1 is autoparallel in $M = \Omega_\emptyset$).

LEMMA 6.1.10

Under the above assumptions

(i) E_k is autoparallel and $\dim(E_k) = 1$.

(ii) The distribution $E_k^\perp$ is integrable.

PROOF Part (i) was just proved. Let us then show part (ii): let $\tilde{\Omega}$ be the subset of M where η_k is not parallel. Then, by condition ex_2, $\tilde{\Omega}$ is open and dense. By part (i) and Lemma 4.4.2 we have that for $q \in \tilde{\Omega}$,

$$\nabla^{\perp}_{Z_q}\eta_k = 0 \qquad \text{if and only if} \qquad Z_q \in E_k^{\perp}(q) .$$

Let X, Y be arbitrary vector fields on M that lie in $E_k^{\perp}$. Since $(\nu M)_0$ is flat

$$0 = R^{\perp}(X,Y)\eta_k = \nabla^{\perp}_X \nabla^{\perp}_Y \eta_k - \nabla^{\perp}_Y \nabla^{\perp}_X \eta_k - \nabla^{\perp}_{[X,Y]}\eta_k = -\nabla^{\perp}_{[X,Y]}\eta_k .$$

Then, if $q \in \tilde{\Omega}$, $[X,Y]_q \in E_k^{\perp}(q)$. So $E_k^{\perp}$ is involutive in $\tilde{\Omega}$ and hence in M, implying $E_k^{\perp}$ is an integrable distribution. □

With the same assumptions and notations throughout this section, since M is simply connected, we have that $E_k = \mathbb{R}\, X$, for some globally defined unit vector field X. Observe that the integral curves of X are unit speed geodesics, since E_k is totally geodesic. Let ϕ_t denote the flow of X. Then, for all $t \in \mathbb{R}$,

$$(I) \qquad (\phi_t)_*(E_k) = E_k, \qquad (\phi_t)_*(E_k^{\perp}) = E_k^{\perp} .$$

The first equality is clear. Let us then show the second one. If $c(s)$ is a curve that lies in $E_k^{\perp}$, then $(\phi_t)_*(c'(0)) = \dfrac{\partial h}{\partial s}_{|s=0} := J(t)$, where $h(s,t) = \phi_t(c(s))$. But $t \mapsto J(t)$ is a Jacobi field along the geodesic $\gamma_{X(c(0))}(t) = \phi_t(c(0))$ with initial conditions $J(0) = c'(0) \perp X(c(0))$ and

$$J'(0) = \frac{D}{\partial t}_{|t=0}\frac{\partial h}{\partial s}_{|s=0} = \frac{D}{\partial s}_{|s=0}\frac{\partial h}{\partial t}_{|t=0} = \frac{D}{ds}_{|s=0} X(c(s)) \perp X(c(0)) ,$$

since $||X|| = 1$. Then $J(t)$ is always perpendicular to $\gamma'_{X(c(0))}(t)$, which generates $E_k(\gamma_{X(c(0))}(t))$. This shows $(\phi_t)_*(E_k^{\perp}) = E_k^{\perp}$.

Let $p \in M$ and let $\mathcal{L}_p$ be the integral leaf through p of $E_k^{\perp}$. Then there exists an open neighbourhood V of p in $\mathcal{L}_p$ and $\varepsilon > 0$ such that $g : [-\varepsilon, \varepsilon] \times V \to M$ defined by

$$g(s,q) = \phi_s(q)$$

is a diffeomorphism onto its image, where ϕ_s is the flow associated to X.

Let $\tilde{c} : [0,1] \to M$ be a piecewise differentiable loop at p contained in $g([-\varepsilon,\varepsilon] \times V)$. If we write $\tilde{c}(t) = g(h(t), c(t)) = \phi_{h(t)}(c(t))$, then both h and c are closed curves starting at 0 and p respectively. Let $g^c : [-\varepsilon, \varepsilon] \times [0,1] \to M$ be defined by

$$g^c(s,t) = g(s, c(t)).$$

From (I) we get that

$$(II) \qquad R^{\perp}\left(\frac{\partial g^c}{\partial s}, \frac{\partial g^c}{\partial t}\right) = 0$$

(we have used the Ricci equation and the fact that E_k is invariant under all shape operators).

Observe now that $c_1(t) = (0,t)$ and $c_2(t) = (h(t),t)$ are both curves in $[-\varepsilon, \varepsilon] \times [0,1]$ from $(0,0)$ to $(0,1)$. Then, by (II) and Exercise 6.3.2,

$$\tau^{\perp}_{g^c \circ c_1} = \tau^{\perp}_{g^c \circ c_2} = \tau^{\perp}_{\tilde{c}}$$

We have shown the following:

LEMMA 6.1.11

Given $p \in M$ there exists a neighbourhood U of p such that for any loop c at p contained in U there exists another loop $\bar{c}$ at p contained in the integral leaf $\mathcal{L}_p$ of $E_k^{\perp}$ and such that $\tau_c^{\perp} = \tau_{\bar{c}}^{\perp}$

For any $p \in M$, $\nu_p M \oplus \mathbb{R}\, X(p) = \nu_p(\mathcal{L}_p)$, regarding the leaf $\mathcal{L}_p$ of $E_k^{\perp}$ as a submanifold of the ambient space. Moreover, the restriction of X to $\mathcal{L}_p$ defines a parallel normal field to this leaf. In fact, let Z be a tangent field to M that lies in $E_k^{\perp}$ and let $\bar{\nabla}$ be the Levi-Civita connection of the ambient space. Then $\langle \bar{\nabla}_Z X, X \rangle = 0$, since X is of unit length, and the projection of $\bar{\nabla}_Z X$ to the normal space of M is $\alpha(X, Z) = 0$, where α is the second fundamental form of M.

The above observation, together with Lemma 6.1.11, implies that, for any $p \in M$,

$$(III) \qquad (\nu_p M)_0 \oplus \mathbb{R}\, X(p) = (\nu_p(\mathcal{L}_p))_0 \,.$$

Let $p \in M$ and let $\gamma : \mathbb{R} \to M$ be an integral curve of X with $\gamma(0) = p$ (observe that $\gamma(t) = \phi_t(p)$ is a geodesic with initial condition $X(p)$). Then, for all $t \in \mathbb{R}$,

$$f_*(E_k^{\perp}(p)) = f_*(E_k^{\perp}(\gamma(t)))$$

or equivalently,

$$(IV) \qquad f_*(T_p \mathcal{L}_p) = f_*(T_{\phi_t(p)} \mathcal{L}_{\phi_t(p)}) \,,$$

where f is the immersion of M into Euclidean space. In fact, let $\bar{v} \in E_k^{\perp}(p)$ and let $v(t)$ be the parallel transport in M of $\bar{v}$ along the geodesic γ. Identifying $w \in TM$ with $f_*(w)$, $\bar{\nabla}_{\gamma'(t)} v(t) = 0$ (using the fact that $\gamma'(t)$ is parallel along $\gamma(t)$ and that $\alpha(\gamma'(t), v(t)) = 0$). Thus $v(t)$ is constant in the ambient space, giving the above equalities.

Let us now fix $t \in \mathbb{R}$ and the leaf $\mathcal{L}_p$ of $E_k^{\perp}$. Let $\xi_t : \mathcal{L}_p \to \mathbb{R}^n$ be the map defined by

$$\xi_t(q) = f(\phi_t(q)) \,, \qquad q \in \mathcal{L}_p \,.$$

Equalities (I) and (IV) imply that ξ_t can be regarded as a parallel normal field to $\mathcal{L}_p$. Now, (I) gives $\phi_{t_0}(\mathcal{L}_p) = \mathcal{L}_{\phi_t(p)}$, hence $\mathcal{L}_p$ and $\mathcal{L}_{\phi_t(p)}$ are parallel submanifolds of Euclidean space. More precisely,

$$(V) \qquad (\mathcal{L}_p)_{\xi_t} = \mathcal{L}_{\phi_t(p)} \,.$$

REMARK 6.1.12 Let $\gamma : \mathbb{R} \to M$ be the integral curve of X starting at p and let ξ be a parallel normal field to M. Identifying $w \in TM$ with $f_*(w)$

$$0 = \frac{d}{dt}_{|0} \langle \xi(\gamma(t)), \gamma'(t) \rangle = \langle -A_{\xi(p)} X(p), X(p) \rangle + \langle \xi(p), \alpha(X(p), X(p)) \rangle ,$$

so

$$\langle \xi(p), \eta_k(p) \rangle = \langle \xi(p), \alpha(X(p), X(p)) \rangle$$

for any $\xi(p) \in (\nu_p(M))_0$. Thus, the curvature normal $\eta_k(p)$ coincides with the projection to $(\nu_p(M))_0$ of $\alpha(X(p), X(p))$. Moreover, one has the equality

$$\eta_k = \alpha(X, X) .$$

In fact, since for any $t \in \mathbb{R}$, $q \mapsto \xi_t(q) = f(\phi_t(q)) - q$ $(q \in \mathcal{L}_p)$ is a parallel normal field to $\mathcal{L}_p$, we obtain that $\frac{d^2}{dt^2}_{|0} \xi_t = \alpha(X, X)$ must be also a parallel normal field to $\mathcal{L}_p$ (X is restricted to $\mathcal{L}_p$). In order to prove this, observe that the parallel transport $\tau^{\perp}_{p,q}$ from p to q in $(\nu(\mathcal{L})_p))_0$ is an isometry mapping the curve $t \mapsto \xi_t(p)$ into $t \mapsto \xi_t(q)$ and so $\tau^{\perp}_{p,q}(\frac{d^2}{dt^2}_{|0} \xi_t(p)) = \frac{d^2}{dt^2}_{|0} \xi_t(q)$ defines a global parallel normal field to $\mathcal{L}_p$. But $\frac{d^2}{dt^2}_{|0} \xi_t(p) = \gamma''(0)$, and this is perpendicular to $\gamma'(0) = X(p)$, since $||X|| = 1$. By (III) $\gamma''(0) \in (\nu_p M)_0$. This gives the desired equality. $\square$

The following lemma relates the curvature normals of the isometric immersion $f : M \to \mathbb{R}^N$ to the curvature normals of the leaves of ${E_k}^{\perp}$.

LEMMA 6.1.13
With the usual assumptions, let $\gamma(t)$ be the integral curve of X starting at p.

(i) The eigenspaces of the simultaneous diagonalization of the shape operators of $(\nu_p(\mathcal{L}))_0$ coincide with the eigenspaces $E_1(p), \ldots, E_{k-1}(p), E_{k+1}(p), \ldots, E_g(p)$.

(ii) Let $\tilde{\eta}_1(p), \ldots, \tilde{\eta}_{k-1}(p), \tilde{\eta}_{k+1}(p), \ldots, \tilde{\eta}_g(p)$ be the curvature normals at p of the leaf $\mathcal{L}_p$ of $E_k^{\perp}$ (regarded as a submanifold of the ambient space) associated to decomposition $T_p\mathcal{L}_p = E_1(p) \oplus \cdots \oplus E_{k-1}(p) \oplus E_{k+1}(p) \oplus \cdots \oplus E_g(p)$. Then

$$\eta_i(\gamma_p(t)) = \frac{\tilde{\eta}_i(p)}{1 - \langle \tilde{\eta}_i(p), \tilde{\gamma}_p(t) \rangle} - \left\langle \frac{\tilde{\eta}_i(p)}{1 - \langle \tilde{\eta}_i(p), \tilde{\gamma}_p(t) \rangle}, \tilde{\gamma}'_p(t) \right\rangle \tilde{\gamma}'_p(t) ,$$

where $i = 1, \ldots, k-1, k+1, \ldots, g$ and γ_p is an integral curve of E_k starting at p and $\tilde{\gamma}_p = f \circ \gamma_p - f(p)$.

PROOF (i): Let $\tilde{A}$ denote the shape operator of the leaves of $E_k^\perp$, regarded as submanifolds of the ambient space, and let X be the unit tangent field to M generating E_k. By (III), it suffices to show that $\tilde{A}_X$ leaves invariant E_i and that it is a multiple of the identity $(i \neq k)$. Let X_i, Y_j be tangent fields to M in E_i and E_j respectively. Let ξ be a (locally defined) parallel normal section of M distinguishing all the different eigenvalues $\lambda_1 = \langle \eta_1, \ \rangle, \cdots, \lambda_g = \langle \eta_g, \ \rangle$ $(i, j \neq k)$. The Codazzi equation $\langle (\nabla_{Y_j} A_\xi)(X), X_i \rangle = \langle (\nabla_{X_i} A_\xi)(X), Y_j \rangle$ straightforwardly implies

$$(VI) \qquad \langle \xi, \eta_k - \eta_i \rangle \langle \nabla_{Y_j} X, X_i \rangle = \langle \xi, \eta_k - \eta_j \rangle \langle \nabla_{X_i} X, Y_j \rangle .$$

But $\langle \nabla_{X_i} X, Y_j \rangle = -\langle \tilde{A}_X X_i, Y_j \rangle = \langle \nabla_{Y_j} X, X_i \rangle$, so (VI) implies $\langle \tilde{A}_X X_i, Y_j \rangle = 0$ if $i \neq j$. Then E_i is invariant under the shape operators of $\mathcal{L}_p$. Now let $i = j$ and assume $X_i \perp Y_i$. A direct computation shows that $\langle (\nabla_X A_\xi)(X_i), Y_i \rangle = 0$. By the Codazzi equation, $0 = \langle (\nabla_{X_i} A_\xi)(X), Y_i \rangle = \langle \xi, \eta_k - \eta_i \rangle \langle \nabla_{X_i} X, Y_i \rangle$. This implies $\langle \tilde{A}_X(X_i), Y_i \rangle = 0$. Then $\tilde{A}_{X|E_i}$ must be a multiple of the identity. This implies part (i).

(ii): from the "tube formula", relating the shape operators of parallel manifolds, the curvature normals of $\mathcal{L}_{\gamma_p(t)}$ are $\dfrac{\tilde{\eta}_i(p)}{1 - \langle \tilde{\eta}_i(p), \tilde{\gamma}_p(t) \rangle}$, $i \neq k$.

If π^t denotes the orthogonal projection to $(\tilde{\gamma}_p'(t))^\perp$, it is not hard to see that $\eta_i = \pi^t \left(\dfrac{\tilde{\eta}_i(p)}{1 - \langle \tilde{\eta}_i(p), \tilde{\gamma}_p(t) \rangle} \right)$, which implies part (ii). □

REMARK 6.1.14 Using the Codazzi equation with $||X_i|| = 1$

$$X(\langle \xi, \eta_i \rangle) = \langle \xi, \eta_i - \eta_k \rangle \langle \tilde{A}_X(X_i), X_i \rangle$$

where $\nabla^\perp \xi = 0$. □

PROOF of Theorem 6.1.7 *(under the assumptions* ex_1 *and* ex_2*)* By part (ii) of Lemma 6.1.13, curvature normals, for each $i = 1, 2, \ldots, k-1, k+1, \ldots, g$, satisfy the equation:

$$\langle \eta_i(\gamma(t)), \eta_i(\gamma(t)) \rangle = \left\langle \frac{\tilde{\eta}_i(p)}{1 - \langle \tilde{\eta}_i(p), \tilde{\gamma}_p(t) \rangle}, \frac{\tilde{\eta}_i(p)}{1 - \langle \tilde{\eta}_i(p), \tilde{\gamma}_p(t) \rangle} \right\rangle - \\ - \left\langle \frac{\tilde{\eta}_i(p)}{1 - \langle \tilde{\eta}_i(p), \tilde{\gamma}_p(t) \rangle}, \tilde{\gamma}_p'(t) \right\rangle^2 .$$

Put $c_i := \langle \eta_i, \eta_i \rangle$, $\tilde{c}_i := \langle \tilde{\eta}_i(p), \tilde{\eta}_i(p) \rangle$. The function $f_i(t) := 1 - \langle \tilde{\eta}_i(p), \tilde{\gamma}_p(t) \rangle$ satisfies

$$f_i(t)^2 c_i = \tilde{c}_i - f_i'(t)^2$$

where c_i and $\tilde{c}_i$ are constants. By taking derivatives it is not hard to conclude that: (i) $f_i'(t) \equiv 0$ or (ii) $f_i(t) c_i + f_i''(t) \equiv 0$. Then either $f_i(t) \equiv 1$ or

$f_i(t) = \sin(\sqrt{c_i}(t + t_0))/\sin(\sqrt{c_i}t_0)$, where t_0 satisfies $\cot^2(\sqrt{c_i}t_0) = \tilde{c}_i - c_i/c_i$ (observe that $(f_i')^2(0) = \tilde{c}_i - c_i f_i(0)^2/c_i$ and that $f_i(0) = 1$). This last case cannot occur because it would imply that we cannot pass to a parallel leaf when $\sqrt{c_i}(t + t_0)$ is a root of $\sin(x) = 0$ (recall that M is complete). So $\langle \tilde{\eta}_i(p), \tilde{\gamma}_p(t) \rangle$ vanishes. Differentiating twice, $\langle \eta_i, \eta_k \rangle = 0$ on M, for $i \neq k$, since $\eta_k(\tilde{\gamma}_p(t)) = \tilde{\gamma}_p''(t)$ by Remark 6.1.12.

We will now prove that $f : M \to \mathbb{R}^n$ splits. Note that E_k is invariant by the shape operators of M. Since M is simply connected, it suffices to show that $E_k^\perp$ is autoparallel (recall that if the orthogonal complement of an autoparallel distribution is autoparallel then both distributions must be parallel). Since $\pi_1(M) = 0$, M must split intrinsically and we can apply Moore's Lemma 2.7.1 to split the immersion.

Let us show that $E_k^\perp$ is an autoparallel distribution of M. Let $\tilde{A}$ be the shape operator of the leaves of $E_k^\perp$, regarded as submanifolds of the ambient space. Observe that $\tilde{A}_X$ coincides with the shape operator of the leaves of $E_k^\perp$ regarded as hypersurfaces of M. We claim that $\tilde{A}_X \equiv 0$. In fact, let $q \in M$ be fixed and let ξ^q be a parallel normal section to M such that $\xi^q(q) = \eta_k(q)$. Then the right hand side of the equation in Remark 6.1.14 vanishes at q, because the function $\langle \xi^q, \eta_k \rangle$ has a maximum at q (using Cauchy-Schwarz inequality, since ξ^q and η_k have both constant length). The other side of the equality of Remark 6.1.14 implies that $\langle \tilde{A}_X(X_i), X_i \rangle_q = 0$, for $\langle \xi^q, \eta_i(q) \rangle = 0$. Since q is arbitrary, we obtain that $\tilde{A}_X \equiv 0$. In summary, we have shown that if some curvature normal is not $\nabla^\perp$-parallel we can globally split the immersion $f : M \to \mathbb{R}^n$. This completes Theorem 6.1.7's proof. □

REMARK 6.1.15 Let $f : M \to \mathbb{R}^n$ be an immersed submanifold with flat normal bundle. The inverse of the length of any nonzero curvature normal $\eta(p)$ coincides with the distance in $\nu_p M$ to the focal hyperplane given by the equation $\langle \eta(p), \ \rangle = 1$. Therefore, M has curvature normals of constant length if and only if the distances to the focal hyperplanes are constant on M (this is always the case if M has, in addition, algebraically constant second fundamental form). □

REMARK 6.1.16 Let $f : M \to \mathbb{R}^n$ be an immersed submanifold with flat normal bundle. Assume that the curvature normals have all the same length l^{-1}. This is equivalent to saying that the Gauss map is homothetic, i.e., the metric induced by the Gauss map is a constant multiple of the Riemannian metric of M. In this case, S. Nölker [168] proved that M must be a product of spheres of radius l and curves with curvature l. This is also true locally. Roughly speaking, the proof goes like this: any eigendistribution on M is autoparallel since it is associated to a curvature normal of maximal length. If M is not isoparametric, there exists a nonparallel curvature normal, so the immersion must split off a curve. If M is an irreducible isoparametric submanifold different from a sphere, then the curvature normals cannot all have the same length. In fact, l must equal the distance to any focal hyperplane. □

d) Local counterexamples

Theorem 6.1.7 fails without the completeness assumption on M. Namely, there exist non-isoparametric (non-complete) submanifolds of Euclidean space with flat normal bundle and algebraically constant second fundamental form [70]. We outline the construction of such examples: one begins with a non-full unit sphere S^n of $\mathbb{R}^{n+k}$. In the affine normal space $p + \nu_p M$ at a fixed point $p \in S^n$, a curve c_p is constructed starting at p and satisfying certain requirements (in particular, the curvature of c_p has to be constant). By means of the parallel transport of the normal connection of S^n one constructs a curve c_q in the affine normal space at any q. The union of the images of such curves gives the desired submanifold. This submanifold has only two eigendistributions. One of them, say E_1, is autoparallel with integral curves c_q. If one starts with a circle in $\mathbb{R}^4$, the simplest nontrivial example produced is a surface in $\mathbb{R}^4$. All examples constructed in such a way are intrinsically $(n+1)$-spheres. Observe that the two curvature normals must satisfy $\langle \eta_1, \eta_2 \rangle = ||\eta_2||^2$, otherwise E_2 would be autoparallel and M would split.

e) Global formulation of isoparametricity

Theorem 6.1.7 and Remark 6.1.15 allow us to give a global (equivalent) definition of an isoparametric submanifold: a complete immersed and irreducible submanifold $f : M^m \to \mathbb{R}^n$, $m \geq 2$ with flat normal bundle is *isoparametric* if the distances to their focal hyperplanes are constant on M.

6.2 Normal holonomy of orbits

To simplify the explanation, we will assume in this section that the submanifolds are embedded, but everything can be carried out for immersed submanifolds.

a) Transvections

Let M be a submanifold of $\mathbb{R}^n$ and let $g \in I(\mathbb{R}^n)$ be such that $g(M) = M$. We will be concerned with the group of transvections $\mathrm{Tr}(M, \nabla^\perp)$ of the normal connection $\nabla^\perp$ we defined in Section 5.4.

Recall that $\mathrm{Tr}(M, \nabla^\perp)$ is the group of isometries of $\mathbb{R}^n$, leaving the submanifold invariant and preserving any normal holonomy subbundle. More explicitly, g is a transvection of M with respect to the normal connection $\nabla^\perp$ if, for any $p \in M$, there exists a piecewise differentiable curve $c : [0,1] \to M$ with $c(0) = p, c(1) = g(p)$

such that $g_{*p|\nu_p M} = \tau_c^\perp$, where $\tau_c^\perp$ is the parallel transport along c with respect to $\nabla^\perp$.

In a similar way, we define $\mathrm{Tr}_0(M, \nabla^\perp)$ (respectively $\mathrm{Tr}_s(M, \nabla^\perp)$) by replacing the above condition by $g_{*p|(\nu_p M)_0} = \tau^\perp_{c|(\nu_p M)_0}$ (respectively $g_{*p|(\nu_p M)_s} = \tau^\perp_{c|(\nu_p M)_s}$). Recall that $(\nu M)_s = ((\nu M)_0)^\perp$ is the subbundle of the normal bundle on which the normal holonomy group acts as an s-representation.

If M is a full submanifold of Euclidean space with constant principal curvatures, then the associated curvature normals $\eta_1, \cdots, \eta_g$ (with respect to $(\nu M)_0$) are parallel and generate $(\nu M)_0$ (since the first normal space coincides with the normal space). If g is an isometry of the ambient space with $g(M) = M$, then g_* maps curvature normals into curvature normals. Namely, if η is a curvature normal, then $g_*(\eta)$ is a curvature normal, where $g_*(\eta)(q) := g_{*g^{-1}q}(\eta(g^{-1}q))$. If g can be continuously deformed to the identity (through extrinsic isometries of M), then $g_*(\eta) = \eta$ (since there are finitely many curvature normals). This observation, together with Corollary 6.1.8, implies the following:

REMARK 6.2.1 Let $M = G \cdot p$, $\dim M \geq 2$, be a full and irreducible homogeneous submanifold of $\mathbb{R}^n$, where $G \subset I(\mathbb{R}^n)$ is connected. Then $G \subseteq \mathrm{Tr}_0(M, \nabla^\perp)$. ∎

We will see in next theorem that the inclusion $G \subseteq \mathrm{Tr}_s(M, \nabla^\perp)$ is a general fact. It depends on the following well-known result, for which we include a proof based on Riemannian holonomy standard theory.

LEMMA 6.2.2
Let H be a connected Lie subgroup of $SO(n)$, acting on $\mathbb{R}^n$ as an s-representation and let $N(H)^o$ be the connected component of the normalizer of H in $SO(n)$. Then $H = N(H)^o$.

PROOF For an irreducible symmetric space X, holonomy and isotropy representations coincide. Moreover, local and global holonomy coincide as well. Then the proof follows from the next proposition, for X cannot be Ricci-flat. ∎

PROPOSITION 6.2.3 (cf. [60])
Let M be a Riemannian manifold irreducible at $q \in M$ and let $\mathfrak{g}$ be the Lie algebra of the local (Riemannian) holonomy group Hol_q^{loc} at q. Let $\mathfrak{n}$ be the normalizer of $\mathfrak{g}$ in $\mathfrak{so}(T_q M)$. Then $\mathfrak{n}$ contains $\mathfrak{g}$ properly if and only if M is Kähler and Ricci-flat near q.

PROOF Let us endow $\mathfrak{so}(T_q M)$ with the usual scalar product $\langle A, B \rangle = -\mathrm{tr}(A.B)$.

Assume $\mathfrak{n} \neq \mathfrak{g}$. If we decompose $\mathfrak{n} = \mathfrak{g} \oplus \mathfrak{k}$ orthogonally, then $\mathfrak{g}$ and $\mathfrak{k}$ are ideals of $\mathfrak{n}$ and so $[\mathfrak{g}, \mathfrak{k}] = 0$. Now choose $0 \neq J_q \in \mathfrak{k}$. Then J_q^2 is a symmetric endomorphism that commutes with $\mathfrak{g}$. Thus, J_q^2 commutes with Hol_q^{loc} and each eigenspace of J_q^2 defines a parallel distribution near q. Since M is locally irreducible at q we conclude, by de Rham decomposition theorem, that $J_q^2 = -c^2\mathrm{id}$. By rescaling J_q, we can assume $J_q^2 = -\mathrm{id}$. Extending J_q by parallelism we obtain a parallel almost complex structure J on M. Thus, M is Kähler near q. The Ricci tensor ric of a Kähler manifold M satisfies (see Appendix A.1)

$$\mathrm{ric}\,(X, JY) = \frac{\langle R(X,Y), J\rangle}{2} .$$

If γ is any curve in a small neighbourhood of q joining q to p and τ_γ is the parallel transport along γ, then

$$\langle R(X_p, Y_p), J_p\rangle = \langle \tau_\gamma^{-1} R(X_p, Y_p)\tau_\gamma, J_q\rangle = 0$$

since $J_q \perp \mathfrak{g}$, and M is Ricci-flat near q.

The above two formulas, together with the Ambrose-Singer holonomy theorem, show that the converse is true. □

THEOREM 6.2.4

Let $M = G \cdot p$ be a homogeneous submanifold of $\mathbb{R}^n$, where $G \subset I(\mathbb{R}^n)$ is connected. Then:

1. $G \subseteq \mathrm{Tr}_s(M, \nabla^\perp)$.
2. $G \subseteq \mathrm{Tr}(M, \nabla^\perp)$ *if M is an irreducible full submanifold of $\mathbb{R}^n$ and* $\dim M \geq 2$.

PROOF Let $g \in G$, and $\tilde{g} : [0,1] \to G$ be a differentiable curve in G with $\tilde{g}(0) = \mathrm{id}$, $\tilde{g}(1) = g$. Let $p \in M$ and $\gamma(t) := \tilde{g}(t) \cdot p$. The restricted normal holonomy groups at p and $\gamma(t)$, are conjugated by the differential of $\tilde{g}$, i.e.,

$$\Phi^*_{\gamma(t)} = \tilde{g}(t)_* \Phi^*_p (\tilde{g}(t)_*)^{-1}.$$

But also the normal holonomy groups are conjugated under parallel transport

$$\Phi^*_p = (\tau^\perp_{\gamma_t})^{-1} \Phi^*_{\gamma(t)} \tau^\perp_{\gamma_t} = h_t \Phi^*_p h_t^{-1},$$

where $\tau^\perp_{\gamma_t}$ is the $\nabla^\perp$ parallel transport along $\gamma_t := \gamma_{|[0,t]}$ and $h_t = (\tau^\perp_{\gamma_t})^{-1}\tilde{g}(t)_*$. Hence $(\tau^\perp_\gamma)^{-1} g_* \in (N(\Phi^*_p))^o$, where $(N(\Phi^*_p))^o$ is the connected component of the normalizer (in the orthogonal group) of Φ^*_p. By the Normal Holonomy Theorem the restricted normal holonomy group Φ^*_p acts on $(\nu_p M)_s$ as an s-representation. Then, by Lemma 6.2.2, there exists $\tau^\perp_c \in \Phi^*_p$, where c

is a null-homotopic loop at p such that $g_{*\,|(\nu_p M)_s} = \tau_\gamma^\perp \tau^\perp_{c\,|(\nu_p(M))_s}$. Hence $g_{*\,|(\nu_p(M))_s} = \tau^\perp_{c*\gamma\,|(\nu_p(M))_s}$, which proves 1.

Let us prove part 2. From Remark 6.2.1

$$\tau^\perp_{\gamma|(\nu_p M)_0} = g_{*p|(\nu_p M)_0}$$

Moreover $\tau^\perp_{c\,|(\nu_p M)_0} = \mathrm{id}$, since $(\nu M)_0$ is flat and c is null-homotopic,. Then $g_{*\,|(\nu_p M)_0} = \tau^\perp_{c*\gamma\,|(\nu_p(M))_0}$ and so $g_* = \tau^\perp_{c*\gamma}$. ∎

Part 2 of Theorem 6.2.4 can be restated as follows:

THEOREM 6.2.5
Let $M = G \cdot v$, $\dim M \geq 2$, be a full irreducible homogeneous submanifold of $\mathbb{R}^n$, where G is a (connected) Lie subgroup of the full group of isometries of $\mathbb{R}^n$. Let $g \in G$ and $p \in M$. Then, there exists $c : [0,1] \to M$ piecewise differentiable with $c(0) = p$, $c(1) = g \cdot p$ such that

$$g_{*p\,|\nu_p M} = \tau_c^\perp\,,$$

where $\tau_c^\perp$ denotes $\nabla^\perp$-parallel transport along c.

The following corollary of Theorem 6.2.4 has an analogue in the Riemannian holonomy: *Let M be a Riemannian homogeneous manifold without flat de Rham factor. Then the isotropy subgroup is contained in the holonomy group* (cf. [90, 4.5, page 110]).

COROLLARY 6.2.6
Let $M = G \cdot p$ be a full homogeneous submanifold of $\mathbb{R}^n$, where $G \subset I(\mathbb{R}^n)$ is connected. Then the image under the slice representation of the isotropy subgroup G_p is contained in the normal holonomy group at p.

We have not assumed that M is irreducible or that $\dim M \geq 2$. This is because the corollary holds for homogeneous curves. (By Remark 3.1.4 the connected component of the extrinsic group of isometries is the product of the connected component of the extrinsic group of isometries of each factor.) In fact, if $M = G \cdot v$ with G connected and $\dim M = 1$, then any element g in the isotropy G_p acts trivially on $T_p M$. Since g is an intrinsic isometry of M then g is the identity map on M. Then the fixed set of g in $\mathbb{R}^n$ is an affine subspace that contains M. Thus, g is the identity if M is full.

b) Computation of the normal holonomy of orbits

The holonomy group $\mathrm{Hol}(M)$ of a locally irreducible homogeneous Riemannian manifold $M = G/H$ can be computed from G. Indeed, Kostant [121] proved that if

M is not Ricci-flat, the Lie algebra of the holonomy group of M is (algebraically) generated by the skew symmetric endomorphisms given by the Nomizu operators (see Section 2.3) ∇X^*, $X \in \mathfrak{g}$. Actually, the assumption that M is not Ricci-flat can be dropped, since Alekseevsky and Kimel'fel'd [4] proved that a homogeneous non-flat Riemannian manifold cannot be Ricci-flat (see also [7, page 553]).

Let us turn now to the case of a full irreducible orbit $M = G \cdot p$ of a representation $G \subseteq I(\mathbb{R}^n)$. We have the following analogous result of C. Olmos and M. Salvai [177] for the computation of the normal holonomy group in terms of G.

THEOREM 6.2.7 [177]
Let $G \subseteq I(\mathbb{R}^n)$ and $M = G \cdot p$ be full and irreducible as a submanifold of $\mathbb{R}^n$ with $\dim M \geq 2$. Then the Lie algebra of the normal holonomy group Φ_p is (algebraically) generated by the orthogonal projection on the affine subspace $p + \nu_p M$ of the Killing vector fields on $\mathbb{R}^n$ induced by G. Moreover $\Phi_p = \bar{G}_p \Phi_p^*$, where Φ_p^* is the restricted normal holonomy group and $\bar{G}_p$ is the isotropy regarded as the subgroup of $O(\nu_p M)$ via the slice representations (i.e., $\bar{G}_p = \{g_{*|\nu_p M} \; : \; g \in G_p\}$).

We introduce some notation first. Let X belong to the Lie algebra $\mathfrak{g}$ of G and consider the curve defined on $[0, 1]$

$$\gamma_p^X(s) := \operatorname{Exp} sX \cdot p.$$

Observe that $\gamma_p^{tX}(1) = \gamma_p^X(t)$. Define the operator

$$\mathcal{A}_X : \nu_p M \to \nu_p M$$

by $\mathcal{A}_X \xi = \left(\frac{d}{dt}_{|t=0} (\operatorname{Exp} tX)_{*p} \, \xi \right)^{\perp} = (X \cdot \xi)^{\perp} = \frac{D^{\perp}}{dt}_{|t=0} (\operatorname{Exp} tX)_{*p} \, \xi$ (where $(\ )^{\perp}$ denotes the orthogonal projection on $\nu_p M$).

REMARK 6.2.8 $\xi \mapsto \mathcal{A}_X \xi$ is the Killing vector field of the normal space $\nu_p M$ obtained by projecting the Killing vector field defined by X. $\square$

Let $\tau_{p,X}^{\perp}$ denote $\nabla^{\perp}$-parallel transport along γ_p^X. ($\tau_{p,tX}^{\perp}$ is the parallel transport along γ_p^X from 0 to t.) Then we have

$$\mathcal{A}_X \xi = \frac{d}{dt}_{|t=0} (\tau_{p,tX}^{\perp})^{-1} \circ (\operatorname{Exp} tX)_* \, \xi \, . \tag{6.2}$$

(We omit the point in the subscript of $(\operatorname{Exp} tX)_*$ here and in the sequel for the sake of simplicity.) Note that $\mathcal{A}_X$ is skew symmetric with respect to the induced inner product on $\nu_p M$, i.e. $\langle \mathcal{A}_X \xi, \eta \rangle + \langle \xi, \mathcal{A}_X \eta \rangle = 0$. Thus $\mathcal{A}_X \in \mathfrak{so}(\nu_p M)$. Moreover, *$\mathcal{A}_X$ belongs to the normalizer of the normal holonomy algebra*, since $(\tau_{p,tX}^{\perp})^{-1} \circ (\operatorname{Exp} tX)_*$ belongs to the normalizer of the normal holonomy group.

REMARK 6.2.9 Let $M = G \cdot p$ ($\dim M \geq 2$) be a full and irreducible submanifold of Euclidean space and let $\xi \in (\nu_p M)_0$. Then $(\mathrm{Exp}\, tX)_* \xi$ is parallel in the normal connection, for any $X \in \mathfrak{g}$ (see Remark 6.2.1). Then $(\tau^{\perp}_{p,tX})^{-1} \circ (\mathrm{Exp} tX)_* \, \xi = \xi$ and so, $\mathcal{A}_X \xi = 0$. But $\mathcal{A}_X$ belongs to the normalizer of the normal holonomy algebra at p. So, by Lemma 6.2.2, $\mathcal{A}_X$ belongs to the normal holonomy algebra at p. □

Let $g = \mathrm{Exp}\, hX$. Then $g \circ \gamma_p^X = \gamma_{g.p}^X$ and hence

$$g_* \circ \tau^{\perp}_{p,tX} \circ g_*^{-1}{}_{|\nu_p M} = \tau^{\perp}_{g.p,tX}$$

(we omit the point in the subscript of g_* here and in the sequel for the sake of simplicity). From this it is not hard to see that

$$(\tau^{\perp}_{p,tX})^{-1} \circ (\mathrm{Exp}\, tX)_{*\,|\nu_p M}$$

is a one-parameter group of linear isometries of $\nu_p M$. By (6.2) $(\mathrm{Exp}(-tX))_* \circ \tau^{\perp}_{p,tX} = e^{-t\mathcal{A}_X}$, and therefore

$$\tau^{\perp}_{p,tX} = (\mathrm{Exp} tX)_* \circ e^{-t\mathcal{A}_X} \tag{6.3}$$

is an explicit formula for computing $\nabla^{\perp}$-parallel transport along γ_p^X from 0 to t.

c) Parallel transport along broken Killing lines.

Let $p \in M$ be fixed $g \in G$ and $Y \in \mathfrak{g}$. Using the notation of previous paragraphs, consider the curve $\gamma^Y_{g \cdot p}(t) = (\mathrm{Exp}\, tY) \cdot (g \cdot p)$. This is the integral curve of the Killing vector field Y^*, where $Y^*(q) = \frac{d}{dt}_{|t=0} (\mathrm{Exp}\, tY) \cdot q$ with initial condition $\gamma^Y_{g \cdot p}(0) = g \cdot p$. One has that $\gamma^Y_{g \cdot p} = g \circ \gamma_p^{Ad(g^{-1})Y}$ and so, by formula (5.7), page 163,

$$\tau^{\perp}_{g \cdot p, Y} \circ g_* = g_* \circ \tau^{\perp}_{p, Ad(g^{-1})Y} \tag{6.4}$$

Let now $X_1, \cdots, X_r \in \mathfrak{g}$ and $g_i = (\mathrm{Exp}\, X_{i-1}) \cdots (\mathrm{Exp}\, X_1) \in G$, for $i = 1, \cdots, r+1$ ($g_1 = \mathrm{id}$).

Consider the *broken Killing line* β obtained by gluing the integral curves $\beta_i = \gamma^{X_i}_{g_i \cdot p}$, $t \in [0,1]$, $i = 1, \cdots r$. Namely, $\beta = \beta_1 * \cdots * \beta_r$ and

$$\beta_i(t) = (\mathrm{Exp}\, tX_i)(\mathrm{Exp}\, X_{i-1}) \cdots (\mathrm{Exp}\, X_1) \cdot p$$

Using formulas (6.3) and (6.4) it is straightforward to compute $\nabla^{\perp}$-parallel transports along β (which is the composition of the parallel transport along β_i). Namely,

$$\tau^{\perp}_{\beta} = (g_{r+1})_* \circ e^{-\mathcal{A}_{Z_r}} e^{-\mathcal{A}_{Z_{r-1}}} \cdots e^{-\mathcal{A}_{Z_1}} \tag{6.5}$$

where $Z_i = Ad(g_i^{-1}) X_i$.

A broken Killing line whose pieces are integral curves of Killing vector fields induced by G is called *G-broken Killing line*.

The following lemma asserts that $\nabla^{\perp}$-parallel transport along loops which are broken Killing lines gives the normal holonomy group. The proof will only be sketched since it requires a background on connections on principal bundles.

LEMMA 6.2.10

Let M be an extrinsic homogeneous submanifold of Euclidean space, orbit of the Lie group G and let $p \in M$. Let $\tilde{\Phi}_p$ be the group obtained by parallel transporting along loops based at p that are G-broken Killing lines. Then $\tilde{\Phi}_p$ coincides with the normal holonomy group Φ_p at p.

PROOF (sketch) Let $\mathcal{P}$ be the family of loops based at p that are G-broken Killing lines. Let c_0, $c_1 \in \mathcal{P}$ be homotopic loops (by means of a family of loops based at p). Then it is standard to show that there exists a piecewise smooth homotopy $c_s \in \mathcal{P}$, $s \in [0,1]$. $s \mapsto \tau^{\perp}_{c_s}$ is a piecewise smooth curve that lies in $\tilde{\Phi}_p$, a Lie subgroup of $O(\nu_p M)$ (cf. [117, vol. I, Appendix 5]). Let ϕ_s and ψ_t be the flows associated to two arbitrary Killing vector fields X, Y induced by G. Then, for any $q \in M$, $h(s,t) = \phi_s \circ \psi_t \cdot q$ is a parametrized surface whose coordinates lines are integral curves of Killing vector fields induced by G. Then, the same argument as in Section 4.1 *(e)*, shows that $R(X_q, Y_q)$ belongs to the Lie algebra of $\tilde{\Phi}_q$.

Let now $\mathcal{B}(\nu M)$ be the bundle of orthonormal bases of the normal space. Then the tangent spaces to the orbits of the groups $\tilde{\Phi}_q$ define an integrable subdistribution $\mathcal{V}$ of the vertical one. $\mathcal{V}$ is invariant under the flow of the horizontal lifting of any Killing vector field (since $\tilde{\Phi}$ conjugates under parallel transport along integral curves of Killing vector fields induced by G). Let $\mathcal{H}$ be the horizontal distribution of $\mathcal{B}(\nu M)$. The horizontal part of the bracket of any two basic fields is the curvature (applied to the corresponding basis). But the curvature lies in $\mathcal{V}$. This implies that $\mathcal{D} = \mathcal{H} \oplus \mathcal{V}$ is integrable. Thus, $\mathcal{D}$ contains the distribution given by the holonomy subbundles, so $\tilde{\Phi}_p$ contains the normal holonomy group at p. The Lemma follows because inclusion $\tilde{\Phi}_p \subseteq \Phi_p$ is trivial. ∎

We return to the operators $\mathcal{A}_X$. Let $\mathfrak{k}$ be the smallest Lie subalgebra of $\mathfrak{so}(\nu_p M)$ which contains $\mathcal{A}_X$, for any $X \in \mathfrak{g}$. Let K be the connected Lie subgroup of $SO(\nu_p M)$ with Lie algebra $\mathfrak{k}$. Let $g \in G_p$. Since K is a geometric object, we see that $\bar{g} K \bar{g}^{-1} = K$, where $\bar{g} = g_{|\nu_p M}$. So $\bar{G}_p K$ is a group.

REMARK 6.2.11 Let X belong to the Lie algebra of the isotropy subgroup G_p. Then $(\operatorname{Exp} tX) \cdot p = p$ and so, by formula (6.3) $(\operatorname{Exp} tX)_{*p|\nu_p M} = e^{t\mathcal{A}_X}$.

This shows that $\bar{\mathfrak{g}}_p \subset \mathfrak{k}$, where $\bar{\mathfrak{g}}_p$ is the Lie algebra of $\bar{G}_p = \{g_{|\nu_p M} \mid g \in G_p\}$. ∎

We can supply a proof of Theorem 6.2.7:

PROOF of Theorem 6.2.7 By formula (6.5) and Lemma 6.2.10 the group $\bar{G}_p K$ contains the normal holonomy group Φ_p. Remark 6.2.9 and Corollary 6.2.6 give the opposite inclusion, so $\Phi_p = \bar{G}_p K$. Moreover, by Remark 6.2.11, the normal holonomy algebra at p coincides with $\mathfrak{k}$. This finishes the proof. □

REMARK 6.2.12 Suppose $M = K \cdot p$ is an orbit of an s-representation, and take the reductive decomposition $\mathfrak{k} = \mathfrak{k}_p \oplus \mathfrak{m}$, with $\mathfrak{m} = \mathfrak{k}_p^{\perp}$. By Lemma 4.1.5 we know

$$([\mathfrak{k}_p, \mathfrak{m}])^{\perp} = 0 .$$

This means that $\mathcal{A}_X = 0$ if $X \in \mathfrak{m}$. Thus, Theorem 6.2.7 gives an alternative proof of Theorem 4.1.7. Namely, the normal holonomy representation of M coincides with the slice representation, i.e., the (effective made) action of the isotropy group K_p on the normal space $\nu_p M$. □

REMARK 6.2.13 Theorem 6.2.7 is used in [91] for the geometric characterization of orthogonal representations with copolarity 1 (and for their classification in the irreducible case, cf. Remark 3.2.11 for the definition of copolarity). This is the only known result that uses that theorem. It would be interesting to find further applications. □

We end by formulating the following:

Conjecture 6.2.14 (cf. [175]) *Let M, $\dim M \geq 2$, be a full homogeneous submanifold of the sphere S^{n-1}, which is not an orbit of an s-representation. Then the normal holonomy group of M acts transitively on the sphere of the normal space (in particular the normal holonomy group acts irreducibly).*

If $\dim M = 2$ the above conjecture is true by the results on surfaces explained in Section 4.5 c).

6.3 Exercises

Exercise 6.3.1 Let M be an isoparametric hypersurface of the sphere S^n and let M_i $i = 1, \ldots, g$ their focal submanifolds. Then,

$$(\nu M_i)_0 = \{0\}$$

for $i = 1, \ldots, g$.

Exercise 6.3.2 Let $g : [a,b] \times [c,d] \to M$ be a piecewise differentiable map of the variables s, t, M is an immersed submanifold of a Riemannian manifold N. Assume that $R^\perp(\frac{\partial g}{\partial s}, \frac{\partial g}{\partial t}) \equiv 0$. Let, for $i = 1, 2$, $c_i : [0,1] \to [a,b] \times [c,d]$ be two piecewise differentiable curves with $c_1(0) = c_2(0)$ and $c_1(1) = c_2(1)$. Prove that

$$\tau^\perp_{g \circ c_1} = \tau^\perp_{g \circ c_2}$$

where $\tau^\perp$ denotes the parallel transport in the normal connection of M. Prove a similar result for the Levi-Civita parallel transport.

These are indeed special cases of flat connections induced on pull back vector bundles (cf. [189]).

Exercise 6.3.3 Let $f : M \to \mathbb{R}^n$ be a connected Riemannian submanifold. Then $\mathrm{rank}^{loc}_f(M)_q$ is constant on M if and only if $\mathrm{rank}^{loc}_f(M) = \mathrm{rank}_f(M)$.

Exercise 6.3.4 Prove Formula (6.1).

Exercise 6.3.5 Let $f : M \to \mathbb{R}^n$ be an immersed simply connected submanifold with $\mathrm{rank}_f(M) \geq 1$ and assume that the number of curvature normals is constant on M. Prove that the curvature normals are globally defined $\mathcal{C}^\infty$ normal fields (assuming the local version). *Hint:* Let H be the subset of the normal space to M that consists of all the curvature normals at any point. Prove that H is a differentiable manifold and that the projection $\pi : \nu M \to M$, restricted to H, is a covering map. Thus, π restricted to any connected component of H is a diffeomorphism. The inverse map is a globally defined curvature normal field.

Exercise 6.3.6 Prove that Corollary 6.2.6 is not true if M is not full.

Exercise 6.3.7 Let M be a compact full submanifold of Euclidean space with parallel second fundamental form. Then any Killing vector field on M extends uniquely to a Killing vector field on the ambient space. *Hint:* Since M is locally symmetric, any bounded Killing vector field lies in the Lie algebra of the transvections. Let $p \in M$ be a fixed point and let g be a transvection of M. Set $\tau_\gamma = g_{*p}$. Then the isometry $\tilde{g}$ of the ambient space, defined by $\tilde{g}(p) = g(p)$, $\tilde{g}_{*p|T_pM} = \tau_\gamma$, $\tilde{g}_{*p|\nu_pM} = \tau^\perp_\gamma$, leaves M invariant.

Exercise 6.3.8 (based on an unpublished proof of Ferus' Theorem by E. Hulett and C. Olmos). Let M be a compact full submanifold of $\mathbb{R}^n$. For $v \in \mathbb{R}^n$, let h_v be the restriction to M of the height function in the direction of v (i.e. $h_v(x) = \langle x, v \rangle$, where $x \in M$). Let X_v be the gradient of h_v (M is endowed with the induced metric). Prove

(i) The second fundamental form of M is parallel if and only if $[X_v, X_w]$ is a Killing vector field on M, for all $v, w \in \mathbb{R}^n$.

(ii) Assume that the second fundamental form of M is parallel. Let $\mathfrak{k}$ be the Lie algebra of Killing vector fields on $\mathbb{R}^n$ that are tangent to M. Define on the vector space $\mathfrak{k} \oplus \mathbb{R}^n$ the following bracket: $[X, v] = X \cdot v$, if $X \in \mathfrak{k}$ and $v \in \mathbb{R}^n$; $[X, Y]$ is the bracket of $\mathfrak{k}$, if $X, Y \in \mathfrak{k}$; $[v, w]$ is the extension to the ambient space of the Killing vector field $[X_v, X_w]$ (see Exercise 6.3.7).

(a) Prove that $\mathfrak{k} \oplus \mathbb{R}^n$ is an orthogonal involutive Lie algebra (see [243]).

(b) M is (orthogonally) equivalent to an orbit of an s-representation.

Chapter 7

Homogeneous structures on submanifolds

In the late '20s, E. Cartan gave a local characterization of Riemannian symmetric spaces by the differential condition

$$\nabla R = 0.$$

For any point p, R_p determines uniquely (up to isometries) the symmetric space. Actually, the curvature tensor R_p of a locally symmetric space M determines a globally symmetric Riemannian manifold $(\widetilde{M}, g)$ whose curvature tensor at p is the same as the one of M. Indeed, R_p, together with the integrability condition $(R_p)_{xy} \cdot R_p = 0$, allows construction of an orthogonal symmetric Lie algebra $\mathfrak{g}$, which uniquely determines $\widetilde{M}$. In terms of holonomy systems (T_pM, R_p, K) is a symmetric holonomy system, where K is the simply connected Lie group with Lie algebra $\mathfrak{k}$. As we did in part b) of Section 4.3, we can set $\mathfrak{g} := \mathfrak{k} \oplus T_pM$, with $\mathfrak{k} = \text{span}\{(R_p)_{xy}\}_{x,y \in T_pM}$ (which is the Lie algebra of the holonomy group, as a consequence of the Ambrose-Singer holonomy Theorem) and define a Lie bracket on $\mathfrak{g}$ by

$$\begin{array}{ll} [B, C] = BC - CB, & B, C \in \mathfrak{k}, \\ [x, y] = (R_p)_{xy}, & x, y \in T_pM, \\ [A, z] = Az, & A \in \mathfrak{k}, z \in T_pM. \end{array}$$

Taking symmetric spaces as models, starting from the '70s, D. Ferus [83], [84], [86], W. Strübing [204], E. Backes and H. Reckziegel [6] studied the symmetric submanifolds of a space form $\bar{M}$. Recall that M is a symmetric submanifold of $\bar{M}$ if it is invariant with respect to reflections at its normal spaces. The symmetry of submanifolds is locally characterized by

$$\nabla \alpha = 0.$$

If $p \in M$, α_p and the integrability condition $(R_p^{\perp})_{xy} \cdot \alpha_p = 0$ (where $R^{\perp}$ is the normal curvature) permits recovery of the submanifold completely (and, possibly, its extension to a complete one). These constructions were provided in Section 3.7.

To come back to the intrinsic case, there is a similar framework for locally homogeneous spaces. A differential characterization is given by the Ambrose-Singer theorem, which asserts that a Riemannian manifold (M, g) is locally homogeneous if and only if there exists on M a metric connection $\hat{\nabla}$ such that

$$\hat{\nabla} S = 0$$
$$\hat{\nabla} R = 0$$

with $S = \nabla - \hat{\nabla}$. $\hat{\nabla}$ is called Ambrose-Singer connection or canonical connection. The tensor S is called a homogeneous structure.

In this chapter we want to describe the analogous setup for homogeneous submanifolds. First we will examine the case of submanifolds of space forms. Using this framework, we will characterize orbits of s-representations and we will study isoparametric submanifolds, giving a proof of Thorbergsson's Theorem 5.4.5.

7.1 Homogeneous structures and homogeneity

Let M be an orbit of a faithful representation ρ of a Lie group G into the isometry group $I(\bar{M})$ of a space form $\bar{M}$ of constant curvature κ. Since ρ is faithful, we will assume $G \subseteq I(\bar{M})$.

The orbit $M = G \cdot p$ through any point $p \in \bar{M}$ is an immersed submanifold of $\bar{M}$ and, at the same time, a reductive homogeneous space G/K. Let $\mathfrak{g} = \mathfrak{k} \oplus \mathfrak{m}$ be a reductive decomposition of the Lie algebra $\mathfrak{g}$ of G, with $\mathfrak{k}$ the Lie algebra of the isotropy group at p. Recall that $\mathfrak{m}$ is isomorphic to T_pM, via the isomorphism sending $X \in \mathfrak{m}$ to the value at p, X^*_p, of the Killing vector field X^*. We denote by $\Gamma : T_pM \to \mathfrak{m}$ the inverse of the above isomorphism.

a) Homogeneity determines homogeneous structures.

As a start, we recall the definition of intrinsic canonical connection on G/K, associated with the reductive decomposition $\mathfrak{g} = \mathfrak{k} \oplus \mathfrak{m}$. G is a principal fibre bundle over G/K with structure group K. The reductive complement $\mathfrak{m}$ of $\mathfrak{k}$ in $\mathfrak{g}$ defines a left-invariant distribution on G, which is right-invariant under K and determines a connection on the bundle G. This connection induces a canonical connection $\hat{\nabla}$ on the tangent bundle $T(G/K)$ of G/K. $\hat{\nabla}$ can be characterized geometrically as the one whose geodesics through $p = eK$ are the one-parameter subgroups $\gamma(t) = \mathrm{Exp}(tX) \cdot p$, for any $X \in \mathfrak{m}$ and such that the $\hat{\nabla}$-parallel transport along γ is given by $\tau_\gamma v = \mathrm{Exp}(tX) \cdot v$. Note that the torsion of $\hat{\nabla}$ at p is given by $\mathrm{pr}([\mathfrak{m}, \mathfrak{m}])$ on $\mathfrak{m}$, so that it vanishes if and only if G/K is a Riemannian symmetric space.

On the other hand, the vector bundle $E = T\bar{M}_{|M} = TM \oplus \nu M$ is a homogeneous vector bundle, i.e., the action of G on E is covered by the action of G on M. We will denote the former action together with its differential by $g \cdot V$ and $X \cdot V$, for $g \in G$, $X \in \mathfrak{g}$ and $V \in E$. Thus, the reductive complement $\mathfrak{m}$ of $\mathfrak{k}$ in $\mathfrak{g}$ determines an extrinsic canonical connection $\tilde{\nabla}$ on the homogeneous vector bundle E. Indeed, $\mathfrak{m}$ determines a unique G-invariant metric connection on E such that the horizontal subspaces along the fibre $\pi^{-1}(p)$ are $\{X \cdot p \mid X \in \mathfrak{m}\}$. More geometrically, $\tilde{\nabla}$ can be characterized as the metric connection whose geodesics through p are $\gamma(t) = \mathrm{Exp}(tX) \cdot p$, $X \in \mathfrak{m}$ and whose parallel displacement along the geodesics coincides with $X \cdot V$, for $V \in E$.

Let $\tilde{\Phi}$ be the holonomy of the connection $\tilde{\nabla}$. We identify $\tilde{\Phi}$ with its representation on $T_pM \oplus \nu_p M$ and K with its action on $E_p = T_pM \oplus \nu_p M$ via the isotropy and the slice representations. The above implies that $\tilde{\Phi} \subseteq K$, as representations. Hence, all G-invariant tensor fields on E are parallel with respect to the canonical connection $\tilde{\nabla}$.

In particular, $S := \nabla \oplus \nabla^\perp - \tilde{\nabla}$ and the second fundamental form are G-invariant. Thus, we get

LEMMA 7.1.1
$\tilde{\nabla} S = 0$ *and* $\tilde{\nabla}\alpha = 0$.

Note that TM and νM are $\tilde{\nabla}$-parallel subbundles of E. This suggests the following

DEFINITION 7.1.2 *A homogeneous structure on a submanifold M of a space form $\bar{M}$ is a tensor field $S \in T^*M \otimes (TM \oplus \nu M)^* \otimes (TM \oplus \nu M)$ such that, if $\tilde{\nabla}$ is the connection, called canonical connection, on $E = TM \oplus \nu M$ given by*

$$\tilde{\nabla} := \nabla \oplus \nabla^\perp - S,$$

(∇ is the Levi-Civita connection on M and $\nabla^\perp$ is the normal connection), then

(1) TM is a $\tilde{\nabla}$-parallel subbundle of E.

(2) $\tilde{\nabla}$ is a metric connection.

(3) $\tilde{\nabla}\alpha = 0$, i.e.

$$(\tilde{\nabla}_X \alpha)(Y, Z) := \tilde{\nabla}_X \alpha(Y, Z) - \alpha(\tilde{\nabla}_X Y, Z) - \alpha(Y, \tilde{\nabla}_X Z) = 0\,,$$

where X, Y, Z are vector fields on M.

(4) $\tilde{\nabla} S = 0$, i.e.,

$$(\tilde{\nabla}_X S)_Y U = \tilde{\nabla}_X (S_Y U) - S_{\tilde{\nabla}_X Y} U + S_Y(\tilde{\nabla}_X U) = 0\,,$$

where X, Y are vector fields on M, U is a vector field on $TM \oplus \nu M$.

REMARK 7.1.3 Let S be a homogeneous structure on M and let $p \in M$. We set $S := S_p$. For any $v \in T_pM$, S_v is an endomorphism of $T_pM \oplus \nu_p M$ which leaves this decomposition invariant. Moreover, by property (2) of the definition, S_v is skew-symmetric. Then S_v can be regarded as $S_v \in \mathfrak{so}(T_pM) \oplus \mathfrak{so}(\nu_p M)$. So, $S(= S_p)$ can be regarded as an element of $(T_pM)^* \otimes (\mathfrak{so}(T_pM) \oplus \mathfrak{so}(\nu_p M))$. □

REMARK 7.1.4 If TM is a $\widetilde{\nabla}$-parallel subbundle of E then its orthogonal complement νM is parallel as well. ☐

REMARK 7.1.5 One can give an alternative and actually shorter definition of canonical connection [79], by taking, instead of the connection $\nabla \oplus \nabla^\perp$ on E, the restriction to E of the Levi-Civita connection $\bar{\nabla}$ on $T\bar{M}$. Then a canonical connection $\widetilde{\nabla}$ is a metric connection satisfying

(a) TM is a $\widetilde{\nabla}$-parallel subbundle of E,

(b) $\Gamma = \bar{\nabla} - \widetilde{\nabla}$ is $\widetilde{\nabla}$-parallel.

Thus (b) takes the place of (3) and (4) in Definition 7.1.2.

To see the equivalence of these two definitions note that, if X and Y are vector fields on M,

$$\Gamma_X Y = \bar{\nabla}_X Y - \widetilde{\nabla}_X Y = \nabla_X Y + \alpha(X,Y) - \widetilde{\nabla}_X Y = S_X Y + \alpha(X,Y),$$

and, if ξ is a normal vector field

$$\Gamma_X \xi = \bar{\nabla}_X \xi - \widetilde{\nabla}_X \xi = -A_\xi X + \nabla^\perp_X \xi - \widetilde{\nabla}_X \xi = S_X \xi - A_X \xi.$$

If φ is the infinitesimal transvection defined in Section 3.7, $\Gamma_X B = S_X B + \varphi_X B$, for any vector field B on E. The above formulas readily imply that the $\widetilde{\nabla}$-parallelity of Γ is equivalent to the ones of S and α. Recall that α is parallel if and only if A is.

This latter approach is more fit for generalizations to submanifolds of homogeneous spaces, cf. [79]. ☐

Let $M = G \cdot p$ be an open part of a homogeneous submanifold of a space form $\bar{M}$. Define on M a canonical connection $\widetilde{\nabla}$ similar to the beginning of this section. Then the difference tensor $S = \nabla \oplus \nabla^\perp - \widetilde{\nabla}$ is a homogeneous structure. Hence, we have the following:

LEMMA 7.1.6

An open part of a homogeneous submanifold admits a homogeneous structure.

We will soon see that the converse is also true. This gives a differential characterization of homogeneous submanifolds of space forms.

REMARK 7.1.7 By the Ambrose-Singer holonomy theorem, the Lie algebra of the holonomy group of $\widetilde{\nabla}$ is generated by $\widetilde{R}$ (since $\widetilde{R}$ is $\widetilde{\nabla}$-parallel). Thus, the holonomy algebra is generated by the projections on $\mathfrak{k}$ of $-\rho_*[X,Y]$. ☐

REMARK 7.1.8 Since $O(n+1) \subset I(\mathbb{R}^{n+1})$, a homogeneous submanifold $M = G \cdot p$ of S^n can be considered as a homogeneous submanifold of $\mathbb{R}^{n+1}$. Conversely, a homogeneous structure S on a submanifold of S^n can be extended to a homogeneous structure S on $\mathbb{R}^{n+1}$ by setting $\widetilde{\nabla} P = 0$, where P is the position vector field regarded as a normal vector field to M in $\mathbb{R}^{n+1}$. ▯

REMARK 7.1.9 To motivate the expression of Γ we gave in Remark 7.1.5, we describe the inverse Γ of the isomorphism $\mathfrak{m} \to T_pM : X \to X \cdot p$.

For the sake of simplicity, we describe Γ in the case of a homogeneous submanifold M of the sphere $S^n \subset \mathbb{R}^{n+1}$. The cases of $\mathbb{R}^n$ and H^n as ambient spaces are very similar.

In the case of submanifolds of S^n, one can regard the elements of $\mathfrak{m}$ as differentials of isometries of $\mathbb{R}^{n+1} = T_pM \oplus \nu_pM$, where νM is the normal space of M in $\mathbb{R}^{n+1}$. Hence, for any $v \in T_pM$, $\Gamma_v \in \mathfrak{m} \subseteq \mathfrak{so}(n+1) = \mathfrak{so}(T_pM \oplus \nu_pM)$. By definition of canonical connection $\mathrm{Exp}(t\Gamma_v) \cdot p$ is a $\widetilde{\nabla}$-geodesic,

$$w(t) := (\mathrm{Exp}(t\Gamma_v))_* w = \mathrm{Exp}(t\Gamma_v) \cdot w$$

is $\widetilde{\nabla}$-parallel in TM and

$$\xi(t) := (\mathrm{Exp}(t\Gamma_v))_* \xi = \mathrm{Exp}(t\Gamma_v) \cdot \xi$$

is $\widetilde{\nabla}$-parallel in νM. Hence, by the Gauss identities and since $w(t)$ is $\widetilde{\nabla}$-parallel

$$\begin{aligned}
\Gamma_v w = \frac{dw(t)}{dt}_{|t=0} &= \bar{\nabla}_v w(t) = \bar{\nabla}_v w(t) - \widetilde{\nabla}^c_v w(t) \\
&= \nabla_v w(t) + \alpha(v,w) - \widetilde{\nabla}_v w(t) \\
&= S_v w + \alpha(v,w),
\end{aligned}$$

where $\bar{\nabla}$ is the Levi-Civita connection of $\mathbb{R}^{n+1}$, ∇ the Levi-Civita connection of M and α is the second fundamental form of M regarded as a submanifold of $\mathbb{R}^{n+1}$. Similarly, using Weingarten identities,

$$\Gamma_v \xi = S_v \xi - A_\xi v,$$

where A is the shape operator of M as a submanifold of $\mathbb{R}^{n+1}$. Hence, the inverse $\Gamma : T_pM \to \mathfrak{m}$ of the isomorphism $\mathfrak{m} \to T_pM : X \to X \cdot p$ has the following expression

$$\Gamma_v X = S_v X + \alpha(v, X^\top) - A_{X^\perp} v = S_x Y + \varphi_x Y\,,$$

where $v \in T_pM$, $X \in \mathbb{R}^{n+1} = T_pM \oplus \nu_pM$, $X^\top$ the orthogonal projection of X on T_pM, $X^\perp$ the orthogonal projection of X on ν_pM, α, A and φ are respectively the second fundamental form, the shape operator and the infinitesimal transvection of M, regarded as a submanifold of $\mathbb{R}^{n+1}$.

Observe that an important special case of homogeneous structures occurs when $S = 0$. It corresponds to the condition that $\nabla\alpha = 0$. We have already examined this situation in Section 3.7. In this case, the operator Γ coincides with the infinitesimal transvection φ. □

b) Existence of homogeneous structures and homogeneity.

Next we prove that the converse of Lemma 7.1.6 holds. For its proof, it is crucial to write down the differential equations of the $\widetilde{\nabla}$-geodesics and of the Darboux frames along it. It is the same type of argument as in [204] (cf. also Section 3.7 and [178]).

For notational convenience, we will use the operator Γ (which entails the pieces of information of second fundamental form, shape operator and canonical connection), given by

$$\Gamma_v X = S_v X + \alpha(v, X^\top) - A_{X^\perp} v = S_x Y + \varphi_x Y,$$

where $v \in T_pM$, $X \in T\bar{M}_{|M} = T_pM \oplus \nu_pM$.

Let γ be a unit speed $\widetilde{\nabla}$-geodesic in M with $\gamma(0) = x$ and

$$B(t) = (B_1(t) = \dot{\gamma}(t), \dots, B_n(t))$$

be a $\widetilde{\nabla}$-parallel Darboux frame at $\gamma(t)$, i.e., the first $m = \dim M$ vectors are tangent to M. Since $\widetilde{\nabla}_{\frac{d}{dt}} B_j = 0$, for any $j = 1, \dots, n$, we have

$$\bar{\nabla}_{\frac{d}{dt}} B_j(t) = \widetilde{\nabla}_{\frac{d}{dt}} B_j(t) + \Gamma_{\frac{d}{dt}} B_j(t) = \Gamma_{\frac{d}{dt}} B_j(t).$$

But

$$\Gamma_{\frac{d}{dt}} B_j(t) = \sum_i c_{ij} B_i(t),$$

for some constant matrix $C = (c_{ij}) = \langle \Gamma_{\dot{\gamma}(0)} B_i(0), B_j(0) \rangle$, because Γ as well as B_j and $\dot{\gamma}$ are $\widetilde{\nabla}$-parallel. Thus, in matrix notation, we have the differential equations

$$\bar{\nabla}_{\frac{d}{dt}} B = BC. \tag{7.1}$$

LEMMA 7.1.10
Let $M \to \bar{M}$ be a connected submanifold of a space form admitting a homogeneous structure S. Let $p, q \in M$. Then there exists an isometry $F : \bar{M} \to \bar{M}$ such that

$$F(p) = q, \qquad F(M) \subseteq M$$

PROOF Let $c(t)$ be a piecewise differentiable curve from p to q and let $\tau(t)$ be the $\widetilde{\nabla}$-parallel transport along c. Define F as the unique isometry of $\bar{M}$ such that

$$F(p) = q, \qquad F_{*p} = \tau(1).$$

The aim is to show that $F(M) \subseteq M$. To this end, let

$$\begin{aligned} J = \{\, x \in M \mid\ & F(x) \in M, F_{*x}(T_xM) = T_{F(x)}M, \\ & F_{*x}(S_x)_v Y = (S_{F(x)})_{F_{*x}v} F_{*x} Y, F_{*x}\alpha_x(v, w) = \alpha_{F(x)}(F_{*x}v, F_{*x}w), \\ & v, w \in T_xM, Y \in T_xM \oplus \nu_xM \}. \end{aligned}$$

Note that the conditions

$$F_{*x}(S_x)_v X = (S_{F(x)})_{F_{*x}v} F_{*x} X \qquad \text{and} \qquad F_{*x}\alpha_x(v,w) = \alpha_{F(x)}(F_{*x}v, F_{*x}w)$$

can be expressed more concisely by $F_{*x}(\Gamma_x)_v X = (\Gamma_{F(x)})_{F_{*x}v} F_{*x} X$. Clearly, $p \in J$, so J is not empty. Moreover, J is closed. Since M is connected, to prove the assertion, it suffices to see that J is open.

Let $x \in J$. Let η be a $\widetilde{\nabla}$ geodesic at x, and $(B_1 = \dot{\eta}(0), \ldots, B_n)$ be a Darboux basis at x. Extend B_j $(j = 1, ..., n)$ to $\widetilde{\nabla}$-parallel Darboux frame $B_j(t)$ along η. Let $\widetilde{\eta}$ be the $\widetilde{\nabla}$ geodesic at $F(x)$, with $\dot{\widetilde{\eta}}(0) = F_{*x}B_1$ and extend the Darboux basis $(F_{*x}B_1, \ldots, F_{*x}B_n) = (\widetilde{B}_1, \ldots, \widetilde{B}_n)$ to a $\widetilde{\nabla}$-parallel Darboux frame $\widetilde{B}_j(t)$ along $\widetilde{\eta}$. Then $(\widetilde{\eta}, \widetilde{B}_j(t))$ satisfies the system of differential equations (7.1), with $c_{jh} = \langle (\Gamma_{F(q)})_{\widetilde{B}_1} \widetilde{B}_j, \widetilde{B}_h \rangle$ and initial conditions $\widetilde{\eta}(0) = F(x)$, $\widetilde{B}_j(0) = \widetilde{B}_j$.

Consider now $F \circ \eta$ and $F_{*\eta(t)}B_j(t)$. The pair $(F \circ \eta, F_{*\eta(t)}B_j(t))$ satisfies the system (7.1) with the same initial conditions as $(\widetilde{\eta}, \widetilde{B}_j(t))$. Thus,

$$(F \circ \eta, F_{*\eta(t)}B_j(t)) = (\widetilde{\eta}, \widetilde{B}_j(t)).$$

If W is a normal neighbourhood of x, then $U \cap W \subseteq J$. This shows that J is open and completes the proof. ∎

REMARK 7.1.11 Lemma 7.1.10 is a generalization of Lemma 3.7.13. ∎

Let $I(M, \bar{M})$ be the group of isometries of the ambient space $\bar{M}$ leaving the submanifold M invariant. The *transvection group* $\mathrm{Tr}(M, \widetilde{\nabla})$ of $\widetilde{\nabla}$ is the group of isometries $g \in I(M, \bar{M})$ such that for any $p \in M$ g_{*p} coincides with the $\widetilde{\nabla}$-parallel transport along some piecewise differentiable curve in M from p to $g(p)$. Then the above result implies that $G = \mathrm{Tr}(M, \widetilde{\nabla})$ is transitive on M. So, for any $p \in M$, the G-orbit of p contains M as an open subset. Thus, Lemma 7.1.6 and Lemma 7.1.10 give the proof of the following theorem [174]

THEOREM 7.1.12
A connected submanifold of $\bar{M}$ is an open part of a globally homogeneous submanifold of $\bar{M}$ if and only if it admits a homogeneous structure.

REMARK 7.1.13 The above theorem can also be stated as "*A closed submanifold of $\bar{M}$ is a globally homogeneous submanifold of $\bar{M}$ if and only if it admits a homogeneous structure.*" ∎

It can be shown that, given a homogeneous structure S on M, a homogeneous submanifold $\widetilde{M}$, which contains M as an open subset, is uniquely determined by the punctual data (S_p, α_p). Indeed, starting from the data (S_p, α_p), one can construct a Lie subalgebra $\mathfrak{g}$ of the Lie algebra of $I(\bar{M})$. Moreover, if G is the connected Lie subgroup of $I(\bar{M})$ having $\mathfrak{g}$ as Lie algebra, $\widetilde{M} := G \cdot p$ is a homogeneous submanifold that contains M as an open subset (see Exercise 7.4.3; cf. [58, 59]).

One can actually get rid of the datum of the homogeneous structure at p, provided that one takes account of some iterated covariant derivative of α at p. One can show [58] that a homogeneous submanifold is uniquely determined by the knowledge of covariant derivatives of α at p up to the order $k + 2$, where k is an integer, which is the extrinsic analogue of the Singer invariant [201].

If M is a homogeneous isoparametric submanifold orbit of the isotropy representation of a symmetric space with reduced root system, then M is uniquely determined by the data $(\alpha_p, \nabla\alpha_p)$ (Exercise 7.4.5).

7.2 Examples of homogeneous structures

a) The space of homogeneous structures

We begin with some remarks on the space of homogeneous structures. Let S be a homogeneous structure on a submanifold and $p \in M$. Set $V := T_pM$ and $W := \nu_pM$. Then $S_p \in V^* \otimes (\mathfrak{so}(V) \oplus \mathfrak{so}(W))$ (see Remark 7.1.3). By using the inner products of V and W, we identify $\mathfrak{so}(V)$ with Λ^2V^* (where Λ^2 is the second exterior power of V^*) and $\mathfrak{so}(W)$ with Λ^2W^*.

We will omit the subscript p, writing simply S for S_p in the sequel.

Hence,

$$S \in (V^* \otimes \Lambda^2V^*) \oplus (V^* \otimes \Lambda^2W^*) \tag{7.2}$$

In other terms, one identifies S with the tensor $S_{xYZ} := \langle S_xY, Z\rangle$, $x \in V$, $Y, Z \in V \oplus W$.

Let us denote by $\mathcal{S}(V, W)$ the module $(V^* \otimes \Lambda^2V^*) \oplus (V^* \otimes \Lambda^2W^*)$ under the (natural) action of $O(m) \times O(h)$ (diagonally immersed subgroup of $O(m + h)$, $m = \dim V$, $h = \dim W$). This action is the following: $g \in O(m) \times O(h)$ acts on the tensors S in $\mathcal{S}(V, W)$ according to the formula

$$(g \cdot S)_{xYZ} := S_{g^{-1}xg^{-1}Yg^{-1}Z}. \tag{7.3}$$

$O(h)$ acts trivially on $V^* \otimes \Lambda^2V^*$ which can be thus considered as a $O(m)$ module. Let us set

$$\mathcal{T}(V) := V^* \otimes \Lambda^2V^*,$$
$$\mathcal{N}(V, W) := V^* \otimes \Lambda^2W^*.$$

Then

$$\mathcal{S}(V, W) = \mathcal{T}(V) \oplus \mathcal{N}(V, W).$$

First we split $\mathcal{T}(V)$ into irreducible factors.

THEOREM 7.2.1
$\mathcal{T}(V)$ decomposes into three irreducible factors.

These three irreducible factors are usually denoted by $\mathcal{T}_1$, $\mathcal{T}_2$ and $\mathcal{T}_3$.
A proof of Theorem 7.2.1 can be found in [223] page 37. Another proof is the following:

PROOF Consider the following $O(m)$-equivariant maps

$$\Lambda : V^* \otimes \Lambda^2 V^* \longrightarrow \Lambda^3 V^*,$$
$$c_{12} : V^* \otimes \Lambda^2 V^* \longrightarrow V^*,$$

where Λ is the antisymmetrization and c_{12} is the contraction given by the trace on the first two components in $V^* \otimes \Lambda^2 V^*$. Note that $\Lambda^3 V^*$ and V^* are irreducible. There are exact sequences of $O(m)$-modules

$$0 \longrightarrow \ker c_{12} \longrightarrow V^* \otimes \Lambda^2 V^* \xrightarrow{c_{12}} V^* \longrightarrow 0\,,$$
$$0 \longrightarrow \ker \Lambda \longrightarrow V^* \otimes \Lambda^2 V^* \xrightarrow{\Lambda} \Lambda^3 V^* \longrightarrow 0\,.$$

The first exact sequence induces the isomorphism of $O(m)$-modules

$$V^* \otimes \Lambda^2 V^* \cong V^* \oplus \ker c_{12}. \tag{7.4}$$

Moreover, one has the following commutative diagram

$$\begin{array}{ccccccc}
 & & 0 & & 0 & & \\
 & & \downarrow & & \downarrow & & \\
 & & \ker c_{12} \cap \ker \Lambda & \hookrightarrow & \ker \Lambda & & \\
 & & \downarrow & & \downarrow & & \\
0 & \longrightarrow & \ker c_{12} & \longrightarrow & V^* \otimes \Lambda^2 V^* & \xrightarrow{c_{12}} & V^* \longrightarrow 0 \\
 & & \downarrow & & \downarrow & & \\
 & & \Lambda^3 V^* & = & \Lambda^3 V^* & & \\
 & & \downarrow & & \downarrow & & \\
 & & 0 & & 0 & &
\end{array}$$

where the left vertical sequence is exact (since $\Lambda_{|\ker c_{12}}$ is surjective). This implies that

$$\ker c_{12} \cong (\ker c_{12} \cap \ker \Lambda) \oplus \Lambda^3 V^*. \tag{7.5}$$

From (7.4) and (7.5) we get the $O(m)$-modules isomorphism

$$V^* \otimes \Lambda^2 V^* \cong V^* \oplus (\ker \Lambda \cap \ker c_{12}) \oplus \Lambda^3 V^*. \tag{7.6}$$

One can verify (e.g., computing the weights of the representation corresponding to the $O(m)$-module $\ker c_{12} \cap \ker \Lambda$) that $\ker c_{12} \cap \ker \Lambda$ is irreducible as well, so that (7.6) is the desired splitting into irreducible factors. Comparing with [223], we have the identifications

$$\mathcal{T}_1 \cong V^*, \qquad \mathcal{T}_2 \cong \ker c_{12} \cap \ker \Lambda, \qquad \mathcal{T}_3 \cong \Lambda^3 V^*.$$ □

LEMMA 7.2.2
$\mathcal{N}(V, W)$ is irreducible.

PROOF V^* is an irreducible $O(m)$-module since it is the standard representation of $O(m)$. $\Lambda^2 W^*$ is an irreducible $O(h)$-module because $\Lambda^2 W^*$ is the adjoint representation of $O(h)$. Lemma 7.2.2 then follows from the following standard result of representation theory: If G and H are compact Lie groups, A is an irreducible G module and B is an irreducible H module, then $A \otimes B$ is an irreducible $G \times H$ module (where we regard A both and B as $G \times H$ modules). (See, for instance, [29, Prop. 4.14] for a proof.) □

b) Examples

We can now give examples of homogeneous structures, dividing them into some classes in accordance with the scheme given by the above algebraic decomposition.

TABLE 7.1: Main classes of homogeneous structures

type of homogeneous structure	class of submanifolds
0	symmetric submanifolds
$\mathcal{T} = \mathcal{T}_1 \oplus \mathcal{T}_2 \oplus \mathcal{T}_3$	orbits of s-representations
$\mathcal{N}$	2-symmetric submanifolds

The examples are indicated in the second column of Table 7.1. Thus, the algebraic decomposition of the previous paragraph corresponds to geometric properties of the submanifolds.

Symmetric submanifolds of space forms.

In Section 3.7, we studied symmetric submanifolds of space forms. We proved that $M \to \bar{M}$ is locally symmetric if and only if its second fundamental form is parallel (Theorem 3.7.2). In terms of homogeneous structures, this property can be restated as follows: *M is a symmetric submanifold of a space form $\bar{M}$ if and only if M admits the homogeneous structure $S = 0$* (where 0 is the null tensor).

Recall that any symmetric submanifolds of space forms split locally as an extrinsic product of extrinsic spheres and of full irreducible connected compact extrinsic symmetric submanifold of Euclidean space $\mathbb{R}^n$, which is minimal in a hypersphere of $\mathbb{R}^n$. Then, as proved in Theorem 3.7.8

THEOREM 7.2.3
A full irreducible connected compact extrinsically symmetric submanifold of

Euclidean space is an orbit of an s-representation.

REMARK 7.2.4 It is possible to give a direct proof of this result using Dadok's Theorem 3.2.15 and results on normal holonomy [80]. Let $K \subset O(\mathbb{R}^n)$ be the group of orthogonal transformations leaving M invariant. Since M is extrinsically symmetric, K contains affine reflections at normal spaces. By the Normal Holonomy Theorem, the slice representation of the isotropy subgroup K_x at $x \in M$ on the normal space $\nu_x M$ is an s-representation (in particular a polar action). Let $\Sigma \subset \nu_x M$ be a section. We claim that Σ is a section for the action of K on $\mathbb{R}^n$ as well. Indeed, every K-orbit meets $\nu_x M$ and thus Σ. Let us prove that Σ meets K-orbits orthogonally. Let $Y = x + \xi \in \Sigma$. Then $\mathfrak{k}_x \cdot Y \perp \Sigma$, for Σ is a section for the slice representation of K_x on $\nu_x M$. Take a Cartan decomposition $\mathfrak{k} = \mathfrak{k}_x \oplus \mathfrak{p}_x$, where $\mathfrak{p}_x$ is the space of infinitesimal transvections at x. For any $X \in \mathfrak{p}_x$ the parallel transport along the geodesic $\mathrm{Exp}tX \cdot x$ both on tangent and normal space is provided by the action of $\mathrm{Exp}tX$. Hence $\mathrm{Exp}tX \cdot Y = \mathrm{Exp}tX \cdot (x + \xi)$ is $\nabla^\perp$-parallel along $\mathrm{Exp}tX \cdot x$, so

$$(X \cdot Y)^\perp = (X \cdot (x + \xi))^\perp = \left(\frac{d}{dt}_{|t=0} \mathrm{Exp}tX \cdot (x + \xi)\right)^\perp = 0 \, .$$

Therefore $X \cdot Y = X \cdot (x + \xi)$ is perpendicular to $\nu_x M$ and hence to Σ. This shows that $\mathfrak{p}_x \cdot Y \perp \Sigma$ and the K-action is polar with section Σ. The result then follows from the Dadok's Theorem 3.2.15. ☐

Orbits of s-representations. In [178] the orbits of the s-representations are characterized as the submanifolds admitting a homogeneous structure S with $S^\perp = 0$ (i.e., $S \in \mathcal{T} = \mathcal{T}_1 \oplus \mathcal{T}_2 \oplus \mathcal{T}_3$). It is proved the following

THEOREM 7.2.5
Let M be a full connected compact submanifold of $\mathbb{R}^n$. Then the following are equivalent:

1. *There exists a homogeneous structure S on M of type $\mathcal{T}$.*
2. *M is orbit of an s-representation.*

Recall that by Theorem 5.4.9 being an orbit of an s-representation is equivalent to having extrinsically homogeneous normal holonomy bundle.

PROOF of Theorem 7.2.5 We first assume that statement 1 holds, i.e., M admits a homogeneous structure S with $S^\perp = 0$. Let c be a piecewise differentiable curve in M joining p and q. By Lemma 7.1.10 there is an isometry F of $\mathbb{R}^n$ such that $F(p) = q$, $F(M) \subseteq M$ and $F_{*p|\nu_p M} = \tau_c^{\tilde{\nabla}^\perp} = \tau_c^{\nabla^\perp}$. So M has extrinsically homogeneous normal holonomy bundle, which is equivalent to being an orbit of an s-representation.

For the converse, suppose that M is an orbit of an s-representation. We use the same notation as in the proof of Proposition 4.1.6. Suppose that $M = K \cdot X = \mathrm{Ad}(K)X$, $X \in \mathfrak{p}$ and that $\mathfrak{k} = \mathfrak{k}_X \oplus \mathfrak{m}$ is a reductive decomposition of M. Let ∇^c be the canonical connection associated to this reductive decomposition. Define a connection on $TM \oplus \nu M$ as $\nabla^c := \nabla^c \oplus \nabla^\perp$.

From the following lemma we then readily see that $\bar{S} := \nabla \oplus \nabla^\perp - \bar{\nabla}$ is a homogeneous structure on M of type $\mathcal{T}$. ∎

LEMMA 7.2.6
Let $M = K \cdot X$ be an orbit of an s-representation. Then $\nabla^c \alpha = 0$.

PROOF Let $Z \in \mathfrak{m}$, then $\gamma(t) := \mathrm{Exp}tZ \cdot X$ is a ∇^c-geodesic and $(\mathrm{Exp}tZ)_{*X}$ is the ∇^c-parallel transport along γ. Then, using Lemma 4.1.5 we have, for any $v, w \in T_X M$,

$$(\nabla^c_{Z \cdot X}\alpha)(v,w) = \nabla^\perp_{Z \cdot X}\alpha((\mathrm{Exp}tZ)_{*X} v, (\mathrm{Exp}tZ)_{*X} w) =$$
$$= \left(\frac{d}{dt}_{|t=0} \alpha((\mathrm{Exp}tZ)_{*X} v, (\mathrm{Exp}tZ)_{*X} w) \right)^\perp =$$
$$= \frac{d}{dt}_{|t=0} (\mathrm{Exp}tZ \cdot \alpha(v,w))^\perp = [Z, \alpha(v,w)]^\perp = 0$$

∎

k-symmetric submanifolds.

DEFINITION 7.2.7 [124] *Let $f : M \to \bar{M}^n(\kappa)$ be a submanifold. A regular s-structure on M is a family of isometries $\{\sigma_x\}_{x \in M}$ of $\bar{M}$ such that:*

(1) $\sigma_x(M) \subseteq M$.

(2) x is an isolated fixed point of $\sigma_{x|M}$.

(3) For any $x, y \in M$, $\sigma_x \cdot \sigma_y = \sigma_z \cdot \sigma_x$, $z = \sigma_x(y)$.

If $\{\sigma_x\}_{x \in M}$ is of finite order $k \geq 2$ (i.e. $\sigma^k = \mathrm{id}$), M is called a k-symmetric submanifold.

A k-symmetric submanifold is extrinsically homogeneous. Indeed, if M is a k-symmetric submanifold, let $\mathrm{Tr}(M, \{\sigma_{x|M}\})$ be the group of the transvections, i.e., the group generated by the isometries $\sigma_{x|M}\sigma_{y|M}^{-1}$ of M. Then one can define the representation

$$F : \mathrm{Tr}(M, \{\sigma_{x|M}\}) \to I(\bar{M})$$
$$\sigma_{x|M}\sigma_{y|M}^{-1} \mapsto \sigma_x \sigma_y^{-1}.$$

M is an orbit of $\mathrm{Tr}(M, \{\sigma_{x|M}\})$ in the representation F.

REMARK 7.2.8 Note that the above definition differs from the one of C. Sánchez [197]. Indeed, in the definition of Sánchez, it is assumed in addition that the restriction to the normal spaces of the differential of the k-symmetries are the identity (i.e., for any x, $(\sigma_x)_{*x|\nu_x M} = \mathrm{id}$). Using Exercise 7.4.6, one can show that a k-symmetric submanifold in the sense of Sánchez is an orbit of an s-representation (see also [178]). □

Let us turn to the case $k = 2$, i.e. of the 2-symmetric submanifolds. Any symmetric submanifold is 2-symmetric, but the converse is not true. As a matter of fact, if M is a symmetric submanifold, the set of reflections σ_x with respect to the normal spaces $\nu_x M$ is a regular $s-$structure of order 2. However, not all regular $s-$structures are of this kind. Indeed, one can prove that a s-structure of order 2 is generated by reflections with respect to (generally proper) subspaces of $\nu_p M$ ([124], [31]).

Let M be a submanifold of $\bar{M}$. The *k-th osculating space of M at p* is the space $\mathcal{O}_p^k(M)$ spanned by the first k derivatives in 0 of curves $\gamma : (-\epsilon, \epsilon) \to M$ with $\gamma(0) = p$. Note that $\mathcal{O}_p^1(M) = T_pM$. The *$k$-th normal space of M at p* is the orthogonal complement $\mathcal{N}_p^k(M)$ of $\mathcal{O}_p^k(M)$ in $\mathcal{O}_p^{k+1}(M)$. So, for instance, $\mathcal{N}_p^1(M) = \mathrm{im}\alpha_p$ (cf. page 22). M is called *nicely curved* if the dimension of any $\mathcal{O}_p^k(M)$ does not depend on p. In this case, one can define the k-th osculating bundle (respectively: k-th normal bundle) as the vector bundle, whose fibres at p is $\mathcal{O}_p^k(M)$ (respectively $\mathcal{N}_p^k(M)$). A metric connection on $\mathcal{N}^k(M)$ is given by

$$\nabla_X^{\mathcal{N}^k} \xi := \mathrm{proj}_{\mathcal{N}_p^k(M)} \bar{\nabla}_X \xi, \qquad X \in T_pM,\ \xi \text{ vector field on } \mathcal{N}^k(M),$$

where proj is the orthogonal projection on $\mathcal{N}_p^k(M)$. One has higher order fundamental forms defined by

$$\alpha^k(X, \xi) := \mathrm{proj}_{\mathcal{N}_p^k(M)} \bar{\nabla}_X \xi, \qquad X \in T_pM,\ \xi \text{ vector field on } \mathcal{N}^{k-1}(M).$$

A. Carfagna, R. Mazzocco and G. Romani proved the following characterization of 2-symmetric submanifolds of Euclidean spaces and spheres in [31], which was proved to hold also for hyperbolic space forms in [30].

THEOREM 7.2.9
A submanifold M of $\bar{M}$ is 2-symmetric if and only if $\nabla^{\mathcal{N}^k} \alpha^k = 0$ for any $k \geq 1$.

($\nabla^{\mathcal{N}^k} \alpha^k$ is defined in a natural way, using the Levi-Civita connection on the tangent part.)

Let $\nabla^{\mathcal{N}}$ be the connection on νN defined by $\nabla_x^{\mathcal{N}} \xi = \nabla_x^{\mathcal{N}^k} \xi$, if $\xi \in \mathcal{N}^k$, $k \geq 1$ (and extended by linearity on any $\xi \in \nu M$). Then one can define on $TM \oplus \nu M$ the connection

$$\tilde{\nabla}^{\mathcal{N}} := \nabla \oplus \nabla^{\mathcal{N}} \qquad (k \geq 1)\,,$$

where ∇ the Levi-Civita connection on M. Consider the tensor $S^{\mathcal{N}} := \nabla \oplus \nabla^{\perp} - \tilde{\nabla}^{\mathcal{N}}$.

THEOREM 7.2.10 [59]
A submanifold M of $\bar{M}$ is 2-symmetric if and only if $S^{\mathcal{N}}$ is a homogeneous structure on the submanifold M.

In particular, by Theorem 7.2.10, one gets that a 2-symmetric submanifold M admits a homogeneous structure of type $\mathcal{N}$. Actually, if M is a compact submanifold of $\mathbb{R}^n$, the converse is also true.

THEOREM 7.2.11 [59]
Let M be a compact connected submanifold of $\mathbb{R}^n$. Then M admits a homogeneous structure $S \in \mathcal{N}$ if and only if M is 2-symmetric.

REMARK 7.2.12 (Historical Note) C. Sánchez in the '80s was the first who used a canonical connection on the tangent space of a submanifold. This was related to k-symmetric submanifolds (in the sense of Sánchez, cf. Remark 7.2.8). He proved that a k-symmetric submanifold of Euclidean space can be characterized by the property of having parallel second fundamental form with respect to the canonical connection of a k-symmetric space (generalizing Strübing [204]). This result was published in [198]. □

7.3 Isoparametric submanifolds of higher rank

In Section 5.4 we cited Thorbergsson's Theorem 5.4.5. Namely, *any irreducible full isoparametric submanifold of $\mathbb{R}^n$ of rank at least three is an orbit of an s-representation.*

The original proof of Thorbergsson [219] uses Tits buildings and the Homogeneous Slice Theorem 5.3.6. There is an alternative proof of Thorbergsson's result using the theory of homogeneous structures on submanifolds [174] and normal holonomy. The idea of the proof is the following: We know that, by Theorem 7.2.5, if there exists on a submanifold M of $\mathbb{R}^n$ a canonical connection ∇^c of type $\mathcal{T}$, then M is an orbit of an s-representation. Given an irreducible full isoparametric submanifold of Euclidean space of codimension at least three one can focalize at the same time any two eigendistributions. The corresponding fibres are, by the Homogeneous Slice Theorem, orbits of s-representations. A canonical connection ∇^c on M is constructed by gluing together the canonical connections that occur naturally on these fibres. The proof of the compatibility between these canonical connections is based on the relation between the normal holonomy groups of the different focal manifolds. The common eigendistributions of the shape operator of M are parallel with respect to the canonical connection. This implies readily that $\nabla^c\alpha = 0$. To show that $\nabla^c(\nabla - \nabla^c) = \nabla^c S^c = 0$ one has to use the geometric fact that the ∇^c

parallel transport along a horizontal curve with respect to some focalization equals the $\nabla^\perp$ parallel displacement in the focal manifold along the projection of the curve.

a) The canonical connection on orbits of s-representations

To motivate the definition of canonical connection we now look at what properties it must have. To this goal, we start with an isoparametric submanifold M that is homogeneous. In this case, M is a principal orbit of an s-representation, i.e., of the isotropy representation of a symmetric space $X = G/K$. Suppose $M = K \cdot v$. Recall from Section 3.2 that $\mathfrak{k} = \mathfrak{k}_v \oplus \mathfrak{k}_+$, (with $\mathfrak{k}_+$ given by the sum of the positive eigenspaces of $\{\mathrm{ad}^2(A), A \in \mathfrak{a}\}$) is an orthogonal direct sum (with respect to the opposite of the Killing form of K) and a reductive decomposition of $\mathfrak{k}$. Recall also from Section 3.6 that the common eigendistributions of the shape operator of M are

$$E_\lambda = \mathfrak{p}_\lambda + \mathfrak{p}_{2\lambda}, \qquad \text{with } \mathfrak{p}_{2\lambda} = \{0\} \text{ if } 2\lambda \text{ is not a root.}$$

In particular, they correspond to positive roots λ of $\{\mathrm{ad}^2(X), X \in \mathfrak{a}\}$.

Consider a focal orbit $M = K \cdot u$, with $u = v + \xi(v)$ and let

$$\bar{E} := \ker(A_\xi - \mathrm{id}).$$

Note that $\bar{E}$ corresponds to a sum of restricted root vectors of $\{\mathrm{ad}^2(A), A \in \mathfrak{a}\}$. Denote this sum by $\mathfrak{f}$. Then $\mathfrak{k}_u = \mathfrak{k}_v \oplus \mathfrak{f}$ and $\mathfrak{k}_u \oplus \mathfrak{n}$ is a reductive decomposition of the homogeneous space M_ξ, where $\mathfrak{n}$ is the orthogonal complement of $\mathfrak{k}_u$ in $\mathfrak{k}$.

PROPOSITION 7.3.1

Let M be an orbit of an s-representation, ∇^c the canonical connection on M associated with the above reductive decomposition of M and $\bar{\nabla}^c$ the analogue on M_ξ. Then ∇^c and $\bar{\nabla}^c$ have the following properties:

1. *The eigendistributions of the shape operator on M are ∇^c-parallel. (This is due to the fact that $\nabla^c \alpha = 0$.)*

2. *If X lies in an eigenspace or a sum of eigenspaces of $\{\mathrm{ad}^2(A) \mid A \in \mathfrak{a}\}$ corresponding to an eigendistribution $E_i \neq \bar{E}$, then*

   $\gamma(t) = \mathrm{Exp}(tX) \cdot v$ is a ∇^c-geodesic in M,

   $\bar{\gamma}(t) = \mathrm{Exp}(tX) \cdot u$ is a ∇^c-geodesic in M_ξ,

   if $w \in \bar{E}$, then

   $\mathrm{Exp}(tX)_* w$ is ∇^c-parallel in M along γ,

   $\mathrm{Exp}(tX)_* w$ is $\nabla^\perp$-parallel in M_ξ along $\bar{\gamma}$.

Next, we suppose that $u =: u_i = v + \xi_i(v)$, where ξ_i is a parallel normal field focalizing only E_i (so that $E_i = \ker(A_{\xi_i} - \mathrm{id})$. Let n_i be the curvature normal relative to E_i and $V_i(v)$ be the affine subspace through v parallel to the vector space spanned by $n_i(v)$ and $E_i(v)$. Set

$$\tilde{K}^i := \left(\{k_{|V_i(v)} \mid k \in K \text{ and } k \cdot S_i(v) = S_i(v)\}\right)_o,$$

and

$$K^i := \left(\{k_{|V_i(v)} \mid k \in K_{u_i}\}\right)_o.$$

Clearly, $\tilde{K}^i \subseteq K^i$. We claim that $K^i \subseteq \tilde{K}^i$, so $\tilde{K}^i = K^i$. In fact, if M_{u_i} is full, K^i is the restricted normal holonomy group of M_{u_i} at u_i and so $K^i \subseteq \tilde{K}^i$ follows from the Homogeneous Slice Theorem. If M_{u_i} is not full, it is a factor of M and obviously in this case $K^i = \tilde{K}^i$.

By the Normal Holonomy Theorem, the representation of K^i on $V_i(v)$ is an s-representation. One can thus construct, as above, a canonical connection ∇^i on $S_i(v)$. This connection is associated to the reductive decomposition of the Lie algebra of K^i with reductive complement given by a sum of restricted root vectors of $\{\mathrm{ad}^2(A), A \in \mathfrak{a}\}$ corresponding to E_i.

Thus ∇^i coincides with the connection induced by ∇^c in the autoparallel submanifold $S_i(v)$.

b) The canonical connection on isoparametric submanifolds of rank at least three

To begin, we show how one can focalize at the same time any two eigendistributions on an irreducible full isoparametric submanifold M of Euclidean space of codimension at least three. We start with some notation. Suppose $E_1, \ldots, E_g$ are the common eigendistributions of the shape operator and $n_1, \ldots, n_g$ the corresponding curvature normals. Let ξ_i be a parallel normal vector field that focalizes only the eigendistribution E_i, and L_{ij} be the span of n_i and n_j. Observe that L_{ij} is a parallel subbundle of the normal bundle νM.

If we choose a parallel normal vector field ξ_{ij} with the property that

$$\langle \xi_{ij}, n_k \rangle = 1 \qquad \text{if and only if } n_k \in L_{ij},$$

then ξ_{ij} focalizes both E_i and E_j. In other words,

$$\mathcal{S}_{ij} = \ker(A_{\xi_{ij}} - \mathrm{id})$$

is an autoparallel distribution that contains $E_i \oplus E_j$, if $i \neq j$. Let $S_{ij}(p)$ be the leaf through p of $\mathcal{S}_{ij}$, which, we know, can be regarded both as a totally geodesic submanifold of M and as a compact full isoparametric submanifold of affine space

$$V_{ij}(p) = p + \mathcal{S}_{ij} \oplus L_{ij}(p),$$

with curvature normals $\{n_{k|S_{ij}(p)} \mid n_k \in L_{ij}\}$. The rank of $S_{ij}(p)$ is one, if $i = j$, and two, if $i \neq j$. By the Homogeneous Slice Theorem, $S_{ij}(p)$ is homogeneous under the normal holonomy group of the focal manifold. So, $S_{ij}(p)$ is an orbit of an s-representation.

Consider now the submersions $\pi_{ij} : M \to M_{\xi_{ij}}$ and $\pi_i : M \to M_{\xi_i}$. Observe that $\xi_{ij} - \xi_i$ is constant on $S_i(q)$, since

$$A^i_{\xi_{ij}-\xi_i} = A_{\xi_{ij}-\xi_i \,|TS_i(q)} = 0 \, ,$$

where A^i is the shape operator of $S_i(q)$. Thus $\xi_{ij} - \xi_i$ defines locally (on some neighbourhood $U \subseteq M_{\xi_i}$) a parallel normal vector field on M_{ξ_i}, which we will denote by η_{ij}.

Let us fix $\eta_{ij}(\pi_i(p))$ and consider the holonomy tube $(M_{\xi_i})_{\eta_{ij}(\pi_i(p))}$ in M_{ξ_i} relative to the singular normal vector $\eta_{ij}(\pi_i(p))$. Because a full focal manifold of an irreducible isoparametric submanifold determines the foliation (cf. Corollary 5.3.7 of Homogeneous Slice Theorem 5.3.6 and Exercise 7.4.1) we have that $M_{\xi_{ij}} = (M_{\xi_i})_{\eta_{ij}(\pi_i(p))}$ (and also $M_{\xi_i} = (M_{\xi_{ij}})_{-\eta_{ij}(\pi_i(p))}$).

Locally (on U) we also have a submersion $p_{ij} : U \subseteq M_{\xi_i} \to p_{ij}(U) \subseteq M_{\xi_{ij}}$ given by $s \mapsto s + \eta_{ij}(s)$. So we have

$$p_{ij} \circ \pi_i = \pi_{ij} \, .$$

REMARK 7.3.2 If γ is a horizontal curve in M with respect to π_{ij} then γ is also horizontal with respect to π_i and $\pi_i \circ \gamma$ is horizontal with respect to p_{ij}. □

We can now give the definition of connection on M, which will turn out to be canonical.

Let ∇^{ij}_q be the canonical connection on $S_{ij}(q)$ naturally induced by the restricted normal holonomy group of $M_{\xi_{ij}}$ (which acts as an s-representation). Denote by $D^{ij} := \nabla - \nabla^{ij}_q$ the corresponding homogeneous structure of type $\mathcal{T}$. For the sake of simplicity, we will still denote by D^{ij} the value of the tensor field D^{ij} at q. Recall that D^{ij}_X, $(X \in T_qS_{ij}(q))$ is a skew-symmetric endomorphism.

Decompose $X, Y \in T_qM$ as

$$X = \sum_{i=1}^{g} X_i \, , \qquad Y = \sum_{j=1}^{g} Y_j \, , \qquad X_h, Y_h \in E_h(q) \, .$$

Define the tensor field D on M by

$$D_X Y = \sum_{ij} D^{ij}_{X_i} Y_j \, ,$$

(again, we write D for the value of the tensor field D at q). Observe that possibly $i = j$ in the above sum.

If $i = j$ (for simplicity $i = j = 1$) then one operates on a curvature sphere $S_1(q)$ that is a totally geodesic submanifold of the s-representation orbit $S_{1k}(q)$ for any k. Take, for instance, $k = 2$. By what we remarked in part (a) of this section (where we discussed the canonical connection on orbits of s-representations), the connection on $S_1(q)$ induced by the canonical connection ∇^{12} on $S_{12}(q)$ coincides with the connection relative to the isotropy representation of the isotropy group Φ^{12}_z, $z = \xi_1(q) - \xi_{12}(q)$ on $\nu_z(\Phi^{12} \cdot z)$, where Φ^{12} is the restricted normal holonomy group of $M_{\xi_{12}}$.

On the other hand, M_{ξ_1} can be regarded as holonomy tube of $M_{\xi_{12}}$ with respect to $z = \xi_1(q) - \xi_{12}(q)$ and by Theorem 5.4.12, the normal holonomy of M_{ξ_1} at z is the image in the slice representation of Φ^{12}_z in $\nu_z(\Phi^{12} \cdot z)$. Thus, the canonical connection $\nabla^{11} = \nabla^1$ on $S_1(q)$, regarded as orbit of the normal holonomy of M_{ξ_1}, coincides with ∇^{12}. This is the key point for the proof of the following

LEMMA 7.3.3
If $X, Y \in T_q S_{ij}(q)$, then $D_X Y = D^{ij}_X Y$.

PROOF If one decomposes X, Y as sums of vectors in eigendistributions, then $D_X Y$ splits as a sum of terms of type $D^{ij}_{X_i} Y_j$ $(i \neq j)$ and of type $D^{ii}_{X_i} Y_i$. For the first kind of term, there is clearly no problem. For the second, by what we remarked above, $D^{ii}_{X_i} Y_i = D^{ij}_{X_i} Y_i$. So we are finished. ∎

As a consequence, we have that D_X is a skew-symmetric endomorphism of $T_p M$. So, it determines a metric connection on M

$$\nabla^c := \nabla - D.$$

Consider the connection on $TM \oplus \nu M$ given by the sum ∇^c and $\nabla^\perp$. We still denote this connection with ∇^c.

REMARK 7.3.4 By the above Lemma, we actually have that $S_{ij}(q)$ is an autoparallel submanifold of M with respect to the connection ∇^c. ∎

Now, Thorbergsson's Theorem 5.4.5 follows from Theorem 7.2.5, the results on homogeneous isoparametric submanifolds of Section 5.4 and the following:

THEOREM 7.3.5
∇^c is a canonical connection of type $\mathcal{T}$ on M.

PROOF We must prove that (1) $\nabla^c \alpha = 0$ and (2) $\nabla^c D = 0$.
(1) $\nabla^c \alpha = 0$. By Exercise 7.4.2, it suffices to show that any eigendistribution of M is ∇^c-parallel. By the above Remark, $S_{ij}(q)$ is an autoparallel submanifold of M with respect to ∇^c. Moreover, E_i is ∇^c-parallel in $S_{ij}(q)$ because the second fundamental form of $S_{ij}(q)$ is ∇^c-parallel. Thus $\nabla^c_{E_j} E_i \subseteq E_i$ and, since j is arbitrary, $\nabla^c E_i \subseteq E_i$, i.e., E_i is ∇^c-parallel.
(2) $\nabla^c D = 0$. If $X, Y, Z \in T_q S_{ij}(q)$ then $\nabla^c_X D_Y Z = \nabla^{ij}_X D^{ij}_Y Z = 0$, because $S_{ij}(q)$ is an orbit of an s-representation and ∇^c is its canonical connection.

Let us assume that $i \neq j$ and k be such that $E_k(q)$ is not contained in $T_qS_{ij}(q)$.

We first remark that $(\nabla^c_X D)_Y Z = 0$ (with $X, Y \in T_qS_{ij}(q)$, $Z \in E_k(q)$) is equivalent to the property that for any ∇^c-geodesic γ_k in $S_k(q)$, with $\gamma_k(0) = q$, $\dot{\gamma}_k(0) = Z$, if τ^c denotes the ∇^c-parallel transport along γ_k from q to $p = \gamma_k(1)$,

$$D_{\tau^c(X)}\tau^c(Y) = \tau^c D_X Y.$$

Now, to prove $\nabla^c D = 0$ using the above reformulation, it is crucial to have relation between parallel transport in a normal direction on a focal manifold and parallel transport (with respect to the canonical connection) on an orbit of an s-representation (cf. Proposition 7.3.1) and also comparing parallel transport in normal directions (Lemma 4.4.6).

Let us consider the focalizations $\pi_i : M \to M_{\xi_i}$, $\pi_{ij} : M \to M_{\xi_{ij}}$ and the local submersion p_{ij} of M_{ξ_i} on $M_{\xi_{ij}}$ (regarded as holonomy tube of M_{ξ_i}).

Let $X_i \in E_i(q)$ and observe that its ∇^c-parallel transport $\widetilde{X}_i(t)$ along γ_k can be done in $S_{ki}(q)$ (which contains $S_k(q)$). $S_{ki}(q)$ focalizes in M_{ξ_i} into $(S_{ki}(q))_{\xi_i}$ and, by Proposition 7.3.1, $\widetilde{X}_i(t)$ is $\nabla^\perp$-parallel in $(S_{ki}(q))_{\xi_i}$ along $\pi_i\gamma_k(t)$. Moreover, by Lemma 4.4.6, $\widetilde{X}_i(t)$ is still $\nabla^\perp$-parallel along $p_{ij} \circ (\pi_i \circ \gamma_k)(t) = \pi_{ij} \circ \gamma_k(t)$ in $M_{\xi_{ij}}$. We can repeat this argument for any $E_h(q) \subseteq T_qS_{ij}(q)$, getting that for any $X \in T_qS_{ij}(q)$, $\tilde{X}(t)$ is ∇^c-parallel along $\gamma_k(t)$ if and only if it is $\nabla^\perp$-parallel along $\pi_{ij} \circ \gamma_k(t)$. In other words, on $S_{ij}(q)$, the ∇^c-parallel transport τ^c along γ_k from q to $p = \gamma_k(1)$ agrees with the $\nabla^\perp$-parallel transport $\tau^\perp$ along $\pi_{ij} \circ \gamma_k(t)$. By the Homogeneous Slice Theorem and the fact that $\tau^\perp_*$ sends the normal holonomy of $M_{\xi_{ij}}$ at q to the normal holonomy of $M_{\xi_{ij}}$ at p, we have that $\tau^\perp_*(D_q) = D_p$.

Thus, if $X, Y \in T_qS_{ij}(q)$,

$$D_{\tau^c(X)}\tau^c(Y) = D_{\tau^\perp(X)}\tau^\perp(Y) = \tau^\perp D_X Y = \tau^c D_X Y.$$

Therefore $\nabla^c D = 0$. $\square$

REMARK 7.3.6 A proof of Thorbergsson's theorem that relies on normal holonomy but avoids the use of homogeneous structures, follows from the remarkable result of Ernst Heintze and Xiaobo Liu that yields homogeneity of infinite dimensional isoparametric submanifolds [94]. $\square$

7.4 Exercises

Exercise 7.4.1 Let $\mathcal{F}$ be the parallel singular foliation of $\mathbb{R}^n - \{0\}$ induced by a compact irreducible isoparametric submanifold. Let M and N be two submanifolds of the family $\mathcal{F}$, $p \in M$ and choose $q \in N$ such that $\xi := q - p$ is normal to M at p. Prove that N coincides with the holonomy tube $(M)_\xi$. Deduce that $-\xi = q - p$ is normal to N at q and M is the holonomy tube $(N)_{-\xi}$.

Exercise 7.4.2 Let M be an isoparametric submanifold. Prove that a connection ∇^c on M with $(\nabla^c)^\perp = \nabla^\perp$ is canonical if and only if any eigendistribution of the operator of M is parallel with respect to ∇^c.

Exercise 7.4.3 (cf. [58,59]) Let S be a homogeneous structure of a submanifold of $\mathbb{R}^n$. Fix $p \in M$ and consider the pair (S_p, α_p). The purpose of this exercise is to describe how one can associate with (S_p, α_p) (for any $p \in M$) a Lie subalgebra $\mathfrak{g}$ of the Lie algebra of $I(\mathbb{R}^n)$ and further a homogeneous submanifold that contains M as an open subset.

To simplify the notation, we set $S := S_p$, $\alpha := \alpha_p$. Let $\gamma(t)$ be a curve in M, with $\gamma(0) = p$, $\dot\gamma(0) = x$. Denote by $\tau_{\gamma(t)}$ the isometry of $T_p\bar{M} = T_pM \oplus \nu_pM$ into $T_{\gamma(t)}\bar{M} = T_{\gamma(t)}M \oplus \nu_{\gamma(t)}M$ determined by the parallel transport with respect to $\widetilde{\nabla}$ along γ. For any t, there exists a unique isometry $F_t \in I(\bar{M})$ such that

$$F_t(p) = \gamma(t), \qquad F_{t*p} = \tau_{\gamma(t)}.$$

(a) Prove that the tangent vector at id to the curve F_t gives an element Ψ_x of the Lie algebra of $I(\bar{M})$ that has the following expression in terms of S and α: Let $Y \in \mathbb{R}^n \simeq T_pM \oplus \nu_pM$ and set $Y(t) := \tau_{\gamma(t)}Y$. The transformation Ψ_x has linear part $L\Psi_x$ given by

$$L\Psi_x Y = S_x Y + \alpha(x, Y^\top) - A_{Y^\perp}x = \Gamma_x Y$$

where $Y^\top$ ($Y^\perp$) is the tangential (normal) component of Y. The translational part of Ψ_x is $p \mapsto x$.

(b) Let $\mathfrak{k}$ be the Lie algebra spanned by the $\widetilde{R}_{xy}$ and $\mathfrak{m}$ the span of the Ψ_x. Set

$$\mathfrak{g} = \mathfrak{k} \oplus \mathfrak{m}.$$

Prove that $\mathfrak{g}$ is a Lie subalgebra of the Lie algebra of $I(\mathbb{R}^n)$ with Lie brackets

$$\begin{aligned}
[\Psi_x, \Psi_y] &= \Psi_{S_xy - S_yx} + \widetilde{R}_{xy}, \\
[\widetilde{R}_{xy}, \Psi_z] &= \Psi_{\widetilde{R}_{xy}z}, \\
[\widetilde{R}_{xy}, \widetilde{R}_{zw}] &= \widetilde{R}_{\widetilde{R}_{xy}zw} + \widetilde{R}_{z\widetilde{R}_{xy}w}.
\end{aligned}$$

(Remark that $[\widetilde{R}_{xy}, \widetilde{R}_{zw}] = \widetilde{R}_{\widetilde{R}_{xy}zw} + \widetilde{R}_{z\widetilde{R}_{xy}w}$ follows from $\widetilde{R}_{xy} \cdot \widetilde{R} = 0$, which is a consequence – or better an integrability condition – of the equations $\widetilde{\nabla}\alpha = 0$, $\widetilde{\nabla}S = 0$.)

(c) Let G be the connected Lie subgroup of $I(\bar{M})$ having $\mathfrak{g}$ as Lie algebra. Prove that $\widetilde{M} := G \cdot p$ is a homogeneous submanifold that contains M as an open subset. *Hint:* modify the proof of Lemma 7.1.10.

(d) Generalize the above to a submanifold of a space form.

Exercise 7.4.4 (cf. [125]) Let $M = K \cdot v$ be a principal orbit of an s-representation (i.e., a homogeneous isoparametric submanifold) corresponding to a symmetric
space $X = G/K$ with a reduced root system (i.e., if λ is a positive root, 2λ is not).

Prove that the canonical connection ∇^c associated with the reductive decomposition $\mathfrak{k} = \mathfrak{k}_v \oplus \mathfrak{k}_+$ (see Section 3.2 and part (a) of Section 7.3) agrees with the *projection connection* $\hat{\nabla}$ defined by projecting the Levi-Civita connection on the various curvature distributions, i.e.,

$$\hat{\nabla}_X Y := \sum_{i=1}^{g} (\nabla_X Y_i)_i,$$

where $(..)_i$ denotes the projection on E_i.

Exercise 7.4.5 Let $M = K \cdot v$ be a principal orbit of an s-representation (i.e., a homogeneous isoparametric submanifold) corresponding to a symmetric space $X = G/K$ with a reduced root system. Let $\hat{S}$ be the homogeneous structure relative to the projection connection $\hat{\nabla}$ of Exercise 7.4.4.

Show that

$$\hat{S}_x v = -(id - A_\xi)^{-1} \left((\nabla_x^\perp A)_\xi v \right), \qquad x \in (E_i)_p, v \in (E_j)_p, i \neq j$$

and that

$$\hat{S}_x v = 0, \qquad x, v \in (E_i)_p.$$

Using Exercise 7.4.3, deduce that M is uniquely determined by the values at p of the second fundamental form and its covariant derivative.

Exercise 7.4.6 Let M be an embedded connected submanifold of $\mathbb{R}^n$ and let $G = \{g \in I(\mathbb{R}^n) : g(M) = M\}$ be its family of extrinsic isometries. Assume that G acts transitively on M. Let $p \in M$ and assume that the subgroup H of the isotropy group given by

$$H = \{g \in G_p : g_{*p\,|\nu_p M} = \mathrm{id}_{\nu_p M}\}$$

has no fixed points in T_pM. Let

$$\mathfrak{g} = \mathfrak{k} \oplus \mathfrak{m}$$

be a reductive decomposition of the Lie algebra $\mathfrak{g}$ of G, where $\mathfrak{k}$ is the Lie algebra of G_p and $\mathfrak{m}$ an $\mathrm{Ad}(G_p)$-invariant subspace (G is not assumed to be connected). Let S be the homogeneous structure on M associated to this decomposition (observe that $\mathfrak{g}$ is also the Lie algebra of the connected component of the identity G^o of G, but it is important for the applications to extrinsic k-symmetric submanifolds in the sense of Sánchez not to assume that G is connected). Prove that S is of type $\mathcal{T}$ and so M is an orbit of an s-representation. *Hint:* if $h \in H$ then $\langle S_X\xi, \eta\rangle = \langle S_{hX} h\xi, h\eta\rangle = \langle S_{hX}\xi, \eta\rangle$ for all $X \in T_pM, \xi, \eta \in \nu_pM$. Hence, $\langle S_{(h-\mathrm{id})X}\xi, \eta\rangle = 0$. So, $\langle S_X\xi, \eta\rangle$ is zero if X is perpendicular to the fixed vectors of h in T_pM.

Chapter 8

Submanifolds of Riemannian manifolds

In this chapter, we present basic material about submanifolds of Riemannian manifolds. The core of submanifold geometry is formed by the first order equations of Gauss and Weingarten and the second order equations of Gauss, Codazzi and Ricci. Since we already discussed these equations in detail for space forms, we present them in Section 8.1 without proof.

In Section 8.2, we discuss an important method for the study of submanifolds based on Jacobi vector fields. The basic idea is to study the geometric behaviour of a submanifold by "pushing it" in certain normal directions.

One of the fundamental objects on a Riemannian manifold is geodesics. If all geodesics in a submanifold are geodesics in the ambient manifold as well, then the submanifold is called totally geodesic. In a sense, totally geodesic submanifolds are the simplest examples of submanifolds. For space forms, we discussed this topic in 2.4. The fundamental problem in a general Riemannian manifold is the existence problem. It has been solved by E. Cartan and, in Section 8.3, we present a proof of his result. We also show that the fixed points of isometries form totally geodesic submanifolds. Finally, we discuss the issue of rigidity for totally geodesic submanifolds.

In Section 8.4, we discuss the relation between totally umbilical submanifolds and extrinsic spheres. A submanifold is totally umbilical if the second fundamental form is proportional to the mean curvature vector field and it is an extrinsic sphere if, in addition, the mean curvature vector field is nonzero and parallel in the normal bundle. Both concepts generalize the idea of Euclidean spheres. As geodesics can be used to characterize totally geodesic submanifolds, we will use circles to characterize extrinsic spheres.

Symmetric submanifolds in submanifold geometry are the analogues of symmetric manifolds in Riemannian geometry. Roughly speaking, a submanifold is symmetric if the reflections in the normal spaces leave the submanifold invariant. In 8.5, we show that the second fundamental form of a symmetric submanifold is parallel and discuss briefly more geometric facts about such submanifolds.

8.1 Submanifolds and the fundamental equations

In this section, we present the basic equations from submanifold theory in Riemannian geometry. Let M be a submanifold of a Riemannian manifold $\bar{M}$. We will use the following notations: The dimensions of M and $\bar{M}$ will be denoted by m and n respectively. The Riemannian metric on $\bar{M}$ and the induced Riemannian metric on M will be denoted by $\langle \cdot, \cdot \rangle$. The Levi Civita covariant derivatives on $\bar{M}$ and M are $\bar{\nabla}$ and ∇, respectively. By $\bar{R}$ and R we denote the Riemannian curvature tensors of $\bar{M}$ and M. We recall that we are using the sign convention $R(X,Y)Z = \nabla_X \nabla_Y Z - \nabla_Y \nabla_X Z - \nabla_{[X,Y]} Z$. The tangent bundle and the normal bundle of M will be denoted by TM and νM. An analogous argument, as in the case of space forms, leads to the following fundamental equations for submanifolds in Riemannian geometry:

THEOREM 8.1.1
Let M be a submanifold of a Riemannian manifold $\bar{M}$. Then the following equations hold for all vector fields X, Y, Z, W tangent to M and all vector fields ξ, η normal to M.
Gauss formula:

$$\bar{\nabla}_X Y = \nabla_X Y + \alpha(X,Y) \; ,$$

Weingarten formula:

$$\bar{\nabla}_X \xi = -A_\xi X + \nabla^\perp_X \xi \; ,$$

Gauss equation:

$$\langle \bar{R}(X,Y)Z, W \rangle = \langle R(X,Y)Z, W \rangle - \langle \alpha(Y,Z), \alpha(X,W) \rangle + \langle \alpha(X,Z), \alpha(Y,W) \rangle \; ,$$

Codazzi equation:

$$(\bar{R}(X,Y)Z)^\perp = (\nabla^\perp_X \alpha)(Y,Z) - (\nabla^\perp_Y \alpha)(X,Z) \; ,$$

Ricci equation:

$$\langle \bar{R}(X,Y)\xi, \eta \rangle = \langle R^\perp(X,Y)\xi, \eta \rangle - \langle [A_\xi, A_\eta] X, Y \rangle \; .$$

Here α denotes the second fundamental form of M and A is the shape operator of M. Since ∇ and $\bar{\nabla}$ are torsion-free connections, α is symmetric, i.e., $\alpha(X,Y) = \alpha(Y,X)$. The second fundamental form and the shape operator are related by

$$\langle \alpha(X,Y), \xi \rangle = \langle A_\xi X, Y \rangle \; .$$

The vector field $\nabla^\perp_X \xi$ is the normal component of $\bar\nabla_X \xi$, and $\nabla^\perp$ is the normal covariant derivative of M. The νM-valued tensor field $\nabla^\perp \alpha$ on M is the covariant derivative of α in the normal bundle of M defined by

$$(\nabla^\perp_X \alpha)(Y, Z) = \nabla^\perp_X \alpha(Y, Z) - \alpha(\nabla_X Y, Z) - \alpha(Y, \nabla_X Z) \ .$$

Finally, $R^\perp$ is the curvature tensor of the normal bundle of M defined by

$$R^\perp(X, Y)\xi = \nabla^\perp_X \nabla^\perp_Y \xi - \nabla^\perp_Y \nabla^\perp_X \xi - \nabla^\perp_{[X,Y]} \xi \ .$$

8.2 Focal points and Jacobi vector fields

The eigenvalues of the shape operator are called the principal curvatures of the submanifold, and corresponding eigenvectors are called principal curvature vectors. A fundamental technique in submanifold geometry is to study the behaviour of submanifolds by displacing them in normal directions at various distances. The main tool for calculating the principal curvatures and principal curvature spaces of the displaced submanifolds is the theory of Jacobi vector fields.

a) M-Jacobi vector fields

Let $\gamma : I \to \bar M$ be a geodesic in $\bar M$ parametrized by arc length and with $0 \in I$, $p := \gamma(0) \in M$, and $\dot\gamma(0) \in \nu_p M$. Suppose $V(s, t) = \gamma_s(t)$ is a smooth geodesic variation of $\gamma = \gamma_0$ with $c(s) = \gamma_s(0) \in M$ and $\xi(s) = \dot\gamma_s(0) \in \nu_{c(s)} M$ for all s. The Jacobi vector field Y along γ induced by this geodesic variation is determined by the initial values

$$Y(0) = \left.\frac{d}{ds}\right|_{s=0} V(s, 0) = \gamma_s(0) = c(s) = \dot c(0) \in T_p M$$

and, using the Weingarten formula,

$$\begin{aligned} Y'(0) &= \left.\frac{\partial}{\partial s}\right|_{s=0} \left.\frac{\partial}{\partial t}\right|_{t=0} V(s, t) \\ &= \left.\frac{d}{ds}\right|_{s=0} \dot\gamma_s(0) = \left.\frac{d}{ds}\right|_{s=0} \xi(s) = \bar\nabla_{Y(0)} \xi = -A_{\xi(0)} Y(0) + \nabla^\perp_{Y(0)} \xi \ . \end{aligned}$$

Thus, the initial values of Y satisfy

$$Y(0) \in T_{\gamma(0)} M \quad \text{and} \quad Y'(0) + A_{\dot\gamma(0)} Y(0) \in \nu_{\gamma(0)} M \ .$$

A Jacobi vector field Y along γ whose initial values satisfy these two conditions is called an *M-Jacobi vector field.* Thus M-Jacobi vector fields correspond to geodesic variations of geodesics intersecting M perpendicularly, roughly speaking. As a Jacobi vector field Y along γ is uniquely determined by its values $Y(0)$ and $Y'(0)$ we easily see that M-Jacobi vector fields along γ form an n-dimensional linear subspace of the $2n$-dimensional vector space of all Jacobi vector fields along γ. Obviously, $Y(t) = t\dot{\gamma}(t)$ is an M-Jacobi vector field along γ. Since this particular Jacobi vector field has no special relevance, we define $\mathfrak{J}(M,\gamma)$ as the $(n-1)$-dimensional vector space consisting of all M-Jacobi vector fields along γ which are perpendicular to the M-Jacobi vector field $t \mapsto t\dot{\gamma}(t)$.

b) Parallel displacement and focal points of hypersurfaces

Let M be a hypersurface of a Riemannian manifold $\bar{M}$ and ξ a unit normal vector field on M. Our aim is to study the displacement M_r of M in direction ξ at distance r. In general, M_r is not a submanifold of M. We will see how one can determine, by means of M-Jacobi vector fields, whether M_r is a submanifold. If M_r is a submanifold, we want to calculate the principal curvatures and the corresponding principal curvature vectors of M_r. Since these are all local objects, there is no loss of generality when we assume that ξ is, in fact, globally defined on M.

Let r be a positive real number and

$$\Phi_r : M \to \bar{M} \ , \ p \mapsto \exp(r\xi_p) \ ,$$

where exp denotes the exponential map of the ambient space $\bar{M}$. The smooth map Φ_r parametrizes the parallel displacement M_r of M in direction ξ at distance r. If $\bar{M}$ is not complete we might have to restrict Φ_r to the subset of M on which Φ_r is defined. Obviously, M_r is an immersed submanifold of $\bar{M}$ if and only if Φ_r is an immersion. But it might happen that M_r is a submanifold of $\bar{M}$ with higher codimension. Just imagine a cylinder with radius r and one-dimensional axis A in $\mathbb{R}^3$. If ξ is the inward unit normal vector field on the cylinder, then parallel displacement of the cylinder in direction ξ at distance r is the axis A, a smooth embedded submanifold of $\mathbb{R}^3$. In general, such submanifolds are called focal manifolds of M and arise when Φ_r has constant rank less than $n-1$, because then, locally, Φ_r is a submersion onto a smooth submanifold of $\bar{M}$ whose dimension is equal to the rank of Φ_r. So the first task is to calculate the differential of Φ_r.

Let $p \in M$ and $\gamma = \gamma_{\xi_p} : t \mapsto \exp(t\xi_p)$. Since ξ has constant length and the normal bundle has rank one, $\mathfrak{J}(M,\gamma)$ consists of Jacobi vector fields Y along γ with initial values $Y(0) \in T_pM$ and $Y'(0) = -A_{\xi_p}Y(0)$. Let $Y \in \mathfrak{J}(M,\gamma)$ and c be a smooth curve in M with $c(0) = p$ and $\dot{c}(0) = Y(0)$. Then

$$V(s,t) = \exp(t\xi_{c(s)})$$

is a smooth geodesic variation of γ consisting of geodesics of $\bar{M}$ intersecting M perpendicularly, and the corresponding M-Jacobi vector field is Y. For the differential Φ_{r*p} of Φ_r at p we get

$$\Phi_{r*p}Y(0) = \Phi_{r*p}\dot{c}(0) = \left.\frac{d}{ds}\right|_{s=0} \exp(r\xi_{c(s)}) = Y(r) \ .$$

Thus, Φ_r is not immersive at p if and only if there exists some nonzero M-Jacobi vector field $Y \in \mathfrak{J}(M, \gamma)$ with $Y(r) = 0$. In such a case, $\Phi_r(p)$ is called a *focal point* of M along γ, and the dimension of the kernel of Φ_{r*p} is called *multiplicity of the focal point*. If $\Phi_r(p)$ is a focal point of M along γ, the multiplicity of it is the dimension of the linear subspace of $\mathfrak{J}(M, \gamma)$ consisting of all $Y \in \mathfrak{J}(M, \gamma)$ with $Y(r) = 0$. The geometric interpretation of Jacobi vector fields in terms of geodesic variations implies that $\Phi_r(p)$ is a focal point of M along γ if and only if there exists a nontrivial geodesic variation of γ, all of whose geodesics intersect M orthogonally and that meet infinitesimally at $\Phi_r(p)$. The most obvious picture of a focal point is that of the center of a sphere, where the geodesics intersecting the sphere orthogonally meet. If there exists a positive integer k such that $\Phi_r(q)$ is a focal point of M along γ_{ξ_q} with multiplicity k for all q in some open neighbourhood U of p, then, if U is sufficiently small, $\Phi_r|U$ parametrizes an embedded $(n-1-k)$-dimensional submanifold F of $\bar{M}$ that is called *focal manifold* of M in $\bar{M}$. If $\Phi_r(p)$ is not a focal point of M along γ, then Φ_{r*} is immersive in some open neighbourhood U of p and we conclude that, if U is sufficiently small, $\Phi_r|U$ parametrizes an embedded hypersurface of $\bar{M}$, which is called *equidistant hypersurface* to M in $\bar{M}$. In both cases, the vector $\dot{\gamma}(r)$ is a unit normal vector of the focal manifold resp. equidistant hypersurface at $\Phi_r(p)$.

Our next aim is to calculate the shape operator of the focal manifold resp. equidistant hypersurface with respect to $\dot{\gamma}(r)$. We denote the focal manifold resp. equidistant hypersurface by M_r (in general only a part of the original M_r).

Let $p \in M$ and Y, c, V be as above. Then $c_r = \Phi_r \circ c$ is a curve in M_r with $\dot{c}_r(0) = \Phi_{r*p}\dot{c}(0) = Y(r)$, and we define a unit normal vector field η_r of M_r along c_r by $\eta_r(s) = \dot{\gamma}_{\xi_{c(s)}}(r)$. We denote by A^r the shape operator of M_r. Then, using the Weingarten formula, we obtain

$$
\begin{aligned}
Y'(r) &= \left.\frac{d}{ds}\right|_{s=0} \left.\frac{d}{dt}\right|_{t=r} V(s,t) = \left.\frac{d}{ds}\right|_{s=0} \eta_r(s) = \eta_r'(0) \\
&= -A^r_{\eta_r(0)} Y(r) + (\eta_r'(0))^\perp ,
\end{aligned}
$$

where $(\cdot)^\perp$ denotes the orthogonal projection onto $\nu_{\gamma(r)} M_r$. Thus, if Y is an M-Jacobi vector field along γ in $\mathfrak{J}(M, \gamma)$, the shape operator A^r of M_r satisfies

$$
A^r_{\dot{\gamma}(r)} Y(r) = -(Y'(r))^T ,
$$

where $(\cdot)^T$ denotes the orthogonal projection onto $T_{\gamma(r)} M_r$. If, in particular, M_r is a hypersurface of $\bar{M}$, then $Y'(r)$ is tangent to M_r because Y is perpendicular to $\dot{\gamma}$ and the normal space of M_r at $\gamma(r)$ is spanned by $\dot{\gamma}(r)$. In summary,

THEOREM 8.2.1

Let M be a hypersurface of a Riemannian manifold $\bar{M}$, $r \in \mathbb{R}_+$, ξ a unit normal vector field on M, and M_r the displacement of M in direction ξ at distance r. Suppose that M_r is a submanifold of $\bar{M}$. Let $p \in M$ and γ the

geodesic in $\bar{M}$ *with* $\gamma(0) = p$ *and* $\dot{\gamma}(0) = \xi_p$. *Then*

$$T_{\gamma(r)} M_r = \{Y(r) \mid Y \in \mathfrak{J}(M, \gamma)\} .$$

The shape operator A^r *of* M_r *with respect to* $\dot{\gamma}(r) \in \nu_{\gamma(r)} M_r$ *is given by*

$$A^r_{\dot{\gamma}(r)} Y(r) = -(Y'(r))^T ,$$

where $(Y'(r))^T$ *is the component of* $Y'(r)$ *tangent to* M_r. *If, in particular,* M_r *has codimension one, then* $Y'(r)$ *is tangent to* M_r.

In case M_r has codimension one, there is another efficient way to describe the shape operator A^r. In the previous situation, we denote by $\gamma^\perp$ the parallel subbundle of the tangent bundle of $\bar{M}$ along γ that is defined by the orthogonal complements of $\mathbb{R}\dot{\gamma}(t)$ in $T_{\gamma(t)}\bar{M}$, and put $\bar{R}^\perp_\gamma := \bar{R}_\gamma|\gamma^\perp = \bar{R}(\cdot, \dot{\gamma})\dot{\gamma}|\gamma^\perp$. Let D be the $\mathrm{End}(\gamma^\perp)$-valued tensor field along γ given by

$$D'' + \bar{R}^\perp_\gamma \circ D = 0 \, , \; D(0) = \mathrm{id}_{T_pM} \, , \; D'(0) = -A_{\xi_p} \, .$$

If $X \in T_pM$ and E_X is the parallel vector field along γ with $E_X(0) = X$, then $Y = DE_X$ is the Jacobi vector field along γ with initial values $Y(0) = X$ and $Y'(0) = -A_{\xi_p}X$. Thus $\gamma(r)$ is a focal point of M along γ if and only if $D(r)$ is singular. If M_r has codimension one, then $D(r)$ is regular, and we obtain

$$A^r_{\dot{\gamma}(r)} D(r) E_X(r) = A^r_{\dot{\gamma}(r)} Y(r) = -Y'(r) = -(DE_X)'(r) = -D'(r) E_X(r) .$$

Therefore

$$A^r_{\dot{\gamma}(r)} = -D'(r) \circ D^{-1}(r) .$$

c) Tubes and focal manifolds of submanifolds with codimension greater than one

We now turn to the case where M is a submanifold of $\bar{M}$ with codimension greater than one. Let $r \in \mathbb{R}_+$ and $\nu^1 M$ the unit normal sphere bundle over M, that is, the sphere bundle over M consisting of unit normal vectors of M. We put

$$M_r := \{\exp(r\xi) \mid \xi \in \nu^1 M\} .$$

In general, M_r is not a submanifold of $\bar{M}$. But if, for instance, M is compact and embedded, then M_r is a compact embedded hypersurface of $\bar{M}$ for sufficiently small r. If M_r is a hypersurface of $\bar{M}$ we call M_r *tube* with radius r around M. And if M_r is a submanifold of $\bar{M}$ with codimension greater than one, we call it *focal manifold* of M.

Let $p \in M$ and $\gamma : I \to \bar{M}$ be a geodesic parametrized by arc length and with $\gamma(0) = p$ and $\dot{\gamma}(0) \in \nu^1 M$. As M has codimension greater than one the vector space $\mathfrak{J}(M, \gamma)$ naturally splits into the direct sum

$$\mathfrak{J}(M, \gamma) = \mathfrak{J}(M, \gamma)^T \oplus \mathfrak{J}(M, \gamma)^\perp$$

of linear subspaces

$$\mathfrak{J}(M,\gamma)^T := \{Y \in \mathfrak{J}(M,\gamma) \mid Y(0) \in T_pM \ , \ Y'(0) = -A_{\dot\gamma(0)}Y(0)\}$$

and

$$\mathfrak{J}(M,\gamma)^\perp := \{Y \in \mathfrak{J}(M,\gamma) \mid Y(0) = 0 \ , \ Y'(0) \in \nu_pM\} \ .$$

M-Jacobi vector fields in $\mathfrak{J}(M,\gamma)^\perp$ arise from geodesic variations all of whose geodesics intersect M perpendicularly at the single point p. We say that $\gamma(r)$ is a *focal point of M along γ* if there exists a nonzero M-Jacobi vector field $Y \in \mathfrak{J}(M,\gamma)$ with $Y(r) = 0$. If $\gamma(r)$ is a focal point of M along γ then the dimension of the linear subspace $\{Y \in \mathfrak{J}(M,\gamma) \mid Y(r) = 0\}$ of $\mathfrak{J}(M,\gamma)$ is called *multiplicity of the focal point.* A focal point arising from $Y \in \mathfrak{J}(M,\gamma)^\perp$ is, in fact, a conjugate point of p in $\bar M$ along γ.

We now assume that M_r is a submanifold of $\bar M$. Let ξ be a smooth curve in ν^1M with $\xi(0) = \dot\gamma(0)$. Then $V(s,t) = \exp(t\xi(s))$ is a smooth geodesic variation of γ consisting of geodesics intersecting M perpendicularly. Let Y be the corresponding M-Jacobi vector field. Y is determined by the initial values $Y(0) = \dot c(0)$ and $Y'(0) = \xi'(0)$, where $c : s \mapsto V(s,0) \in M$ and ξ is considered as a vector field along c. Since ξ is of unit length Y belongs to $\mathfrak{J}(M,\gamma)$. The curve $c_r : s \mapsto \exp(r\xi(s))$ is smooth in M_r and hence $Y(r) = \dot c_r(0) \in T_{\gamma(r)}M_r$. As any tangent vector of M_r at $\gamma(r)$ arises in this manner we have

$$T_{\gamma(r)}M_r = \{Y(r) \mid Y \in \mathfrak{J}(M,\gamma)\} \ .$$

Denote by A^r the shape operator of M_r. A calculation shows that

$$A^r_{\dot\gamma(r)}Y(r) = -(Y'(r))^T$$

for all $Y \in \mathfrak{J}(M,\gamma)$, where $(\cdot)^T$ denotes the component tangent to M_r. In case M_r is a tube, $Y'(r)$ is always tangent to M_r. We summarize this in

THEOREM 8.2.2

Let M be a submanifold of a Riemannian manifold $\bar M$ with codimension greater than one, $r \in \mathbb{R}_+$, and $M_r = \{\exp(r\xi) \mid \xi \in \nu^1M\}$. Suppose that M_r is a submanifold of $\bar M$. Let $p \in M$ and γ a geodesic in $\bar M$ with $\gamma(0) = p$ and $\dot\gamma(0) \in \nu^1_pM$. Then

$$T_{\gamma(r)}M_r = \{Y(r) \mid Y \in \mathfrak{J}(M,\gamma)\} \ ,$$

and the shape operator A^r of M_r with respect to $\dot\gamma(r) \in \nu_{\gamma(r)}M_r$ is given by

$$A^r_{\dot\gamma(r)}Y(r) = -(Y'(r))^T \ ,$$

(where $(Y'(r))^T$ is the component of $Y'(r)$ tangent to M_r). If, in particular, M_r is a tube around M, then $Y'(r)$ is tangent to M_r.

Of particular interest is the case of M consisting of a single point $\{p\}, p \in \bar{M}$. For sufficiently small $r \in \mathbb{R}_+$, the set M_r is a compact embedded hypersurface of $\bar{M}$, a so-called *geodesic hypersphere* of $\bar{M}$. M_r is sometimes called a *distance sphere* in $\bar{M}$ because, at least for r small enough, it consists of points in $\bar{M}$ with distance r to p. When M is a point, $\mathfrak{J}(M,\gamma)$ consists of Jacobi vector fields Y along γ with $Y(0) = 0$ and $Y'(0)$ orthogonal to $\dot{\gamma}(0)$, and the shape operator A^r of M_r is given by $A^r_{\dot{\gamma}(r)}Y(r) = -Y'(r)$.

As before, we present another efficient way of describing the shape operator of a tube. In the set-up of Theorem 8.2.2, suppose M_r is a tube. We decompose $\gamma_p^\perp$ orthogonally $\gamma_p^\perp = T_pM \oplus (\nu_pM \cap \gamma_p^\perp)$.

Let D be the $\mathrm{End}(\gamma^\perp)$-valued tensor field along γ, which is the solution of the Jacobi equation

$$D'' + \bar{R}_\gamma^\perp \circ D = 0\ ,\ D(0) = \begin{pmatrix} \mathrm{id}_{T_pM} & 0 \\ 0 & 0 \end{pmatrix}\ ,\ D'(0) = \begin{pmatrix} -A_{\dot{\gamma}(0)} & 0 \\ 0 & \mathrm{id}_{\nu_pM\cap\gamma_p^\perp} \end{pmatrix}\ .$$

The matrix decomposition corresponds to the above decomposition of $\gamma_p^\perp$. A similar argument as in the previous case shows that the shape operator $A^r_{\dot{\gamma}(r)}$ of M_r with respect to $\dot{\gamma}(r)$ is given by

$$A^r_{\dot{\gamma}(r)} = -D'(r) \circ D^{-1}(r)\ .$$

Note that $D(r)$ is singular if and only if $\gamma(r)$ is a focal point of M along γ.

8.3 Totally geodesic submanifolds

Let M be a submanifold of a Riemannian manifold $\bar{M}$. Recall that M is called totally geodesic in $\bar{M}$ if every geodesic in M is also a geodesic in the ambient manifold $\bar{M}$. This is equivalent to the vanishing of the second fundamental form of M.

a) Existence of totally geodesic submanifolds

Let p be a point in $\bar{M}$ and V a linear subspace of $T_p\bar{M}$. Is there a totally geodesic submanifold M of $\bar{M}$ with $p \in M$ and $T_pM = V$? Suppose there is such a submanifold M. Since the exponential map $\exp_p : T_p\bar{M} \to \bar{M}$ maps straight lines through the origin $0 \in T_p\bar{M}$ to geodesics in $\bar{M}$ there is an open neighbourhood U of 0 in $T_p\bar{M}$ such that $\exp_p$ maps $U \cap V$ diffeomorphically onto some open neighbourhood of p in M. This implies that M is uniquely determined near p and that any totally geodesic submanifold of $\bar{M}$ containing p and tangent to V is contained as an open part in a maximal one with this property. This feature is known as rigidity of totally geodesic submanifolds.

We have seen that a totally geodesic submanifold arises necessarily as the image under the exponential map of some open part of a linear subspace of the tangent space. So, for the existence of a totally geodesic submanifold with given initial data (p, V), one has to investigate whether such an image is totally geodesic. The Gauss formula readily implies that the second fundamental form at p vanishes. When does it vanish at all other points? An answer was given by E. Cartan [36].

THEOREM 8.3.1 (E. Cartan)
Let $\bar{M}$ be a Riemannian manifold, $p \in \bar{M}$ and V a linear subspace of $T_p\bar{M}$. There exists a totally geodesic submanifold M of $\bar{M}$ with $p \in M$ and $T_pM = V$ if and only if there exists a real number $\epsilon \in \mathbb{R}_+$ such that for every geodesic γ in $\bar{M}$ with $\gamma(0) = p$ and $\dot{\gamma}(0) \in V \cap U_\epsilon(0)$ the Riemannian curvature tensor of $\bar{M}$ at $\gamma(1)$ preserves the parallel translate of V along γ from p to $\gamma(1)$.

PROOF If M is totally geodesic, the condition on the parallel translation is an immediate consequence of the Gauss formula and the Codazzi equation.

Conversely, let $p \in \bar{M}$, V a linear subspace of $T_p\bar{M}$, and $\epsilon \in \mathbb{R}_+$ such that for every geodesic γ in $\bar{M}$ with $\gamma(0) = p$ and $\dot{\gamma}(0) \in V \cap U_\epsilon(0)$ the Riemannian curvature tensor of $\bar{M}$ at $\gamma(1)$ preserves the parallel translate of V along γ from p to $\gamma(1)$. Without loss of generality we can assume that $\exp_p |U_\epsilon(0)$ is a diffeomorphism into $\bar{M}$. We set

$$M := \exp_p(V \cap U_\epsilon(0)) \ .$$

By construction $p \in M$, $T_pM = V$, and the second fundamental form of M at p vanishes. It remains to show that the second fundamental form of M vanishes everywhere.

Let $q \in M$ and $v \in V \cap U_\epsilon(0)$ so that $q = \exp_p(v)$. We show that T_qM is obtained by $\bar{\nabla}$-parallel translation of V along γ_v from p to q. By construction, the tangent space of M at q is given by

$$T_qM = \exp_{p*v}(V) \ ,$$

where we identify $T_p\bar{M}$ and $T_v(T_p\bar{M})$ in the canonical way. Let $\xi \in V$ and Y_ξ the Jacobi vector field along γ_v with initial values $Y_\xi(0) = 0$ and $Y_\xi'(0) = \xi$. Then $\exp_{p*v}(\xi) = Y_\xi(1)$.

Since, by assumption, $\bar{R}$ preserves $\bar{\nabla}$-parallel translation of V along γ_v, the equation for Jacobi vector fields implies that Y_ξ takes values in the vector bundle V_v along γ_v obtained by $\bar{\nabla}$-parallel translation of V along γ_v. Since $\exp_p |U_\epsilon(0)$ is a diffeomorphism into $\bar{M}$, a dimension argument shows that

$$T_qM = \exp_{p*v}(V) = \{Y_\xi(1) \mid \xi \in V\} = V_v(1) \ ,$$

which means that T_qM is obtained by $\bar{\nabla}$-parallel translation of V along γ_v from p to q.

It is now sufficient to prove that $\bar{\nabla}$-parallel translation along loops in M based at p leaves V invariant. For the moment, assume this is true. Let $q \in M$ and $v \in V \cap U_\epsilon(0)$ so that $q = \exp_p(v)$. Let c be any loop in M based at q. We get a loop $\hat{c}$ based at p by running along γ_v from p to q first, then along the loop c, and then back to p along γ_v. As already seen, T_qM is the $\bar{\nabla}$-parallel translate of V along γ_v from p to q. Since $\bar{\nabla}$-parallel translation along loops in M based at p preserves V, it follows that $\bar{\nabla}$-parallel translation along c preserves T_qM. Now let $c : [0,1] \to M$ be any curve in M. From each point $c(t)$ we construct a loop by running first from $c(0)$ to $c(t)$ along c, and then along the radial geodesic from $c(t)$ to p, and finally along the radial geodesic from p to $c(0)$. The invariance of the tangent spaces of M with respect to $\bar{\nabla}$-parallel transport along loops and along the radial geodesics implies that $T_{c(t)}M$ is obtained by $\bar{\nabla}$-parallel transport of $T_{c(0)}M$ along c. Hence, we have shown that $\bar{\nabla}$-parallel transport along curves in M leaves the tangent spaces of M invariant. This implies that the induced connection on M coincides with the restriction of $\bar{\nabla}$ to tangent vector fields of M. From the Gauss formula, we finally get that the second fundamental form of M vanishes, and hence M is totally geodesic in $\bar{M}$.

Thus, it remains to prove that $\bar{\nabla}$-parallel translation along loops in M based at p leaves V invariant. We first prove the following:

LEMMA 8.3.2

Let $\bar{M}$ be a Riemannian manifold and $p \in \bar{M}$. Let $f : [0,\delta] \times [0,1] \to \bar{M}$ be a smooth map with $f(s,0) = p$ for all $s \in [0,\delta]$. For each $s \in [0,\delta]$ we define $f_s : [0,1] \to \bar{M}$, $t \mapsto f(s,t)$. Similarily, for each $t \in [0,1]$ we define $f^t : [0,\delta] \to \bar{M}$, $s \mapsto f(s,t)$. For each $s \in [0,\delta]$, we denote by $\tau(s) \in SO(T_p\bar{M})$ the orthogonal transformation of $T_p\bar{M}$ obtained by parallel translation along f_0 from $p = f_0(0)$ to $f_0(1) = f^1(0)$, then along f^1 from $f^1(0)$ to $f^1(s) = f_s(1)$, and finally along f_s from $f_s(1)$ to $f_s(0) = p$. Let $A(s) \in \mathfrak{so}(T_p\bar{M})$ be the skew-symmetric transformation of $T_p\bar{M}$ defined by $A(s) = \tau'(s) \circ \tau(s)^{-1}$ for all $s \in [0,\delta]$. Then, for each $u, w \in T_p\bar{M}$,

$$\langle A(s)u, w\rangle = \int_0^1 \langle \bar{R}\left(\frac{\partial f}{\partial s}(s,t), \frac{\partial f}{\partial t}(s,t)\right) U_s(t), W_s(t)\rangle dt \ ,$$

where $U_s(t)$ and $W_s(t)$ are the parallel vector fields along f_s with $U_s(0) = u$ and $W_s(0) = w$, respectively.

PROOF Let $s \in (0,\delta)$. We define a smooth map $\tilde{f} : [0,\delta - s] \times [0,1] \to \bar{M}$ by $\tilde{f}(\tilde{s},t) = f(s+\tilde{s},t)$. To avoid confusion, we denote the objects associated to $\tilde{f}$ with tilded symbols $\tilde{}$. We have

$$\tau(s+\tilde{s}) = \tilde{\tau}(\tilde{s}) \circ \tau(s)$$

and

$$A(s) \circ \tau(s) = \tau'(s) = \tilde{\tau}'(0) \circ \tau(s) = \tilde{A}(0) \circ \tau(s) \ ,$$

since $\tilde{\tau}(0)$ is the identity transformation of $T_p\bar{M}$. This implies $A(s) = \tilde{A}(0)$.

This shows that it suffices to prove the above formula only for $s = 0$, because then

$$\begin{aligned}\langle A(s)u, w\rangle &= \langle \tilde{A}(0)u, w\rangle \\ &= \int_0^1 \langle \bar{R}\left(\frac{\partial \tilde{f}}{\partial \tilde{s}}(0,t), \frac{\partial \tilde{f}}{\partial t}(0,t)\right)\tilde{U}_0(t), \tilde{W}_0(t)\rangle dt \\ &= \int_0^1 \langle \bar{R}\left(\frac{\partial f}{\partial s}(s,t), \frac{\partial f}{\partial t}(s,t)\right)U_s(t), W_s(t)\rangle dt \ .\end{aligned}$$

For $s = \delta$ the formula follows by a continuity argument.

Let $U(s,t)$ be the vector field along $f(s,t)$ obtained by parallel translation of u along f_0 from $p = f_0(0)$ to $f_0(1) = f^1(0)$, then along f^1 from $f^1(0)$ to $f^1(s) = f_s(1)$, and finally along f_s from $f_s(1)$ to $f_s(t)$. Then we have

$$U(s,0) = \tau(s)u$$

and

$$A(0)u = \tau'(0)u = Z(0) \ ,$$

where Z is the vector field along f_0 defined by

$$Z(t) = \left(\frac{D}{\partial s}U\right)(0,t) \ ,$$

where D is the covariant derivative along curves associated to the connection $\bar{\nabla}$. By construction, the vector field $t \mapsto U(s,t)$ is parallel along f_s, so

$$Z'(t) = \left(\frac{D}{\partial t}\frac{D}{\partial s}U\right)(0,t) = \bar{R}\left(\frac{\partial f}{\partial t}(0,t), \frac{\partial f}{\partial s}(0,t)\right)U_0(t) \ .$$

For the smooth function

$$g(t) = \langle Z(t), W_0(t)\rangle$$

we therefore get

$$g'(t) = \langle Z'(t), W_0(t)\rangle = \langle \bar{R}\left(\frac{\partial f}{\partial t}(0,t), \frac{\partial f}{\partial s}(0,t)\right)U_0(t), W_0(t)\rangle$$

with initial condition

$$g(1) = \langle Z(1), W_0(1)\rangle = 0$$

since $Z(1) = 0$ by construction. Using Barrow's rule

$$\begin{aligned}\langle A(0)u, w\rangle &= g(0) \\ &= g(1) - \int_0^1 \langle \bar{R}\left(\frac{\partial f}{\partial t}(0,t), \frac{\partial f}{\partial s}(0,t)\right)U_0(t), W_0(t)\rangle dt \\ &= \int_0^1 \langle \bar{R}\left(\frac{\partial f}{\partial s}(0,t), \frac{\partial f}{\partial t}(0,t)\right)U_0(t), W_0(t)\rangle dt \ ,\end{aligned}$$

by which the lemma is proved. □

PROOF of Theorem 8.3.1 (continued): We still have to prove that V is invariant by $\bar{\nabla}$-parallel translation along loops in M based at p. Let $c : [0,1] \to M$ be a loop in M with $c(0) = c(1) = p$. Then there exists a unique curve $\xi : [0,1] \to V \cap U_\epsilon(0) \subset T_pM$ so that $c = \exp_p \circ \xi$. We define

$$f : [0,1] \times [0,1] \to M \ , \ (s,t) \mapsto \exp_p(t\xi(s))$$

and use the same notations as in the previous lemma. By assumption, the Riemannian curvature tensor $\bar{R}$ of $\bar{M}$ preserves the $\bar{\nabla}$-parallel translate of V along the geodesics f_s. In the first part, we saw that $T_{f(s,t)}M$ is obtained by $\bar{\nabla}$-parallel translation of V along the geodesic f_s from $p = f_s(0)$ to $f_s(t) = f(s,t)$. Combining these facts with the equation of the previous lemma, we obtain

$$\langle A(s)u, w\rangle = 0$$

for all $s \in [0,1]$, $u \in V$ and $w \in V^\perp$, that is

$$A(s) \in \mathfrak{so}(V) \oplus \mathfrak{so}(V^\perp)$$

for all $s \in [0,1]$. Since $\tau(0)$ is the identity transformation of $T_p\bar{M}$, we conclude that

$$\tau(s) \in O(V) \times O(V^\perp)$$

for all $s \in [0,1]$. In particular, $\tau(1) \in O(V) \times O(V^\perp)$. But $\tau(1)$ is, by construction, the $\bar{\nabla}$-parallel translation along the loop c from $c(0)$ to $c(1)$. This concludes the proof of Theorem 8.3.1. □

If the manifold $\bar{M}$ is real analytic, the assumption on the geodesics can be replaced by the local property that the Riemannian curvature tensor $\bar{R}_p$ and all its covariant derivatives $(\bar{\nabla}^k\bar{R})_p$, $k \geq 1$, at p preserve V.

A global version of the existence of complete totally geodesic immersed submanifolds of complete Riemannian manifolds has been obtained by Hermann [100] using once-broken geodesics. Let $p \in \bar{M}$ and V be a linear subspace of $T_p\bar{M}$. Further, let $\gamma : [0,b] \to \bar{M}$ be a once-broken geodesic starting at p and broken at $t_o \in (0,b)$. Following Hermann, we say that γ is *V-admissible* if $\dot{\gamma}(t)$ lies in the parallel translate of V along γ from p to $\gamma(t)$ for all $t \in [0,b]$, and if $\gamma([t_o,b])$ is contained in some convex neighbourhood of $\gamma(t_o)$. It is convenient to encompass smooth geodesics among once-broken geodesics.

THEOREM 8.3.3 (Hermann)
Let $\bar{M}$ be a complete Riemannian manifold, $p \in \bar{M}$ and V a linear subspace of $T_p\bar{M}$. Then there exists an immersed complete totally geodesic submanifold M of $\bar{M}$ with $p \in M$ and $T_pM = V$ if and only if for each V-admissible

once-broken geodesic $\gamma : [0, b] \to \bar{M}$ the Riemannian curvature tensor of $\bar{M}$ at $\gamma(b)$ preserves the parallel translate of V along γ from p to $\gamma(b)$.

PROOF The "only if" part of the theorem is trivial. Let us assume that, for each V-admissible once-broken geodesic $\gamma : [0, b] \to \bar{M}$, the Riemannian curvature tensor of $\bar{M}$ at $\gamma(b)$ preserves the parallel translate of V along γ from p to $\gamma(b)$. According to Cartan's Local Existence Theorem 8.3.1 there exists a connected totally geodesic submanifold M of $\bar{M}$ with $p \in M$ and $T_pM = V$. Without loss of generality, we can assume that M is maximal in the sense that each connected totally geodesic submanifold N of $\bar{M}$ with $q \in N \cap M \neq \emptyset$ and $T_qN = T_qM$ is contained in M. The only point we have to keep in mind here is M may not be embedded but only immersed in $\bar{M}$. We now assume that M is not complete and derive a contradiction.

If M is not complete, there exists a geodesic $\beta : [0, 1) \to M$ for which $\lim_{t \to 1} \beta(t)$ does not exist in M. Since $\bar{M}$ is complete and M is totally geodesic in $\bar{M}$ there exists a geodesic $\alpha : [0, 1] \to \bar{M}$ such that $\alpha|[0, 1) = \beta$. Let W be the $\bar{\nabla}$-parallel translate of V along α from p to $q = \alpha(1)$. It follows from the assumption and Theorem 8.3.1 that there exists a connected totally geodesic submanifold N of $\bar{M}$ with $q \in N$ and $T_qN = W$. Now consider a once-broken geodesic $\gamma : [0, 1 + \epsilon] \to \bar{M}$ for which $\gamma(t) = \beta(t) = \alpha(t)$ for $t \in [0, 1)$, $\gamma(1) = q = \alpha(1)$ and $\gamma(t) = \beta(1 - t) = \alpha(1 - t)$ for $t \in (1, 1 + \epsilon)$, $\epsilon \in \mathbb{R}_+$ sufficiently small. By construction, the $\bar{\nabla}$-parallel translate of V along β from p to $\beta(1 - \delta)$ coincides with the parallel translate of W along γ from q to $\gamma(1 + \delta) = \beta(1 - \delta)$ for all $\delta \in (0, \epsilon)$. It follows that the tangent spaces of M and N coincide at all points on $\gamma((1, 1 + \epsilon)) = \beta((1 - \epsilon, 1))$. Since we assumed that M is maximal totally geodesic, rigidity of totally geodesic submanifolds implies that N is contained in M. But this is a contradiction since q is in N but not in M. It follows that M is complete. $\square$

We will discuss the existence problem for totally geodesic submanifolds of symmetric spaces in Section 9.1.

b) Fixed point sets of isometries

An important class of totally geodesic submanifolds is given by fixed point sets of isometries.

PROPOSITION 8.3.4

Let $f : \bar{M} \to \bar{M}$ be an isometry of a Riemannian manifold $\bar{M}$ and $\bar{M}_f := \{p \in \bar{M} \mid f(p) = p\}$ the set of fixed points of f. If $\bar{M}_f$ is non-empty, then each connected component M of $\bar{M}_f$ is a totally geodesic submanifold of $\bar{M}$ and for each $p \in M$ we have $T_pM = \{X \in T_p\bar{M} \mid f_{*p}X = X\}$.

PROOF Let M be a connected component of $\bar{M}_f$, $p \in M$, and $V_p = \{X \in$

$T_p\bar{M} \mid f_{*p}X = X\}$. Since isometries map geodesics to geodesics we have $f \circ \gamma_X = \gamma_X$ for all $X \in V_p$, where γ_X is the maximal geodesic in $\bar{M}$ with $\gamma_X(0) = p$ and $\dot{\gamma}_X(0) = X$. This implies $\exp_p(V_p) \subset M$, where $\exp_p$ is the exponential map of $\bar{M}$ at p. If $q \in M$ is sufficiently close to p, say in some convex open neighbourhood of p in $\bar{M}$, there exists a unique geodesic in this neighbourhood connecting them. As p and q are fixed by f and the geodesic between them is unique, the entire geodesic is fixed by f. Hence, its tangent vector at p is fixed, and the geodesic is of the form γ_X with some $X \in V_p$. Thus, there exists an open neighbourhood of p in M contained in $\exp_p(V_p)$. As $\exp_p$ is a local diffeomorphism near $0 \in T_p\bar{M}$ we can now conclude that M is a submanifold of $\bar{M}$ and $T_pM = V_p$ (recall that p was arbitrary). Next, let $\gamma : I \to M$ be a geodesic in M with $\gamma(0) = p$ and $\dot{\gamma}(0) = X \in T_pM$. As $T_pM = V_p$ and $\exp_p(V_p) \subset M$, uniqueness of geodesics implies that $\gamma = \gamma_X|I$, that is, γ is a geodesic in $\bar{M}$. □

This proposition is of particular interest when f is an isometric involution on $\bar{M}$. If $\bar{M}_f$ is non-empty, then f is the reflection of $\bar{M}$ in each connected component of $\bar{M}_f$. An interesting example is given by the geodesic symmetry s_p of a Riemannian symmetric space $\bar{M}$ at a given point $p \in \bar{M}$. The point p is an isolated fixed point of s_p. Each other connected component of the fixed point set of s_p is called a *polar* of $\bar{M}$ (with respect to p). Polars contain deep information about the geometry and topology of a symmetric space, see, for instance, [147–151].

c) The congruence problem for totally geodesic submanifolds

Another fundamental problem concerning totally geodesic submanifolds is congruence. By this we mean the following: Given two Riemannian manifolds M and $\bar{M}$ and two totally geodesic isometric immersions $f_1, f_2 : M \to \bar{M}$, is there an isometry g of $\bar{M}$ so that $f_1 = g \circ f_2$? If such a g exists, the two immersions are said to be congruent. A basic problem is to determine the congruence classes of totally geodesic isometric immersions from a fixed Riemannian manifold M into another fixed Riemannian manifold $\bar{M}$. This is, in general, a rather difficult problem and has been solved so far only for some particular ambient spaces $\bar{M}$, for instance $\mathbb{R}^n$, rank-one symmetric spaces, real [52] and complex [15] two-plane Grassmannians. Even in Grassmannians, the congruence problem seems to be still open except for the above-mentioned special cases.

8.4 Totally umbilical submanifolds and extrinsic spheres

Recall that a submanifold M of a Riemannian manifold $\bar{M}$ is said to be *umbilical in the direction* ξ if the shape operator A_ξ of M in the direction of the normal vector

ξ is a multiple of the identity. If M is umbilical in any normal direction ξ, then M is called a *totally umbilical submanifold* of $\bar{M}$. The *mean curvature vector field* H of an m-dimensional submanifold M of $\bar{M}$ is defined by

$$H := \frac{1}{m}\mathrm{tr}(\alpha),$$

and $h := ||H||$ is the *mean curvature function* of M. M is totally umbilical if and only if

$$\alpha(X,Y) = \langle X,Y\rangle H \tag{8.1}$$

for all vector fields X, Y on M. A totally umbilical submanifold with nonzero parallel mean curvature vector field is called *extrinsic sphere*. The Codazzi equation provides a criterion to tell when a totally umbilical submanifold is an extrinsic sphere.

a) When is a totally umbilical submanifold an extrinsic sphere?

PROPOSITION 8.4.1
Let M be a totally umbilical and nontotally geodesic submanifold of a Riemannian manifold $\bar{M}$ with dimension $m = \dim M \geq 2$, which is not totally geodesic. Then M is an extrinsic sphere if and only if the tangent bundle of M is curvature-invariant, that is, if $\bar{R}(T_pM, T_pM)T_pM \subset T_pM$ for all $p \in M$.

PROOF Inserting equation (8.1) into the Codazzi equation yields

$$(\bar{R}(X,Y)Z)^{\perp} = \langle Y,Z\rangle \nabla^{\perp}_X H - \langle X,Z\rangle \nabla^{\perp}_Y H .$$

When M is an extrinsic sphere, TM is curvature-invariant. Conversely, suppose that TM is curvature-invariant. As $m \geq 2$ we can choose locally orthonormal vector fields X, Y on M. Putting $Y = Z$, the previous equation implies that H is parallel. □

In a space of constant curvature, the tangent bundle of any submanifold is curvature-invariant. Thus, Proposition 8.4.1 implies

COROLLARY 8.4.2
Every totally umbilical and nontotally geodesic submanifold with dimension ≥ 2 in a space of constant curvature is an extrinsic sphere.

b) Circles

A smooth curve $\gamma : I \to \bar{M}$ parametrized by arc length is called a *circle* in $\bar{M}$ if it parametrizes a one-dimensional extrinsic sphere in $\bar{M}$. Let $\gamma : I \to \bar{M}$ be a smooth curve parametrized by arc length and $M = \gamma(I)$, which is an immersed submanifold of $\bar{M}$. Then

$$H \circ \gamma = \alpha(\dot{\gamma}, \dot{\gamma}) = \bar{\nabla}_{\dot{\gamma}}\dot{\gamma}$$

by the Gauss formula, and the Weingarten formula implies

$$\nabla^{\perp}_{\dot\gamma} H = \bar\nabla_{\dot\gamma} H + A_H \dot\gamma = \bar\nabla_{\dot\gamma}\bar\nabla_{\dot\gamma}\dot\gamma + \langle \bar\nabla_{\dot\gamma}\dot\gamma, \bar\nabla_{\dot\gamma}\dot\gamma\rangle \dot\gamma \,.$$

This shows that a smooth curve $\gamma : I \to \bar M$ parametrized by arc length is a circle in $\bar M$ if and only if it satisfies the third order differential equation

$$\bar\nabla_{\dot\gamma}\bar\nabla_{\dot\gamma}\dot\gamma + \langle \bar\nabla_{\dot\gamma}\dot\gamma, \bar\nabla_{\dot\gamma}\dot\gamma\rangle \dot\gamma = 0 \,.$$

This equation says that a circle has non-vanishing constant first geodesic curvature and vanishing higher geodesic curvatures. Standard arguments from theory of differential equations imply the following existence and uniqueness result about circles, which establishes the classification of one-dimensional extrinsic spheres in Riemannian manifolds.

PROPOSITION 8.4.3
Let $\bar M$ be a Riemannian manifold, $p \in \bar M$, $X \in T_p\bar M$ a unit vector and $Y \in T_p\bar M$ nonzero and orthogonal to X. Then there exists a unique maximal circle $\gamma : I \to \bar M$ with $0 \in I$, $\gamma(0) = p$, $\dot\gamma(0) = X$ and $(\bar\nabla_{\dot\gamma}\dot\gamma)(0) = Y$.

c) Characterization of extrinsic spheres by circles

A submanifold M of a Riemannian manifold $\bar M$ is totally geodesic if and only if any geodesic in M is also a geodesic in $\bar M$. For extrinsic spheres, we have a similar characterization by using circles [171].

THEOREM 8.4.4
A submanifold M of a Riemannian manifold $\bar M$ is an extrinsic sphere if and only if every circle in M is also a circle in $\bar M$.

PROOF We first assume that M is an extrinsic sphere in $\bar M$. The second fundamental form α of M satisfies $\alpha(X,Y) = \langle X, Y\rangle H$, where H is the mean curvature vector field of M that is parallel in the normal bundle of M. Let $\gamma : I \to M$ be a circle in M (parametrized by arc length). The Gauss formula gives $\bar\nabla_{\dot\gamma}\dot\gamma = \nabla_{\dot\gamma}\dot\gamma + \alpha(\dot\gamma,\dot\gamma)$. Applying the formulas of Gauss and Weingarten, then using the facts that M is totally umbilical and $\alpha(\dot\gamma,\dot\gamma) = H \circ \gamma$ is parallel in the normal bundle of M, we get

$$\begin{aligned}
\bar\nabla_{\dot\gamma}\bar\nabla_{\dot\gamma}\dot\gamma &= \nabla_{\dot\gamma}\nabla_{\dot\gamma}\dot\gamma + \alpha(\nabla_{\dot\gamma}\dot\gamma,\dot\gamma) - A_{\alpha(\dot\gamma,\dot\gamma)}\dot\gamma + \nabla^{\perp}_{\dot\gamma}\alpha(\dot\gamma,\dot\gamma)\\
&= \nabla_{\dot\gamma}\nabla_{\dot\gamma}\dot\gamma - \langle A_{\alpha(\dot\gamma,\dot\gamma)}\dot\gamma, \dot\gamma\rangle\dot\gamma\\
&= \nabla_{\dot\gamma}\nabla_{\dot\gamma}\dot\gamma - \langle \alpha(\dot\gamma,\dot\gamma), \alpha(\dot\gamma,\dot\gamma)\rangle\dot\gamma\\
&= \nabla_{\dot\gamma}\nabla_{\dot\gamma}\dot\gamma - \langle \bar\nabla_{\dot\gamma}\dot\gamma - \nabla_{\dot\gamma}\dot\gamma, \bar\nabla_{\dot\gamma}\dot\gamma - \nabla_{\dot\gamma}\dot\gamma\rangle\dot\gamma\\
&= \nabla_{\dot\gamma}\nabla_{\dot\gamma}\dot\gamma + 2\langle \bar\nabla_{\dot\gamma}\dot\gamma, \nabla_{\dot\gamma}\dot\gamma\rangle\dot\gamma - \langle \nabla_{\dot\gamma}\dot\gamma, \nabla_{\dot\gamma}\dot\gamma\rangle\dot\gamma - \langle \bar\nabla_{\dot\gamma}\dot\gamma, \bar\nabla_{\dot\gamma}\dot\gamma\rangle\dot\gamma\\
&= \nabla_{\dot\gamma}\nabla_{\dot\gamma}\dot\gamma + \langle \nabla_{\dot\gamma}\dot\gamma, \nabla_{\dot\gamma}\dot\gamma\rangle\dot\gamma - \langle \bar\nabla_{\dot\gamma}\dot\gamma, \bar\nabla_{\dot\gamma}\dot\gamma\rangle\dot\gamma\\
&= -\langle \bar\nabla_{\dot\gamma}\dot\gamma, \bar\nabla_{\dot\gamma}\dot\gamma\rangle\dot\gamma \,.
\end{aligned}$$

This shows that γ is a circle in $\bar{M}$.

We now assume that every circle in M is also a circle in $\bar{M}$. Let $\gamma : I \to M$ be a circle in M with $0 \in I$ and put $p = \gamma(0)$. The formulas of Gauss and Weingarten imply

$$\bar{\nabla}_{\dot\gamma}\bar{\nabla}_{\dot\gamma}\dot\gamma = \nabla_{\dot\gamma}\nabla_{\dot\gamma}\dot\gamma + \alpha(\nabla_{\dot\gamma}\dot\gamma,\dot\gamma) - A_{\alpha(\dot\gamma,\dot\gamma)}\dot\gamma + \nabla^{\perp}_{\dot\gamma}\alpha(\dot\gamma,\dot\gamma) ,$$

and using the assumption that γ is a circle in $\bar{M}$ and, once again, the Gauss formula, we get

$$\bar{\nabla}_{\dot\gamma}\bar{\nabla}_{\dot\gamma}\dot\gamma = -\langle\bar{\nabla}_{\dot\gamma}\dot\gamma,\bar{\nabla}_{\dot\gamma}\dot\gamma\rangle\dot\gamma = -\langle\nabla_{\dot\gamma}\dot\gamma,\nabla_{\dot\gamma}\dot\gamma\rangle\dot\gamma - \langle\alpha(\dot\gamma,\dot\gamma),\alpha(\dot\gamma,\dot\gamma)\rangle\dot\gamma .$$

As γ is a circle in M, these two equations imply

$$\langle\alpha(\dot\gamma,\dot\gamma),\alpha(\dot\gamma,\dot\gamma)\rangle\dot\gamma + \alpha(\nabla_{\dot\gamma}\dot\gamma,\dot\gamma) - A_{\alpha(\dot\gamma,\dot\gamma)}\dot\gamma + \nabla^{\perp}_{\dot\gamma}\alpha(\dot\gamma,\dot\gamma) = 0 .$$

Taking the normal component

$$\alpha(\nabla_{\dot\gamma}\dot\gamma,\dot\gamma) + \nabla^{\perp}_{\dot\gamma}\alpha(\dot\gamma,\dot\gamma) = 0 .$$

But

$$(\nabla^{\perp}_{\dot\gamma}\alpha)(\dot\gamma,\dot\gamma) = \nabla^{\perp}_{\dot\gamma}\alpha(\dot\gamma,\dot\gamma) - 2\alpha(\nabla_{\dot\gamma}\dot\gamma,\dot\gamma) ,$$

so

$$3\alpha(\nabla_{\dot\gamma}\dot\gamma,\dot\gamma) = -(\nabla^{\perp}_{\dot\gamma}\alpha)(\dot\gamma,\dot\gamma) .$$

According to Proposition 8.4.3, we can realize any pair (X,Y) with $X \in T_pM$ a unit vector and $Y \in T_pM$ nonzero and orthogonal to X as $\dot\gamma(0) = X$ and $(\nabla_{\dot\gamma}\dot\gamma)(0) = Y$ with some circle γ in M. The right-hand side of the previous equation is independent of Y. By replacing Y with $-Y$, we conclude that $\alpha(X,Y) = 0$ whenever $X,Y \in T_pM$ are orthogonal. Let $E_1,\ldots,E_m$ be an orthonormal basis of T_pM. For $i \neq j$ we get

$$0 = \alpha(E_i + E_j, E_i - E_j) = \alpha(E_i,E_i) - \alpha(E_j,E_j) .$$

Let $U,V \in T_pM$ be arbitrary and write $U = \sum u_iE_i$ and $V = \sum v_iE_i$. Then

$$\begin{aligned}\alpha(U,V) &= \textstyle\sum_{i,j=1}^m u_iv_j\alpha(E_i,E_j) = \sum_{i=1}^m u_iv_i\alpha(E_i,E_i) = \\ &= \textstyle\sum_{i=1}^m u_iv_i\alpha(E_1,E_1) = \langle U,V\rangle\alpha(E_1,E_1).\end{aligned}$$

Thus M is totally umbilical and the mean curvature normal vector H_p of M at p satisfies $\alpha(X,X) = H_p$ for any unit vector $X \in T_pM$. Eventually, as γ is parametrized by arc length,

$$\nabla^{\perp}_{\dot\gamma}H = \nabla^{\perp}_{\dot\gamma}\alpha(\dot\gamma,\dot\gamma) = -\alpha(\nabla_{\dot\gamma}\dot\gamma,\dot\gamma) = -\langle\nabla_{\dot\gamma}\dot\gamma,\dot\gamma\rangle H \circ \gamma = 0 .$$

As any unit tangent vector can be realized as the tangent vector of a circle, H is parallel in the normal bundle of M and M is an extrinsic sphere in $\bar{M}$. □

The previous two results show that an extrinsic sphere is locally uniquely determined by its tangent space at a point and the mean curvature normal at that point. In this sense, extrinsic spheres are rigid. The discussion also indicates how to construct a given extrinsic sphere geometrically. Fix a point p in an extrinsic sphere M of a Riemannian manifold $\bar{M}$. The Gauss formula implies that every geodesic in M is a circle in $\bar{M}$. Thus, M can be reconstructed by running along every circle γ in $\bar{M}$ with initial values $\gamma(0) = p$, $\dot{\gamma}(0) \in T_pM$ and $(\bar{\nabla}_{\dot{\gamma}}\dot{\gamma})(0) = H_p$.

8.5 Symmetric submanifolds

Recall that a submanifold M of a Riemannian manifold $\bar{M}$ is called a *symmetric submanifold* if for each $p \in M$ there exists an isometry σ_p of $\bar{M}$ with

$$\sigma_p(p) = p \,,\ \sigma_p(M) = M \text{ and } \sigma_{p*}X = \begin{cases} -X & , X \in T_pM \\ X & , X \in \nu_pM \end{cases} .$$

In this section, we want to study the relation between symmetry of submanifolds and parallelity of the second fundamental form. Let M be a symmetric submanifold of a Riemannian manifold $\bar{M}$. For σ_p as above, the connected component of the fixed point set of σ_p containing p is a totally geodesic submanifold of $\bar{M}$ whose tangent space at p coincides with the normal space of M at p. We call the latter *normal submanifold* of M at p and denote it by $M_p^\perp$. Note that $M_p^\perp$ is just the image under the exponential map of $\bar{M}$ of the normal space ν_pM of M at p. Thus, a necessary condition for a submanifold to be symmetric is that there exists a totally geodesic submanifold of the ambient space tangent to each normal space. This is no restriction in a space of constant curvature, but quite restrictive in more general Riemannian manifolds like symmetric spaces. It is also clear from the very definition that every symmetric submanifold is a Riemannian symmetric space. In particular, symmetric submanifolds are complete. Let q_1, q_2 be two distinct points in M. Then there exists a geodesic γ in M connecting q_1 and q_2. The geodesic symmetry σ_p at the midpoint p on γ between q_1 and q_2 maps $M_{q_1}^\perp$ to $M_{q_2}^\perp$, and vice versa. This shows that any two normal submanifolds $M_{q_1}^\perp$ and $M_{q_2}^\perp$ are congruent to each other under an isometry of $\bar{M}$. Thus, we can talk about the congruence class of normal submanifolds associated to M at points in M, any representative of which we simply denote by $M^\perp$.

Any isometry of $\bar{M}$ is an affine map with respect to the Levi Civita covariant derivative. Using the Gauss formula, we therefore get

$$(\nabla_X^\perp\alpha)(Y,Z) = \sigma_{p*}(\nabla_X^\perp\alpha)(Y,Z) = (\nabla_{\sigma_{p*}X}^\perp\alpha)(\sigma_{p*}Y,\sigma_{p*}Z) = -(\nabla_X^\perp\alpha)(Y,Z)$$

for all $p \in M$ and $X, Y, Z \in T_pM$. Thus, the second fundamental form of a symmetric submanifold is parallel. The Codazzi equation then implies that each tangent space of M is curvature-invariant, that is, $\bar{R}(T_pM, T_pM)T_pM \subset T_pM$ for all $p \in M$. We summarize this in

PROPOSITION 8.5.1
Let M be a symmetric submanifold of a Riemannian manifold $\bar{M}$. Then, the second fundamental form of M is parallel, each tangent space of M is curvature-invariant, and for each $p \in M$ there exists a totally geodesic submanifold $M_p^\perp$ of $\bar{M}$ with $p \in M_p^\perp$ and $T_pM_p^\perp = \nu_pM$. Any two normal submanifolds $M_{q_1}^\perp$ and $M_{q_2}^\perp$, $q_1, q_2 \in M$, are congruent to each other under an isometry of $\bar{M}$.

A natural question arising from this proposition is whether parallelity of the second fundamental form implies symmetry of the submanifold. Since the first condition is local whereas the second one is global, this question makes sense only for some kind of local symmetry. To make this precise, we introduce the notion of a *locally symmetric submanifold* M of a Riemannian manifold $\bar{M}$ by requiring that for each $p \in M$ there exists an isometry σ_p on some open neighbourhood U of p in M with

$$\sigma_p(p) = p \ , \ \sigma_p(U) = U \text{ and } \sigma_{p*}X = \begin{cases} -X & , X \in T_pM \\ X & , X \in \nu_pM \end{cases} .$$

For submanifolds of spaces of constant curvature, local symmetry is equivalent to the parallelity of the second fundamental form [204]. But this result does not generalize to more general Riemannian manifolds. For example, a totally geodesic real projective space $\mathbb{R}P^k$ in complex projective space $\mathbb{C}P^n$ is not a locally symmetric submanifold for $k < n$, but obviously has parallel second fundamental form. We will say more about symmetric submanifolds of symmetric spaces in Section 9.3.

8.6 Exercises

Exercise 8.6.1 Let $f : M \to N$ and $g : N \to P$ be isometric immersions and let ϕ be the composition $g \circ f$. Let α^f, α^g and α^ϕ be the second fundamental forms of f, g and ϕ respectively. Prove that for any $p \in M$, $v, w \in T_pM$

$$\alpha^\phi(v, w) = \alpha^f(v, w) + \alpha^g(f_{*p}v, f_{*p}w) .$$

Exercise 8.6.2 Prove that a totally geodesic $\mathbb{R}P^k$ in $\mathbb{C}P^n$ is not a symmetric submanifold if $k < n$.

Exercise 8.6.3 The present exercise generalizes the well-known fact that the intersection of the fixed point set of an isometry with an invariant submanifold is totally geodesic in the submanifold. Let P be a Riemannian manifold and let P_1, P_2 and $P_1 \cap P_2$ be submanifolds with P_1 totally geodesic. Assume that $T_x P_2$ is invariant under the reflection at $T_x P_1$ for any $x \in P_1 \cap P_2$. Then show that $P_1 \cap P_2$ is totally geodesic in P_2. Furthermore, the shape operator A_η of P_2 leaves $T_x(P_1 \cap P_2)$ invariant for any $x \in P_1 \cap P_2$ and any $\eta \in \nu_x P_2 \cap T_x P_1$.

Exercise 8.6.4 Using Lemma 8.3.2 prove the Ambrose-Singer Theorem on Riemannian holonomy. Adapt the argument for a general connection.

Chapter 9

Submanifolds of Symmetric Spaces

In this chapter, we study submanifolds of Riemannian symmetric spaces: totally geodesic submanifolds (Section 9.1), totally umbilical submanifolds and extrinsic spheres (Section 9.2), symmetric submanifolds (Section 9.3), submanifolds with parallel second fundamental form (Section 9.4) and homogeneous hypersurfaces (Section 9.5). We mainly discuss the classification problems for these submanifolds.

Notation: *Throughout this chapter we will denote real hyperbolic space by $\mathbb{R}H^n$ (and not H^n as in the previous chapters), to distinguish it from its complex and quaternionic analogues.*

9.1 Totally geodesic submanifolds

Throughout this section, we denote a Riemannian symmetric space by $\bar{M}$. The corresponding Riemannian symmetric pair is (G, K) with $G = I^0(\bar{M})$ and K the isotropy group at some point $o \in \bar{M}$. We denote the corresponding Cartan decomposition of $\mathfrak{g}$ by $\mathfrak{g} = \mathfrak{k} \oplus \mathfrak{m}$ and identify $T_o\bar{M}$ with $\mathfrak{m}$ in the usual way.

a) Lie triple systems

Let V be a linear subspace of $T_o\bar{M}$. The Riemannian curvature tensor of $\bar{M}$ is parallel. Thus, from Theorem 8.3.1, we see that there exists a totally geodesic submanifold M of $\bar{M}$ with $o \in M$ and $T_oM = V$ if and only if V is a curvature-invariant subspace of $T_o\bar{M}$, that is, if $\bar{R}(V, V)V \subset V$. Since the Riemannian curvature tensor of $\bar{M}$ at o is given by

$$\bar{R}_o(X, Y)Z = -[[X, Y], Z]$$

for all $X, Y, Z \in \mathfrak{m} = T_o\bar{M}$, we get

COROLLARY 9.1.1
Let $\bar{M}$ be a Riemannian symmetric space, $o \in \bar{M}$ and V a linear subspace of $T_o\bar{M}$. Then the following statements are equivalent:

(i) There exists a totally geodesic submanifold M of $\bar{M}$ with $o \in M$ and $T_oM = V$.

(ii) V is a curvature-invariant subspace of $T_o\bar{M}$, that is, $\bar{R}(V,V)V \subset V$.

(iii) V is a Lie triple system in $\mathfrak{m}$, that is, $[[V,V],V] \subset V$.

Since every Riemannian locally symmetric space is locally isometric to a Riemannian symmetric space, one can apply this criterion also for the local existence of totally geodesic submanifolds in Riemannian locally symmetric spaces. Generalizations to more general classes of homogeneous spaces have been obtained by A.A. Sagle [194] and K. Tsukada [229].

b) Construction of a totally geodesic submanifold from a Lie triple system

How can we actually construct a totally geodesic submanifold from a given Lie triple system? Suppose that $V \subset \mathfrak{m}$ is a Lie triple system and recall that $[V,V]$ denotes the linear subspace of $\mathfrak{k}$, which is spanned by all vectors of the form $[X,Y]$ with $X, Y \in V$. We define

$$\mathfrak{g}' := [V,V] \oplus V$$

and claim that $\mathfrak{g}'$ is a Lie subalgebra of $\mathfrak{g}$. In fact, if $X, Y \in V$, then $[X,Y] \in [V,V] \subset \mathfrak{g}'$. Next, if $X,Y,Z \in V$, then $[[X,Y],Z] \in V \subset \mathfrak{g}'$ since V is a Lie triple system. Since $[V,V]$ is spanned by vectors of the form $[X,Y]$ it follows that

$$[[V,V],V] \subset V \subset \mathfrak{g}' .$$

Eventually, let $X,Y,Z,W \in V$. Then, using the Jacobi identity for the Lie bracket, we obtain

$$\begin{aligned} [[X,Y],[Z,W]] &= -[[Y,[Z,W]],X] - [[[Z,W],X],Y] \\ &\in [[[V,V],V],V] \subset [V,V] \subset \mathfrak{g}' , \end{aligned}$$

and hence

$$[[V,V],[V,V]] \subset [V,V] \subset \mathfrak{g}' .$$

Altogether, it now follows that $\mathfrak{g}'$ is a Lie subalgebra of $\mathfrak{g}$. Let σ be the Cartan involution on $\mathfrak{g}$ corresponding to the Cartan decomposition $\mathfrak{g} = \mathfrak{k} \oplus \mathfrak{m}$. For all $X,Y,Z \in V$ we have

$$\sigma([X,Y]+Z) = \sigma([X,Y]) + \sigma(Z) = [\sigma(X),\sigma(Y)] + \sigma(Z) = [X,Y] - Z \in \mathfrak{g}' .$$

This shows that $\mathfrak{g}'$ is invariant under σ.

Let G' be the connected Lie subgroup of G with Lie algebra $\mathfrak{g}'$. Then G' is invariant under the corresponding Cartan involution on G. Let $M := G' \cdot o$ be the orbit of the action of G' on $\bar{M}$ containing o and denote by Exp the Lie exponential map from $\mathfrak{g}$ to G. For every $X \in V$ the curve $t \mapsto \mathrm{Exp}(tX) \cdot o$ is a geodesic in $\bar{M}$ and contained in M, hence also a geodesic in M. It follows from the Gauss formula that the second fundamental form of M at o vanishes. Since G' acts transitively on M, the second fundamental form of M vanishes everywhere and it follows that M is totally geodesic in $\bar{M}$. This shows that the maximal totally geodesic submanifold of $\bar{M}$ tangent to V is homogeneous and, in particular, complete. Note that M is not embedded in general. For instance, choose for V a line corresponding to a dense geodesic on a flat torus. One can say even more about M. The geodesic symmetry s_o of $\bar{M}$ at o reflects in o each geodesic in $\bar{M}$ through o, hence leaves M invariant and its restriction to M is an isometric geodesic symmetry. Thus M is also a Riemannian symmetric space.

Conversely, suppose that G' is a connected Lie subgroup of G that is invariant under the Cartan involution. Then the Lie algebra $\mathfrak{g}'$ of G' has the direct sum decomposition

$$\mathfrak{g}' = (\mathfrak{g}' \cap \mathfrak{k}) \oplus (\mathfrak{g}' \cap \mathfrak{m}) .$$

We define $V := \mathfrak{g}' \cap \mathfrak{m}$ and claim that V is a Lie triple system. Indeed, let $X, Y, Z \in V$. Then X, Y, Z are in $\mathfrak{g}'$, and since $\mathfrak{g}'$ is a Lie subalgebra of $\mathfrak{g}$, also $[[X, Y], Z] \in \mathfrak{g}'$. Also, X, Y, Z are in $\mathfrak{m}$, and hence

$$[[X, Y], Z] \in [[\mathfrak{m}, \mathfrak{m}], \mathfrak{m}] \subset [\mathfrak{k}, \mathfrak{m}] \subset \mathfrak{m} .$$

Altogether, this implies that $[[X, Y], Z] \in \mathfrak{g}' \cap \mathfrak{m} = V$, which shows that V is a Lie triple system. Let H be the connected Lie subgroup of G that is determined by the Lie triple system V as described above. Then the orbit $H \cdot o$ is the connected, complete, totally geodesic submanifold of $\bar{M}$ with $o \in H \cdot o$ and $T_o(H \cdot o) = V$. The Lie algebra $\mathfrak{h}$ of H satisfies

$$\mathfrak{h} = [V, V] \oplus V = [\mathfrak{g}' \cap \mathfrak{m}, \mathfrak{g}' \cap \mathfrak{m}] \oplus (\mathfrak{g}' \cap \mathfrak{m}) \subset (\mathfrak{g}' \cap \mathfrak{k}) \oplus (\mathfrak{g}' \cap \mathfrak{m}) = \mathfrak{g}' .$$

From this, we conclude that $H \cdot o \subset G' \cdot o$. Since $H \cdot o$ is complete and connected, and as

$$\dim(G' \cdot o) = \dim(\mathfrak{g}' \cap \mathfrak{m}) = \dim V = \dim(H \cdot o) ,$$

we see that $G' \cdot o = H \cdot o$, which implies that $G' \cdot o$ is totally geodesic. We summarize this in

PROPOSITION 9.1.2

Let (G, K) be a Riemannian symmetric pair with Cartan decomposition $\mathfrak{g} = \mathfrak{k} \oplus \mathfrak{m}$ at some point $o \in M$ and corresponding Cartan involution $\sigma : G \to G$.

(i) Let $V \subset \mathfrak{m}$ be a Lie triple system. Then

$$\mathfrak{g}' := [V, V] \oplus V$$

is a Lie subalgebra of $\mathfrak{g}$. Let G' be the connected Lie subgroup of G with Lie algebra $\mathfrak{g}'$. Then G' is a σ-invariant Lie subgroup of G and the orbit $M = G' \cdot o$ is the connected, complete, totally geodesic submanifold of $\bar{M} = G/K$ with $o \in M$ and $T_oM = V$. Moreover, M is intrinsically a Riemannian symmetric space.

(ii) Conversely, let G' be a σ-invariant connected Lie subgroup of G. Then the orbit $G' \cdot o$ is a totally geodesic submanifold of the symmetric space $\bar{M} = G/K$.

The following example illustrates how this result can be used for the construction of totally geodesic submanifolds of compact Lie groups. Recall that any connected compact Lie group admits a biinvariant Riemannian metric, turning it into a symmetric space.

PROPOSITION 9.1.3
Let H be a connected compact Lie group equipped with a biinvariant Riemannian metric. Then each connected Lie subgroup of H is totally geodesic in H.

PROOF The Lie group $G = H \times H$ acts on H isometrically by

$$(h_1, h_2) \cdot h = h_1 h h_2^{-1} \quad (h, h_1, h_2 \in H) ,$$

and

$$G \to G \ , \ (h_1, h_2) \mapsto (h_2, h_1)$$

is an involutive automorphism of G with fixed point set $K = \{(h, h) \mid h \in H\}$, which is canonically isomorphic to the connected Lie group H. The pair (G, K) is the Riemannian symmetric pair of H and the above involution is the corresponding Cartan involution. Now let L be any connected Lie subgroup of H. Then $L \times L$ is a connected Lie subgroup of G, which is obviously invariant under the Cartan involution. It follows from Proposition 9.1.2 that the orbit of $L \times L$ through e, which is precisely the subgroup L, is totally geodesic in H. □

What is the least codimension of a totally geodesic submanifold of a given symmetric space? This problem has been studied by A.L. Onishchik [181], who called this minimal codimension the index of the symmetric space. He classified all symmetric spaces of index one and two. In particular, he proved

THEOREM 9.1.4
An irreducible Riemannian symmetric space contains a totally geodesic hypersurface if and only if it is a space of constant curvature.

c) The classification problem

The classification problem for totally geodesic submanifolds in Riemannian symmetric spaces can therefore be reduced to the classification problem of Lie triple systems or σ-invariant Lie subgroups. Unfortunately, these are very difficult algebraic problems and not helpful to solve the original problem.

Another approach to the classification problem of totally geodesic submanifolds in Riemannian symmetric spaces of compact type has been suggested by B.Y. Chen and T. Nagano [53]. Due to the fact that every totally geodesic submanifold of a Riemannian symmetric space is contained in a complete one, which, in addition, is again a Riemannian symmetric space, it is sufficient to classify the maximal totally geodesic submanifolds. Recall that a totally geodesic submanifold M of $\bar{M}$ is called maximal if there exists no totally geodesic submanifold M' of $\bar{M}$ with $M \subset M'$, $M \neq M'$ and $\bar{M} \neq M'$. So the first question one has to solve is: Given a Riemannian symmetric space $\bar{M}$, which Riemannian symmetric spaces can be realized in $\bar{M}$ as a maximal totally geodesic submanifold?

Chen and Nagano approached this problem by means of the (M_+, M_-)-method. The idea of this method is as follows. According to Proposition 8.3.4, each connected component of the fixed point set of the geodesic symmetry s_o of $\bar{M}$ at o is a totally geodesic submanifold of $\bar{M}$. If $q \in \bar{M}$ is a point different from o and contained in such a component, denote this component by $M_+(q)$, then q is an antipodal point of o and $M_+(q)$ is the orbit through q of the action of the identity component K^o of the isotropy group K at o. Note that two points o and q in a Riemannian manifold $\bar{M}$ are *antipodal* if there exists a closed geodesic γ in $\bar{M}$, say of length l, so that the distance between o and q along γ is $l/2$. Any such orbit $M_+(q)$ is called *polar* of o, or, if $M_+(q)$ consists just of the single point q, *pole* of o. There exists a complete totally geodesic submanifold $M_-(q)$ of $\bar{M}$ tangent to the normal space of $M_+(q)$ at q, namely the connected component containing q of the fixed point set of $s_q \circ s_o$. Consider the set $B(\bar{M})$ of all quadruples $(o, q, M_+(q), M_-(q))$ modulo congruence by isometries of $\bar{M}$. When M is a totally geodesic submanifold of $\bar{M}$, then there is a natural map $B(M) \to B(\bar{M})$, where the relation among the totally geodesic submanifolds in the quadruples is inclusion. In other words, when M is a Riemannian symmetric space and there is no such map $B(M) \to B(\bar{M})$, then M cannot be realized in $\bar{M}$ as a totally geodesic submanifold. So the strategy is to compute all the quadruples $(o, q, M_+(q), M_-(q))$ for Riemannian symmetric spaces and then to compare them. Then eliminate all pairs $(M, \bar{M})$ where there is no natural map $B(M) \to B(\bar{M})$ and investigate the remaining cases whether they can actually be realized by totally geodesic submanifolds. One useful remark for this is that the rank of a totally geodesic submanifold cannot exceed the rank of the ambient space.

We illustrate this with the classification of maximal totally geodesic submanifolds of compact Riemannian symmetric spaces of rank one. These spaces are the spheres S^n and the projective spaces $\mathbb{R}P^n$, $\mathbb{C}P^n$, $\mathbb{H}P^n$ and $\mathbb{O}P^2$. For each of these spaces there is only one such quadruple. This is because all geodesics are closed with the same length and the isotropy subgroup at a point acts transitively on the set of unit tangent vectors at that point. The quadruples are

$$\begin{aligned}
S^n &: (o, q, \{q\}, S^n) \\
\mathbb{R}P^n &: (o, q, \mathbb{R}P^{n-1}, \mathbb{R}P^1 = S^1) \\
\mathbb{C}P^n &: (o, q, \mathbb{C}P^{n-1}, \mathbb{C}P^1 = S^2) \\
\mathbb{H}P^n &: (o, q, \mathbb{H}P^{n-1}, \mathbb{H}P^1 = S^4) \\
\mathbb{O}P^2 &: (o, q, \mathbb{O}P^1 = S^8, \mathbb{O}P^1 = S^8) \ .
\end{aligned}$$

For example, we see that the Cayley projective plane $\mathbb{O}P^2$ cannot be realized as a totally geodesic submanifold in any other compact Riemannian symmetric space of rank one. Following the above strategy we get all maximal totally geodesic submanifolds:

$$\begin{aligned}
S^n &: S^{n-1} \\
\mathbb{R}P^n &: \mathbb{R}P^{n-1} \\
\mathbb{C}P^n &: \mathbb{C}P^{n-1}, \mathbb{R}P^n \\
\mathbb{H}P^n &: \mathbb{H}P^{n-1}, \mathbb{C}P^n \\
\mathbb{O}P^2 &: \mathbb{O}P^1, \mathbb{H}P^2 \ .
\end{aligned}$$

The classification of totally geodesic submanifolds of compact Riemannian symmetric spaces of rank one is originally due to J.A. Wolf [242]. Further lists of maximal totally geodesic submanifolds in Riemannian symmetric spaces of compact type can be found in [53]. The (M_+, M_-)-method is also the starting point for T. Nagano and M. Sumi [152] toward a classification of totally geodesic spheres in Riemannian symmetric spaces of compact type. Their work extends previous results of S. Helgason [98] about totally geodesic spheres of maximal curvature in compact symmetric spaces.

d) Reflective submanifolds

An interesting subclass of the totally geodesic submanifolds is formed by the reflective submanifolds, which are, in general, defined as follows. Let $\bar{M}$ be a Riemannian manifold and M a submanifold of $\bar{M}$. When the geodesic reflection of $\bar{M}$ in M is a globally well-defined isometry of $\bar{M}$, then M is called *reflective submanifold.* Since any reflective submanifold is a connected component of the fixed point set of an isometry, it is totally geodesic. Obviously, any connected component of the fixed point set of an involutive isometry on a Riemannian manifold is a reflective submanifold. In particular, this implies

COROLLARY 9.1.5

Any polar in a Riemannian symmetric space is a reflective submanifold.

We already encountered reflective submanifolds in the framework of symmetric submanifolds, namely any normal submanifold $M^{\perp}$ of a symmetric submanifold is reflective. For symmetric spaces, we have the following useful criterion.

PROPOSITION 9.1.6

Let $\bar{M}$ be a Riemannian symmetric space. If M is a reflective submanifold of $\bar{M}$ then T_pM and ν_pM are Lie triple systems in $T_p\bar{M}$ for each $p \in M$. Moreover, the complete totally geodesic submanifold $M^\perp$ of $\bar{M}$ with $p \in M^\perp$ and $T_pM^\perp = \nu_pM$ is also reflective. Conversely, if $\bar{M}$ is simply connected, $p \in \bar{M}$, and if V is a Lie triple system in $T_p\bar{M}$ so that $V^\perp$ is also a Lie triple system in $T_p\bar{M}$, then there exists a reflective submanifold M of $\bar{M}$ with $p \in M$ and $T_pM = V$.

PROOF Suppose first that M is a reflective submanifold of $\bar{M}$ and let p be any point in M. It follows from Corollary 9.1.1 that T_pM is a Lie triple system. Denote by s_M the reflection of $\bar{M}$ in M and, as usual, by s_p the geodesic symmetry of $\bar{M}$ in p. Then the isometry $s_M \circ s_p$ fixes p, and its differential at p fixes normal vectors of M at p and maps tangent vectors of M at p into its negative. It follows that the connected component $M^\perp$ of the fixed point set of $s_M \circ s_p$ containing p, which is a totally geodesic submanifold of $\bar{M}$, has tangent space equal to ν_pM at p. This implies that ν_pM is also a Lie triple system and, since $M^\perp$ is a connected component of the fixed point set of an involutive isometry, that $M^\perp$ is reflective.

Conversely, assume that $\bar{M}$ is simply connected, $p \in \bar{M}$, and V is a Lie triple system in $T_p\bar{M}$ so that $V^\perp$ is also a Lie triple system. According to Proposition 9.1.2 there exists a totally geodesic submanifold M of $\bar{M}$ with $p \in M$, $T_pM = V$, that is an orbit of some Lie subgroup of $I^o(\bar{M})$. We denote by s_M the reflection of $\bar{M}$ in M, which is a well-defined smooth map, at least in some open tubular neighborhood U in $\bar{M}$ of some open neighborhood of p in M. We will now show that s_M is an isometry on U. Let $o \in M \cap U$ and ξ a unit normal vector of M at o. Then $q = \exp^\perp(r\xi)$ is in U for sufficiently small $r \in \mathbb{R}_+$, where $\exp^\perp$ denotes the normal exponential map of M. We have to show that

$$s_{M*q} : T_q\bar{M} \to T_{\bar{q}}\bar{M}$$

is a linear isometry, where $\bar{q} = \exp^\perp(-r\xi)$.

Let γ_ξ be the geodesic in $\bar{M}$ with $\gamma_\xi(0) = o$ and $\dot{\gamma}_\xi(0) = \xi$. Then we have

$$T_q\bar{M} = \{Y(r) \mid Y \text{ is an } M\text{–Jacobi vector field along } \gamma_\xi\} ,$$

and similarily

$$T_{\bar{q}}\bar{M} = \{Y(-r) \mid Y \text{ is an } M\text{–Jacobi vector field along } \gamma_\xi\}$$

for sufficiently small r. The differential of s_M at q is determined by

$$s_{M*q}Y(r) = Y(-r)$$

for such M-Jacobi vector fields. Hence, we must show that $||Y(r)|| = ||Y(-r)||$ for all M-Jacobi vector fields along γ_ξ.

Now consider the Jacobi operator $\bar{R}_\xi := \bar{R}(\cdot, \xi)\xi$, which is a self-adjoint endomorphism of $T_o\bar{M}$. As $V^\perp = \nu_p M$ is a Lie triple system in $T_p\bar{M}$ and M is an orbit of a Lie subgroup of $I^o(\bar{M})$, also $\nu_o M$ is a Lie triple system in $T_o\bar{M}$. From this, and since $\xi \in \nu_o M$, it follows that

$$\bar{R}_\xi T_o M \subset T_o M \ , \ \bar{R}_\xi \nu_o M \subset \nu_o M \ .$$

Hence, there exists an orthonormal basis $e_1, \ldots, e_m, f_1, \ldots, f_{n-m}$ of $T_o\bar{M}$ so that

$$e_i \in T_o M \ , \ f_i \in \nu_o M \ , \ \bar{R}_\xi e_i = \lambda_i e_i \ , \ \bar{R}_\xi f_i = \mu_i f_i$$

with some real numbers $\lambda_1, \ldots, \lambda_m, \mu_1, \ldots, \mu_{n-m}$. Using the fact that M is totally geodesic in $\bar{M}$ and that the eigenvalues of $\bar{R}(\cdot, \dot{\gamma}_\xi(t))\dot{\gamma}_\xi(t)$ are constant (since $\bar{\nabla}\bar{R} = 0$), we get that each M-Jacobi vector field Y along γ_ξ is of the form

$$Y(t) = \sum_{i=1}^{m} a_i \cos_{\lambda_i}(t) E_i(t) + \sum_{i=1}^{n-m} b_i \sin_{\mu_i}(t) F_i(t) \ ,$$

with some $a_1, \ldots, a_m, b_1, \ldots, b_{n-m} \in \mathbb{R}$ and where E_i resp. F_i is the $\bar{\nabla}$-parallel vector field along γ_ξ with $E_i(0) = e_i$ resp. $F_i(0) = f_i$. For all such vector fields Y we easily see that $||Y(r)|| = ||Y(-r)||$, by which we now conclude that s_M is a local isometry. Since $\bar{M}$ is connected, complete, simply connected and real analytic, it follows that this local isometry extends to a global isometry, which then, by construction, is the reflection of $\bar{M}$ in M. □

A thorough study and classification of reflective submanifolds in simply connected Riemannian symmetric spaces can be found in a series of papers by D.S.P. Leung [126–129]. We shall illustrate now how this is related to the classification of involutions on Lie groups.

Let $\bar{M}$ be a symmetric space, $o \in \bar{M}$ and (G, K) the corresponding Riemannian symmetric pair. The geodesic symmetry s_o of $\bar{M}$ in o determines the Cartan involution

$$\sigma : G \to G \ , \ g \mapsto s_o g s_o$$

on G. Now let M be a reflective submanifold of $\bar{M}$ with $o \in M$ and let $M^\perp$ be the complete totally geodesic submanifold of $\bar{M}$ tangent to the normal space of M at o. Both M and $M^\perp$ are reflective and the reflections s_M and $s_{M^\perp}$ of $\bar{M}$ in M and $M^\perp$ are involutive isometries of $\bar{M}$, respectively. Thus, we get two involutive automorphisms τ and $\tau^\perp$ on G by

$$\tau : G \to G \ , \ g \mapsto s_M g s_M \quad , \quad \tau^\perp : G \to G \ , \ g \mapsto s_{M^\perp} g s_{M^\perp} \ .$$

It is clear from the construction that $s_o, s_M, s_{M^\perp}$ air pairwise commuting involutive isometries on $\bar{M}$ and the product of any two of them is equal to the third one. This implies that $\sigma, \tau, \tau^\perp$ are also pairwise commuting involutive automorphisms on G and the product of any two of them is equal to the third one.

Conversely, suppose we have given three pairwise commuting involutive automorphisms σ, τ, ρ on G such that the product of any two of them is equal to the third one and where σ is still the Cartan involution of the Riemannian symmetric pair (G, K). We denote by G_σ the fixed point set of σ. Then we have $G_\sigma^o \subset K \subset G_\sigma$. The identity component G_τ^o of G_τ acts isometrically on $\bar{M} = G/K$. Since $\sigma\tau = \tau\sigma$, the group G_τ^o is σ-invariant and hence, its orbit $M = G_\tau^o \cdot o$ is totally geodesic in $\bar{M}$. Analogously, the orbit $N = G_\rho^o \cdot o$ is totally geodesic in $\bar{M}$. We denote by

$$\mathfrak{g} = \mathfrak{k}_\sigma \oplus \mathfrak{p}_\sigma \ , \ \mathfrak{g} = \mathfrak{k}_\tau \oplus \mathfrak{p}_\tau \ , \ \mathfrak{g} = \mathfrak{k}_\rho \oplus \mathfrak{p}_\rho$$

the usual decompositions of $\mathfrak{g}$ into the ± 1-eigenspaces of the corresponding involutions on $\mathfrak{g}$. Since $\tau\rho = \rho\tau = \sigma$ and $T_o\bar{M} = \mathfrak{p}_\sigma$, we have

$$T_oM = \mathfrak{p}_\sigma \cap \mathfrak{p}_\tau = \mathfrak{p}_\sigma \cap \mathfrak{k}_\rho = \nu_oN \ , \ T_oN = \mathfrak{p}_\sigma \cap \mathfrak{p}_\rho = \mathfrak{p}_\sigma \cap \mathfrak{k}_\tau = \nu_oM \ .$$

Thus, these three involutive automorphisms on G induce a pair of reflective submanifolds M and N of $\bar{M}$ so that $M^\perp = N$. It follows that the classification of reflective submanifolds in symmetric spaces is equivalent to the classification of triples σ, τ, ρ of pairwise commuting involutive automorphisms on certain Lie groups such that the product of any two of them is equal to the third one.

The reflective submanifolds in the compact Riemannian symmetric spaces of rank one are

$$\begin{aligned}
S^n &: \{o\}, S^1, \ldots, S^{n-1} \\
\mathbb{R}P^n &: \{o\}, \mathbb{R}P^1, \ldots, \mathbb{R}P^{n-1} \\
\mathbb{C}P^n &: \{o\}, \mathbb{C}P^1, \ldots, \mathbb{C}P^{n-1}, \mathbb{R}P^n \\
\mathbb{H}P^n &: \{o\}, \mathbb{H}P^1, \ldots, \mathbb{H}P^{n-1}, \mathbb{C}P^n \\
\mathbb{O}P^2 &: \{o\}, \mathbb{O}P^1, \mathbb{H}P^2 \ .
\end{aligned}$$

Already from this list we see that not every totally geodesic submanifold is reflective. For instance, $\mathbb{R}P^k$ in $\mathbb{C}P^n$ is not reflective for $k = 1, \ldots, n-1$. The motivation for Leung to study these submanifolds was to generalize the classical Schwartz reflection principle for minimal surfaces in $\mathbb{R}^3$ to certain kinds of minimal submanifolds in Riemannian symmetric spaces. Leung established the complete classification of reflective submanifolds in Riemannian symmetric spaces.

e) Duality and totally geodesic submanifolds

A very useful observation is that totally geodesic submanifolds are preserved under duality. To be precise, let (G, K) be a Riemannian symmetric pair so that G/K is a simply connected Riemannian symmetric space of compact type or of noncompact type, respectively. Consider the complexification $\mathfrak{g}^\mathbb{C} = \mathfrak{g} + i\mathfrak{g}$ of $\mathfrak{g}$ and the Cartan decomposition $\mathfrak{g} = \mathfrak{k} \oplus \mathfrak{m}$ of $\mathfrak{g}$. Then $\mathfrak{g}^* = \mathfrak{k} \oplus i\mathfrak{m}$ is a real Lie subalgebra of $\mathfrak{g}^\mathbb{C}$ with respect to the induced Lie algebra structure. Let G^* be the real Lie subgroup of $G^\mathbb{C}$ with Lie algebra $\mathfrak{g}^*$. Then G^*/K is a simply connected Riemannian symmetric space of noncompact type or of compact type, respectively, with Cartan decomposition $\mathfrak{g}^* = \mathfrak{k} \oplus i\mathfrak{m}$. It is straightforward to check that V is a Lie triple system in $\mathfrak{m}$ if and only if iV is a Lie triple system in $i\mathfrak{m}$. In this manner, we get a one-to-one correspondence between the totally geodesic submanifolds of G/K and its dual symmetric space G^*/K. As an application, we get the following list of the maximal totally geodesic submanifolds in noncompact Riemannian symmetric spaces of rank one, namely the hyperbolic spaces over $\mathbb{R}$, $\mathbb{C}$, $\mathbb{H}$ and $\mathbb{O}$:

$$\begin{aligned}
\mathbb{R}H^n &: \mathbb{R}H^{n-1} \\
\mathbb{C}H^n &: \mathbb{C}H^{n-1}, \mathbb{R}H^n \\
\mathbb{H}H^n &: \mathbb{H}H^{n-1}, \mathbb{C}H^n \\
\mathbb{O}H^2 &: \mathbb{O}H^1, \mathbb{H}H^2 .
\end{aligned}$$

9.2 Totally umbilical submanifolds and extrinsic spheres

a) Circles

We already discussed the existence and uniqueness of circles in Riemannian manifolds in Proposition 8.4.3. It is well known that each geodesic in a Riemannian symmetric space is an orbit of a one-parameter group of isometries. It is easy to show that each circle in $\mathbb{R}^n$, S^n, $\mathbb{R}P^n$ and $\mathbb{R}H^n$ is an orbit of a one-parameter group of isometries. S. Maeda and Y. Ohnita [133] proved that this is also true for circles in $\mathbb{C}P^n$ and $\mathbb{C}H^n$. This was extended to all two-point homogeneous spaces by K. Mashimo and K. Tojo [136]. In fact, they proved that this property characterizes two-point homogeneous spaces.

THEOREM 9.2.1 (Mashimo, Tojo)
Let $\bar{M}$ be a Riemannian homogeneous space. Then each circle in $\bar{M}$ is an orbit of a one-parameter group of isometries if and only if $\bar{M}$ is a two-point homogeneous space.

The "only if" part is proved by showing that the isotropy group at some point acts transitively on unit tangent vectors at that point.

b) The classification problem for extrinsic spheres

One step toward the classification of extrinsic spheres of dimension ≥ 2 in Riemannian symmetric spaces is the following result:

THEOREM 9.2.2
Let M be an extrinsic sphere in a Riemannian symmetric space $\bar{M}$ with $\dim M \geq 2$. Then there exists a connected, complete, totally geodesic submanifold N of $\bar{M}$ with constant curvature so that M is contained in N as an extrinsic sphere with codimension one.

PROOF Let M be an extrinsic sphere in a Riemannian symmetric space $\bar{M}$ and $o \in M$. According to Theorem 8.4.4, M is uniquely determined by T_oM and the mean curvature vector H_o by running along circles in $\bar{M}$ tangent to T_oM and whose second derivative at o is H_o. As a consequence, we see that when there is a connected, complete, totally geodesic submanifold N of $\bar{M}$ with $o \in N$ and $T_oM \oplus \mathbb{R}H_o \subset T_oN$, then $M \subset N$. According to Corollary 9.1.1, we thus have to show that $T_oM \oplus \mathbb{R}H_o$ is a curvature-invariant subspace of $T_o\bar{M}$.

From Proposition 8.4.1, we already know that T_oM is a curvature-invariant subspace of $T_o\bar{M}$. If the codimension of M is one, we deduce from Theorem 9.1.4 that $\bar{M}$ has constant curvature. Since we treated this case in Theorem 2.6.2, we assume from now on that the codimension of M is at least two.

Let η be a normal vector of M at o that is perpendicular to the mean curvature vector H_o of M at o. Then the shape operator A_η of M with respect to η vanishes. Further, since the mean curvature vector field H of M is parallel in the normal bundle of M, we have $R^\perp(X,Y)H_o = 0$ for all $X, Y \in T_oM$, where $R^\perp$ is the normal curvature tensor of M. The Ricci equation therefore implies

$$\langle \bar{R}(X,Y)H_o, \eta\rangle = 0$$

for all $X, Y \in T_oM$ and all $\eta \in \nu_oM$ which are perpendicular to H_o, and hence

$$\bar{R}(T_oM, T_oM)H_o \subset T_oM \oplus \mathbb{R}H_o\,. \tag{9.1}$$

So far, all arguments are true in the general case of a Riemannian manifold. We will now use the assumption that $\bar{M}$ is a Riemannian symmetric space. In this case, its curvature tensor $\bar{R}$ is parallel, which means that

$$\begin{aligned} X\langle \bar{R}(Y,Z)W,U\rangle = \langle \bar{R}(\bar{\nabla}_XY,Z)W,U\rangle + \langle \bar{R}(Y,\bar{\nabla}_XZ)W,U\rangle \\ +\langle \bar{R}(Y,Z)\bar{\nabla}_XW,U\rangle + \langle \bar{R}(Y,Z)W,\bar{\nabla}_XU\rangle \end{aligned} \tag{9.2}$$

for all tangent vector fields X, Y, Z, U, W on $\bar{M}$.

We now assume that X, Y, Z, W are tangent to M and $U = \eta$ is a normal vector field of M that is also perpendicular to H and such that $\nabla^\perp\eta$ vanishes at o. Then $\bar{\nabla}_X\eta$ vanishes at o. The left-hand side of equation (9.2) then vanishes because T_oM is a curvature-invariant subspace of $T_o\bar{M}$. Using the equations

$$\bar{\nabla}_XY = \nabla_XY + \langle X,Y\rangle H\ ,\ \ldots\ ,$$

and (9.1), equation (9.2) then implies

$$0 = \langle X, Y\rangle\langle \bar{R}(H,Z)W,\eta\rangle + \langle X, Z\rangle\langle \bar{R}(Y,H)W,\eta\rangle \ .$$

Since $\dim M \geq 2$, we can choose $X = Y$ to be of unit length and Z perpendicular to X. The previous equation then reduces to

$$0 = \langle \bar{R}(H,Z)W,\eta\rangle \ .$$

Since this holds for all Z, W (by varying with X) and all η, this implies

$$\bar{R}(H_o, T_oM)T_oM \subset T_oM \oplus \mathbb{R}H_o \ . \tag{9.3}$$

We now choose X, Y, Z tangent to M, put $W = H$, and let $U = \eta$ be a normal vector field of M that is also perpendicular to H and such that $\nabla^\perp\eta$ vanishes at o. The left-hand side of (9.2) vanishes because of equation (9.1), which holds at each point of M because o was chosen to be arbitrary. For the right-hand side, first observe that

$$\bar{\nabla}_X H = -A_H X = -\langle H, H\rangle X$$

and, once again,

$$\bar{\nabla}_X \eta = -A_\eta X + \nabla^\perp_X \eta = 0 \text{ at } o \ .$$

In a similar way, we then obtain

$$0 = \langle X, Y\rangle\langle \bar{R}(H,Z)H,\eta\rangle + \langle X, Z\rangle\langle \bar{R}(Y,H)H,\eta\rangle \ ,$$

and conclude from this, using again the fact that $\dim M \geq 2$, that

$$\bar{R}(H_o, T_oM)H_o \subset T_oM \oplus \mathbb{R}H_o \ . \tag{9.4}$$

Eventually, equations (9.1), (9.3) and (9.4), the algebraic curvature identities, and the fact that T_oM is a curvature-invariant subspace of $T_o\bar{M}$, imply that $T_oM \oplus \mathbb{R}H_o$ is a curvature-invariant subspace of $T_o\bar{M}$. Thus, we have shown that there exists a connected, complete, totally geodesic submanifold N of $\bar{M}$ so that M is contained in N as an extrinsic sphere with codimension one. It remains to prove that N has constant curvature.

To establish this, we first choose $X = Y, Z = W$ to be orthonormal and tangent and put $U = H$ in equation (9.2). The left-hand side vanishes because T_oM is curvature-invariant and o was chosen arbitrarily, and again using $\bar{\nabla}_X H = -\langle H, H\rangle X$ we get

$$0 = \langle \bar{R}(H,Z)Z,H\rangle - \langle H, H\rangle\langle \bar{R}(X,Z)Z,X\rangle \ .$$

If we denote by $\bar{K}(A,B)$ the sectional curvature of $\bar{M}$ with respect to the two-plane spanned by A and B, this implies

$$0 = \langle H, H\rangle(\bar{K}(H,Z) - \bar{K}(X,Z)) \ .$$

Since H is nonzero everywhere, this implies

$$\bar{K}(X,Z) = \bar{K}(H,Z) \tag{9.5}$$

whenever X and Z are orthonormal and tangent to M. An arbitrary 2-plane tangent to N is spanned by orthonormal vectors of the form X and $\cos(\alpha)Z+\sin(\alpha)H/|H|$ with X,Z orthonormal and tangent to M. A straightforward calculation, using equation (9.5) and once again the fact that T_oM is curvature-invariant, yields

$$\langle \bar{R}(X,\cos(\alpha)Z+\sin(\alpha)H/|H|)(\cos(\alpha)Z+\sin(\alpha)H/|H|),X\rangle = \bar{K}(X,H) .$$

From this we conclude that, at each point of N, the sectional curvature in $\bar{M}$ of 2-planes tangent to N is independent of the 2-plane. As N is totally geodesic in $\bar{M}$, we thus get that the sectional curvature of N depends only on the point. But N is homogeneous, since it is a connected, complete, totally geodesic submanifold of a symmetric space and hence itself a symmetric space. Thus, we eventually conclude that N has constant sectional curvature. □

Therefore, the classification of extrinsic spheres in Riemannian symmetric spaces can be worked out in two steps. First, classify the totally geodesic submanifolds with constant curvature in a Riemannian symmetric space. For symmetric spaces of compact type, one can apply the results of T. Nagano and M. Sumi [152] mentioned in the previous section. Using duality between symmetric spaces of compact and noncompact type, the classification can be transferred to symmetric spaces of noncompact type. In the second step, one has to classify the extrinsic spheres in spaces of constant curvature with codimension one. This has been done explicitly in Theorem 2.6.2.

c) The classification problem for totally umbilical submanifolds

The classification of totally umbilical submanifolds of dimension > 2 in Riemannian symmetric spaces has been achieved by Y.A. Nikolaevskii [166]. Basically, these submanifolds live in totally geodesically embedded products of spaces of constant curvature. A partial classification was previously obtained by B.Y. Chen in [51]. In particular, Chen proved:

THEOREM 9.2.3
If an irreducible Riemannian symmetric space $\bar{M}$ contains a totally umbilical hypersurface M then both M and $\bar{M}$ have constant curvature.

A special case of this result is Theorem 9.1.4.

9.3 Symmetric submanifolds

a) Symmetry versus parallel second fundamental form

In Proposition 8.5.1, we proved that the second fundamental form of a symmetric submanifold is parallel and that tangent to each normal space there exists a totally geodesic submanifold of the ambient space. For simply connected Riemannian symmetric spaces, Naitoh [156] proved that the converse also holds.

PROPOSITION 9.3.1 (Naitoh)
Let M be a complete submanifold of a simply connected Riemannian symmetric space $\bar{M}$. Then M is a symmetric submanifold if and only if the second fundamental form of M is parallel and each normal space $\nu_p M$ of M is a curvature-invariant subspace of $T_p\bar{M}$.

PROOF First note that we have already given the proof of this result in Theorem 3.7.2 in the special case when $\bar{M}$ is the Euclidean space.

Suppose that the second fundamental form of M is parallel and that each normal space of M is curvature-invariant. We fix a point $p \in M$ and define a linear isometry λ on $T_p\bar{M}$ by

$$\lambda : T_p\bar{M} \to T_p\bar{M} \ , \ X \mapsto \begin{cases} -X \ , & \text{if } X \in T_pM \\ X \ , & \text{if } X \in \nu_pM \ . \end{cases}$$

Since the second fundamental form of M is parallel, the Codazzi equation implies that T_pM is a curvature-invariant subspace of $T_p\bar{M}$. By assumption, the normal space ν_pM is also a curvature-invariant subspace of $T_p\bar{M}$. The algebraic curvature identities therefore imply

$$\begin{array}{ll} \bar{R}(T_pM, T_pM)T_pM \subset T_pM \ , & \bar{R}(T_pM, T_pM)\nu_pM \subset \nu_pM \ , \\ \bar{R}(T_pM, \nu_pM)T_pM \subset \nu_pM \ , & \bar{R}(T_pM, \nu_pM)\nu_pM \subset T_pM \ , \\ \bar{R}(\nu_pM, \nu_pM)T_pM \subset T_pM \ , & \bar{R}(\nu_pM, \nu_pM)\nu_pM \subset \nu_pM \ . \end{array} \tag{9.6}$$

From this, we easily derive that λ leaves $\bar{R}$ invariant, that is,

$$\lambda(\bar{R}(X,Y)Z) = \bar{R}(\lambda(X), \lambda(Y))\lambda(Z))$$

for all $X, Y, Z \in T_p\bar{M}$. This implies (cf. [99]) that there exists a local isometry Λ of $\bar{M}$ with $\Lambda(p) = p$ and whose differential at p coincides with the linear isometry λ. But, since $\bar{M}$ is connected, complete, simply connected, and real analytic, Λ can be extended to a global isometry σ_p of $\bar{M}$. By construction, we have

$$\sigma_p(p) = p \ \text{ and } (\sigma_p)_*(v) = \begin{cases} -v, & v \in T_pM \\ v, & v \in \nu_pM \end{cases} .$$

A result by Strübing [204] shows that there exists an open neighborhood U of p in M such that $\sigma_p(U) \subset U$. Completeness of M then eventually implies that $\sigma_p(M) = M$, because s_p reflects in p the geodesics in M through p. □

b) Totally geodesic symmetric submanifolds

The classification of totally geodesic symmetric submanifolds in Riemannian symmetric spaces follows from the one of reflective submanifolds (see Section 9.1).

PROPOSITION 9.3.2

A totally geodesic submanifold M of a simply connected Riemannian symmetric space $\bar{M}$ is symmetric if and only if it is a reflective submanifold.

PROOF Each reflective submanifold has obviously parallel second fundamental form and, by definition, each of its normal spaces $\nu_p M$ is a curvature-invariant subspace of $T_p\bar{M}$. According to Proposition 9.3.1, M is a symmetric submanifold.

Conversely, assume that M is a totally geodesic symmetric submanifold. Again, Proposition 9.3.1 tells us that each normal space $\nu_p M$ of M is a curvature-invariant subspace of $T_p\bar{M}$. Thus, both T_pM and $\nu_p M$ are Lie triple systems for all $p \in M$, which means that M is a reflective submanifold of $\bar{M}$. □

c) Grassmann geometries

The obvious question now is: Are there any non-totally geodesic symmetric submanifolds in a given Riemannian symmetric space? We will discuss this question in the framework of Grassmann geometries.

Let $\bar{M}$ be a Riemannian manifold. The isometry group $I(\bar{M})$ of $\bar{M}$ acts in a natural way on the Grassmann bundle $G_m(T\bar{M})$ of m-planes in the tangent bundle $T\bar{M}$. An m-dimensional connected submanifold M of $\bar{M}$ belongs to the (m-dimensional) *Grassmann geometry* of $\bar{M}$ if all its tangent spaces lie in the same orbit of the action of $I(\bar{M})$ on $G_m(T\bar{M})$. For example, any homogeneous submanifold of $\bar{M}$ belongs to some Grassmann geometry of $\bar{M}$. If M belongs to some Grassmann geometry of $\bar{M}$, the *Grassmann geometry associated to* M is the set $\mathfrak{G}(M, \bar{M})$ of all connected submanifolds of $\bar{M}$ whose tangent spaces lie in the same orbit as those of M. For instance, the Grassmann geometry $\mathfrak{G}(S^m, S^n)$ associated to an m-sphere in S^n is the geometry of all m-dimensional submanifolds of S^n. Also, the Grassmann geometry $\mathfrak{G}(\mathbb{C}P^m, \mathbb{C}P^n)$ associated to an m-dimensional complex projective subspace in $\mathbb{C}P^n$ is the geometry of all m-dimensional complex submanifolds in $\mathbb{C}P^n$.

Now suppose that M is an m-dimensional symmetric submanifold of $\bar{M}$. Let p_1 and p_2 be two different points in M. Connecting p_1 and p_2 by a geodesic in M, the symmetry σ_{p_0} at the midpoint p_0 on γ between p_1 and p_2 is an isometry of $\bar{M}$ leaving M invariant and interchanging p_1 and p_2. This shows in particular that M is a homogeneous submanifold and hence belongs to the m-dimensional Grassmann geometry of $\bar{M}$.

From now on we suppose that $\bar{M}$ is a Riemannian symmetric space. As we have seen above, each tangent space and each normal space of a symmetric submanifold of $\bar{M}$ is a Lie triple system. This implies:

PROPOSITION 9.3.3
Each symmetric submanifold of a Riemannian symmetric space $\bar{M}$ belongs to the Grassmann geometry $\mathfrak{G}(M, \bar{M})$ associated to a suitable reflective submanifold M of $\bar{M}$.

This proposition motivates investigation of the Grassmann geometries associated to reflective submanifolds of Riemannian symmetric spaces in more detail. For simply connected Riemannian symmetric spaces of compact type this was done by H. Naitoh in a series of papers [157–161]. His proof also works for the Riemannian symmetric spaces of noncompact type.

THEOREM 9.3.4 (Naitoh)
All Grassmann geometries associated to reflective submanifolds of simply connected irreducible Riemannian symmetric spaces have only totally geodesic submanifolds with the following exceptions:

1. *$\mathfrak{G}(S^m, S^n)$ and $\mathfrak{G}(\mathbb{R}H^m, \mathbb{R}H^n)$ $(1 \leq m \leq n-1)$, that is, the geometry of m-dimensional submanifolds of S^n resp. $\mathbb{R}H^n$.*
2. *$\mathfrak{G}(\mathbb{C}P^m, \mathbb{C}P^n)$ and $\mathfrak{G}(\mathbb{C}H^m, \mathbb{C}H^n)$ $(1 \leq m \leq n-1)$, that is, the geometry of m-dimensional complex submanifolds of $\mathbb{C}P^n$ resp. $\mathbb{C}H^n$.*
3. *$\mathfrak{G}(\mathbb{R}P^n, \mathbb{C}P^n)$ and $\mathfrak{G}(\mathbb{R}H^n, \mathbb{C}H^n)$, that is, the geometry of n-dimensional totally real submanifolds of $\mathbb{C}P^n$ resp. $\mathbb{C}H^n$.*
4. *$\mathfrak{G}(\mathbb{C}P^n, \mathbb{H}P^n)$ and $\mathfrak{G}(\mathbb{C}H^n, \mathbb{H}H^n)$, that is, the geometry of n-dimensional totally complex submanifolds of $\mathbb{H}P^n$ resp. $\mathbb{H}H^n$.*
5. *$\mathfrak{G}(M, \bar{M})$, where the rank of $\bar{M}$ is greater than one and the isotropy representation of $\bar{M}$ has a symmetric orbit M, that is, the geometries associated with irreducible symmetric R-spaces and their noncompact dual geometries.*

Naitoh also obtained a decomposition theorem that settles the reducible case, see [159].

Still remaining is the classification of the symmetric submanifolds in these Grassmann geometries. This has been carried out by various authors whose results we will now describe (see also [163] for a survey about symmetric submanifolds of symmetric spaces of rank one). We already discussed the symmetric submanifolds of spheres in Section 3.7.

d) Symmetric complex submanifolds of $\mathbb{C}P^n$

In this part, we describe the classification of symmetric complex submanifolds in complex projective spaces. All these submanifolds arise from so-called canonical embeddings of certain Hermitian symmetric spaces.

Let $\mathfrak{g}$ be a complex simple Lie algebra, $\mathfrak{h}$ a Cartan subalgebra of $\mathfrak{g}$ and Δ the corresponding set of roots. We choose a Weyl canonical basis $\{H_\alpha, X_\alpha\}$, $\alpha \in \Delta$, of $\mathfrak{g}$ and define a compact real form $\mathfrak{g}_u$ of $\mathfrak{g}$ by

$$\mathfrak{g}_u := \sum_{\alpha\in\Delta} \mathbb{R}(iH_\alpha) + \sum_{\alpha\in\Delta} \mathbb{R}(X_\alpha + X_{-\alpha}) + \sum_{\alpha\in\Delta} \mathbb{R}i(X_\alpha - X_{-\alpha})\,.$$

Let $\alpha_1, \ldots, \alpha_l \in \Delta$ be a set of simple roots. For each $j \in \{1, \ldots, l\}$ we put

$$\Delta_j := \{\alpha = \sum_{\nu=1}^{l} n_\nu \alpha_\nu \in \Delta \mid n_j < 0\},$$

and define a complex Lie subalgebra $\mathfrak{l}_j$ of $\mathfrak{g}$ by

$$\mathfrak{l}_j := \mathfrak{h} + \sum_{\alpha\in\Delta\setminus\Delta_j} \mathbb{C}X_\alpha$$

and a Lie subalgebra $\mathfrak{h}_{u,j}$ of $\mathfrak{g}_u$ by

$$\mathfrak{h}_{u,j} := \mathfrak{g}_u \cap \mathfrak{l}_j\,.$$

Let G be the simply connected complex Lie group with Lie algebra $\mathfrak{g}$ and L_j the connected complex Lie subgroup of G with Lie algebra $\mathfrak{l}_j$. Then G/L_j is a simply connected compact homogeneous complex manifold. Let G_u and $H_{u,j}$ be the connected real Lie subgroups of G with Lie algebra $\mathfrak{g}_u$ and $\mathfrak{h}_{u,j}$, respectively. The inclusion $G_u \hookrightarrow G$ induces a homeomorphism from $M_j := G_u/H_{u,j}$ onto G/L_j turning M_j into a *C-space*, that is, a simply connected compact complex homogeneous space, on which G_u acts transitively by holomorphic transformations. Note that the second Betti number $b_2(M_j)$ of M_j is one. Conversely, as was shown by H.C. Wang [236], every irreducible C-space M with $b_2(M) = 1$ arises in this manner.

We now describe a family of holomorphic embeddings of M_j into complex projective spaces. Let p be a positive integer and $\rho : G \to \mathfrak{gl}(\mathbb{C}^{n(p)+1})$ the irreducible representation of G with highest weight $p\Lambda_j$, where Λ_j is the fundamental weight corresponding to the simple root α_j. Denote by $V \subset \mathbb{C}^{n(p)+1}$ the one-dimensional eigenspace of ρ corresponding to $p\Lambda_j$. Then the map

$$G \to \mathbb{C}P^{n(p)}\;,\; g \mapsto \rho(g)V$$

induces a full holomorphic embedding of $M_j = G_u/H_{u,i} = G/L_j$ into $\mathbb{C}P^{n(p)}$, which is called the *p-th canonical embedding* of M_j into a complex projective space. The submanifold M_j of $\mathbb{C}P^{n(p)}$ is the unique compact orbit of the action of the complex Lie group G on $\mathbb{C}P^{n(p)}$. The dimension $n(p)$ can be calculated explicitly by means of Weyl's dimension formula. The induced metric on $M_j \subset \mathbb{C}P^{n(p)}$ is Kähler-Einstein. Note that M_j is Hermitian symmetric if and only if $n_j = -1$ for all roots $\alpha \in \Delta_j$, and every Hermitian symmetric space arises in this manner.

It follows from Proposition 9.3.1 that a complete complex submanifold M of $\mathbb{C}P^n$ is symmetric if and only if its second fundamental form is parallel. The complex submanifolds of $\mathbb{C}P^n$ with parallel second fundamental form were classified by H. Nakagawa and R. Takagi [165].

THEOREM 9.3.5 (Nakagawa-Takagi)
A complete complex submanifold M of $\mathbb{C}P^N$ is a symmetric submanifold if and only if it is either a totally geodesic complex projective subspace or congruent to one of the following models:

TABLE 9.1: Symmetric complex submanifolds of $\mathbb{C}P^N$

M	N	embedding	Remarks
$\mathbb{C}P^a \times \mathbb{C}P^b$	$ab + a + b$	$F_1 \otimes F_1$	$1 \leq a \leq b$
$\mathbb{C}P^n$	$\frac{1}{2}(n+1)(n+2) - 1$	F_2	$n \geq 2$
$G_2(\mathbb{C}^n)$	$\frac{1}{2}n(n-1) - 1$	F_1	$n \geq 5$
$G_2^+(\mathbb{R}^{n+1})$	n	F_1	$n \geq 1$
$SO(10)/U(5)$	15	F_1	
$E_6/T \cdot Spin(10)$	26	F_1	

Here, F_1 resp. F_2 denotes the first resp. second canonical embedding, and $F_1 \otimes F_1$ is the embedding that is induced by the exterior tensor product of the two representations associated to the first canonical embedding F_1. The embedding of $\mathbb{C}P^a \times \mathbb{C}P^b$ is also known as the Segre embedding and is explicitly given by

$$([z_0 : \ldots : z_a], [w_o : \ldots w_b]) \mapsto [z_o w_o : \ldots : z_\nu w_\mu : \ldots z_a w_b]$$

(all possible products of the coordinates). The embedding of $\mathbb{C}P^n$ is known as the second Veronese embedding and is explicitly given by

$$[z_0 : \ldots : z_n] \mapsto [z_0^2 : \sqrt{2} z_o z_1 : \ldots : \sqrt{2} z_{n-1} z_n : z_n^2] \,.$$

The embedding of $G_2(\mathbb{C}^n)$ is the Plücker embedding, and the one of $G_2^+(\mathbb{R}^{n+1})$ gives the complex quadric in $\mathbb{C}P^n$ that is determined by the equation $z_0^2 + \ldots + z_n^2 = 0$.

e) Symmetric totally real submanifolds of $\mathbb{C}P^n$

The classification of n-dimensional totally real symmetric submanifolds in $\mathbb{C}P^n$ was established by H. Naitoh [153] (for the irreducible case) and by H. Naitoh and M. Takeuchi [162] (for the general case). The reflective submanifold in the corresponding Grassmann geometry is the totally geodesic real projective spaces $\mathbb{R}P^n$ in $\mathbb{C}P^n$. The crucial observation for the classification is that an n-dimensional totally real symmetric submanifold M of $\mathbb{C}P^n$ is symmetric if and only if its inverse image $\tilde{M}$ under the Hopf map $S^{2n+1} \to \mathbb{C}P^n$ is a symmetric submanifold of the sphere. This shows the relation with symmetric R-spaces. In the irreducible case, the relevant symmetric R-spaces $\tilde{M}$ are $U(n)/SO(n)$, $U(2n)/Sp(n)$, $U(n)$ and $(T \cdot E_6)/F_4$. Among all the standard embeddings of irreducible symmetric R-spaces they are characterized by the property that the dimension of the ambient Euclidean space is twice the dimension of the symmetric R-space. So, if $n + 1$ denotes the dimension of the symmetric R-space $\tilde{M}$, its image lies in the sphere $S^{2n+1} \subset \mathbb{R}^{2n+2} = \mathbb{C}^{n+1}$. It turns out that $\tilde{M}$ is invariant under the canonical S^1-action on S^{2n+1} and hence projects via the Hopf map to an n-dimensional submanifold M of $\mathbb{C}P^n$. Each of these submanifolds is a totally real symmetric submanifold of $\mathbb{C}P^n$ that is the image of an embedding of the following Riemannian symmetric spaces M^n into $\mathbb{C}P^n$:

TABLE 9.2: Symmetric totally real submanifolds of $\mathbb{C}P^N$

M	$N = \dim M$	Remarks
$SU(n)/SO(n)$	$\frac{1}{2}(n-1)(n+2)$	$n \geq 3$
$SU(n)$	$n^2 - 1$	$n \geq 3$
$SU(2n)/Sp(n)$	$(n-1)(2n+1)$	$n \geq 3$
E_6/F_4	26	

These embeddings can be described explicitly in an elementary way. Consider the natural action of $SL(n, \mathbb{C})$ on $J_n(\mathbb{R}) \otimes \mathbb{C}$, the complexification of the real Jordan algebra $J_n(\mathbb{R})$ of all symmetric $n \times n$-matrices with real coefficients, given by

$$(A, X) \mapsto AXA^T$$

for $A \in SL(n, \mathbb{C})$ and $X \in J_n(\mathbb{R}) \times \mathbb{C}$. The complex dimension of $J_n(\mathbb{R}) \otimes \mathbb{C}$ is $n(n+1)/2$, and hence this action induces an action of $SL(n, \mathbb{C})$ on $\mathbb{C}P^N$ with $N = n(n+1)/2 - 1 = (n-1)(n+2)/2$. This action has exactly n orbits that are parametrized by the rank of the matrices. The subgroup of $SL(n, \mathbb{C})$ preserving complex conjugation on $\mathbb{C}P^N$ is $SL(n, \mathbb{R})$. Now fix a maximal compact subgroup $SO(n)$ of $SL(n, \mathbb{R})$. The restriction to $SO(n, \mathbb{C})$ of the action of $SL(n, \mathbb{C})$ on $J_n(\mathbb{R}) \otimes \mathbb{C}$ splits off a one-dimensional trivial factor corresponding to the trace. This means that $SO(n, \mathbb{C})$, and hence $SO(n)$, fixes the point o in $\mathbb{C}P^N$ given by complex scalars of the identity matrix in $J_n(\mathbb{R}) \otimes \mathbb{C}$. The maximal compact subgroup $SO(n)$ of $SL(n, \mathbb{R})$ determines a maximal compact subgroup $SU(n)$ of $SL(n, \mathbb{C})$. The orbit of the action of $SU(n)$ through o gives an embedding of $SU(n)/SO(n)$ in $\mathbb{C}P^N$ as a totally real symmetric submanifold of real dimension N. The other three embeddings can be constructed in a similar fashion by using the real Jordan algebras $J_n(\mathbb{C})$, $J_n(\mathbb{H})$ and $J_3(\mathbb{O})$. The corresponding subgroups are

TABLE 9.3: Some subgroups of some complex Lie groups

$SL(n,\mathbb{C})$	$SL(n,\mathbb{C}) \times SL(n,\mathbb{C})$	$SL(2n,\mathbb{C})$	$E_6(\mathbb{C})$
$SL(n,\mathbb{R})$	$SL(n,\mathbb{C})$	$SL(n,\mathbb{H})$	E_6^{-26}
$SO(n)$	$SU(n)$	$Sp(n)$	F_4
$SU(n)$	$SU(n) \times SU(n)$	$SU(2n)$	E_6

THEOREM 9.3.6 (Naitoh)

A complete irreducible totally real submanifold M^n of $\mathbb{C}P^n$ is symmetric if and only if it is a totally geodesic real projective subspace $\mathbb{R}P^n \subset \mathbb{C}P^n$, or if it is congruent to one of the embeddings described above.

H. Naitoh and M. Takeuchi proved in [162] that each totally real symmetric submanifold M^n of $\mathbb{C}P^n$ is basically a product of the irreducible submanifolds discussed above and a flat torus. A suitable product of $n+1$ circles in S^{2n+1} projects via the Hopf map to a flat torus T^n embedded in $\mathbb{C}P^n$ as a totally real symmetric submanifold. Naitoh and Takeuchi gave in [162] a unifying description of all symmetric submanifolds in the Grassmann geometry $\mathfrak{G}(\mathbb{R}P^n, \mathbb{C}P^n)$ using the Shilov boundary of symmetric bounded domains of tube type.

f) Symmetric totally complex submanifolds of $\mathbb{H}P^n$

The symmetric totally complex submanifolds of $\mathbb{H}P^n$ have been classified by K. Tsukada [226]. The reflective submanifold in the corresponding Grassmann geometry is the totally geodesic $\mathbb{C}P^n \subset \mathbb{H}P^n$. A basic tool for the classification is the twistor map $\mathbb{C}P^{2n+1} \to \mathbb{H}P^n$. Consider $\mathbb{H}^{n+1}$ as a (right) vector space and pick a unit quaternion, say i, which turns $\mathbb{H}^{n+1}$ into a complex vector space $\mathbb{C}^{2n+2}$. The twistor map $\mathbb{C}P^{2n+1} \to \mathbb{H}P^n$ maps a complex line in $\mathbb{C}^{2n+2}$ to the quaternionic line in $\mathbb{H}^{n+1}$ spanned by it. The fiber over each point is a complex projective line $\mathbb{C}P^1 \subset \mathbb{C}P^{2n+1}$. Alternatively, the set of all almost Hermitian structures in the quaternionic Kähler structure of a quaternionic Kähler manifold $\bar{M}$ forms the so-called twistor space Z of $\bar{M}$, and the natural projection $Z \to \bar{M}$ is the so-called twistor map onto $\bar{M}$. In the case of $\mathbb{H}P^n$ the twistor space is just $\mathbb{C}P^{2n+1}$.

Now let M be a non-totally geodesic symmetric totally complex submanifold of $\mathbb{H}P^n$ belonging to the Grassmann geometry $\mathfrak{G}(\mathbb{C}P^n, \mathbb{H}P^n)$. The first step in the classification is to show that M is a Hermitian symmetric space with respect to a Kähler structure that is induced from the quaternionic Kähler structure of $\mathbb{H}P^n$. Then one shows that M can be lifted to a Kähler immersion into the twistor space $\mathbb{C}P^{2n+1}$. The main part of the proof is then to show, using representation theory of complex semisimple Lie algebras, that this lift is one of the following embeddings in $\mathbb{C}P^{2n+1}$:

TABLE 9.4: Symmetric totally complex submanifolds of $\mathbb{H}P^n$

M	$n = \dim_{\mathbb{C}} M$	embedding
$\mathbb{C}P^1 \times G_2^+(\mathbb{R}^m)$ $(m \geq 3)$	$m-1$	$F_1 \otimes F_1$
$Sp(3)/U(3)$	6	F_1
$G_3(\mathbb{C}^6)$	9	F_1
$SO(12)/U(6)$	15	F_1
$E_7/(T \cdot E_6)$	27	F_1

In the first case, the embedding is via the exterior tensor product of the first canonical embedding of each factor; in the other cases, it is the first canonical embedding. Note that, in the first case, the submanifold is isometric to $\mathbb{C}P^1 \times \mathbb{C}P^1$ for $m = 3$ and isometric to $\mathbb{C}P^1 \times \mathbb{C}P^1 \times \mathbb{C}P^1$ for $m = 4$. The embedding of $G_3(\mathbb{C}^6)$ into $\mathbb{C}P^{19}$ is the Plücker embedding. The image of each of these embeddings under the Hopf map $\mathbb{C}P^{2n+1} \to \mathbb{H}P^n$ is indeed an n-dimensional symmetric totally complex submanifold of $\mathbb{H}P^n$. Tsukada proved:

THEOREM 9.3.7 (Tsukada)
A complete n-dimensional totally complex submanifold of $\mathbb{H}P^n$ is symmetric if and only if it is either a totally geodesic $\mathbb{C}P^n$ or congruent to one of the examples described above.

g) Symmetric submanifolds associated with irreducible symmetric R-spaces

The pairs $(M, \bar{M})$ mentioned in part 5 of Theorem 9.3.4 are, for irreducible $\bar{M}$, precisely the pairs $(K \cdot X, G)$ and $(K \cdot X, G/K)$ in the two tables of the classification of irreducible symmetric R-spaces that can be found at the end of the appendix. The embedding of M in $\bar{M}$ can be described as follows. Write $\bar{M} = G/K$ with (G, K) a symmetric pair and put $o = eK \in \bar{M}$. Let $\mathfrak{g} = \mathfrak{k} \oplus \mathfrak{p}$ be the corresponding Cartan decomposition of $\mathfrak{g}$. Then there exists an element $Z \in \mathfrak{p}$ so that the eigenvalues of $ad(Z)$ are $+1, 0, -1$. The element Z determines a closed geodesic γ in $\bar{M}$. The antipodal point q to o on γ is a pole of o, that is, a fixed point of the action of K on $\bar{M}$. The reflective submanifold M is the centrosome of o and q, that is, the orbit of K through the midpoint on γ between o and q (it does not matter which of the two possible midpoint one chooses). The orbits of K through the other points on γ and distinct from o and q are non-totally geodesic symmetric submanifolds of $\bar{M}$ belonging to the Grassmann geometry $\mathfrak{G}(M, \bar{M})$. In this way, we get a one-parameter family of non-congruent symmetric submanifolds of $\bar{M}$, and every symmetric submanifold in $\mathfrak{G}(M, \bar{M})$ arises in this way up to congruence. In particular, any non-totally geodesic symmetric submanifold of $\bar{M}$ arises as an orbit of the action of the isotropy group at a suitable point. It is worthwhile to mention that, among the reflective submanifolds in $\bar{M}$, the symmetric R-spaces are precisely those for which the totally geodesic submanifolds tangent to the normal spaces of M are locally reducible with a one-dimensional flat factor.

h) Symmetric submanifolds of symmetric spaces of noncompact type

In this part, we describe the classification of symmetric submanifolds of Riemannian symmetric spaces of noncompact type. For the real hyperbolic space $\mathbb{R}H^n$, this was already done in Section 3.7. It was shown by Kon [120] resp. Tsukada [226] that every symmetric submanifold in $\mathfrak{G}(\mathbb{C}H^m, \mathbb{C}H^n)$ resp. $\mathfrak{G}(\mathbb{C}H^n, \mathbb{H}H^n)$ is totally geodesic. The classification of symmetric submanifolds in $\mathfrak{G}(\mathbb{R}H^n, \mathbb{C}H^n)$ was obtained by Naitoh [155]. Here we want to describe the classification of symmetric submanifolds in the remaining Grassmann geometry $\mathfrak{G}(M, \bar{M})$ listed in Theorem 9.3.4 (5). This classification has been obtained by Berndt, Eschenburg, Naitoh and Tsukada [18].

We start with recalling the theory of symmetric R-spaces from another viewpoint (see Kobayashi and Nagano [116], Nagano [146], and Takeuchi [211] for details). Let $(\bar{\mathfrak{g}}, \sigma)$ be a positive definite symmetric graded Lie algebra, that is, $\bar{\mathfrak{g}}$ is a real semisimple Lie algebra with a gradation $\bar{\mathfrak{g}} = \bar{\mathfrak{g}}_{-1} + \bar{\mathfrak{g}}_0 + \bar{\mathfrak{g}}_1$ so that $\bar{\mathfrak{g}}_{-1} \neq (0)$ and the adjoint action of $\bar{\mathfrak{g}}_0$ on the vector space $\bar{\mathfrak{g}}_{-1}$ is effective, and a Cartan involution σ satisfying $\sigma(\bar{\mathfrak{g}}_p) = \bar{\mathfrak{g}}_{-p}$ $(p = -1, 0, 1)$. The positive definite symmetric graded Lie algebras have been completely classified (see [116], [211]).

By defining $\tau(X) = (-1)^p X$ for $X \in \bar{\mathfrak{g}}_p$ we obtain an involutive automorphism τ of $\bar{\mathfrak{g}}$ that satisfies $\sigma\tau = \tau\sigma$. Let $\bar{\mathfrak{g}} = \bar{\mathfrak{k}} + \bar{\mathfrak{p}}$ be the Cartan decomposition induced by σ. Then we have $\tau(\bar{\mathfrak{k}}) = \bar{\mathfrak{k}}$ and $\tau(\bar{\mathfrak{p}}) = \bar{\mathfrak{p}}$. Let $\bar{\mathfrak{k}} = \mathfrak{k}_+ + \mathfrak{k}_-$ and $\bar{\mathfrak{p}} = \mathfrak{p}_+ + \mathfrak{p}_-$ be the ± 1-eigenspace decompositions of $\bar{\mathfrak{k}}$ and $\bar{\mathfrak{p}}$ with respect to τ. Obviously, we have $\mathfrak{k}_+ = \bar{\mathfrak{k}} \cap \bar{\mathfrak{g}}_0$, $\mathfrak{k}_- = \bar{\mathfrak{k}} \cap (\bar{\mathfrak{g}}_{-1} + \bar{\mathfrak{g}}_1)$, $\mathfrak{p}_+ = \bar{\mathfrak{p}} \cap \bar{\mathfrak{g}}_0$ and $\mathfrak{p}_- = \bar{\mathfrak{p}} \cap (\bar{\mathfrak{g}}_{-1} + \bar{\mathfrak{g}}_1)$. Since $\bar{\mathfrak{g}}$ is a semisimple Lie algebra, there is a unique element $\nu \in \bar{\mathfrak{g}}_0$ so that

$$\bar{\mathfrak{g}}_p = \{X \in \bar{\mathfrak{g}} \,|\, \mathrm{ad}(\nu)X = pX\} \;,\; p = -1, 0, 1 \,.$$

It can be easily seen that $\nu \in \bar{\mathfrak{p}}$ and hence $\nu \in \mathfrak{p}_+$.

We denote by B the Killing form of $\bar{\mathfrak{g}}$. The restriction of B to $\bar{\mathfrak{p}} \times \bar{\mathfrak{p}}$ is a positive definite inner product on $\bar{\mathfrak{p}}$ and will be denoted by $\langle \cdot, \cdot \rangle$. This inner product is invariant under the adjoint action of $\bar{\mathfrak{k}}$ on $\bar{\mathfrak{p}}$ and under the involution $\tau|\bar{\mathfrak{p}}$. In particular, $\mathfrak{p}_+$ and $\mathfrak{p}_-$ are perpendicular to each other. Let $\bar{G}$ be the simply connected Lie group with Lie algebra $\bar{\mathfrak{g}}$ and $\bar{K}$ be the connected Lie subgroup of $\bar{G}$ corresponding to $\bar{\mathfrak{k}}$, and define the homogeneous space $\bar{M} = \bar{G}/\bar{K}$. Let $\pi : \bar{G} \to \bar{M}$ be the natural projection, and put $o = \pi(e)$, where $e \in \bar{G}$ is the identity. The restriction to $\bar{\mathfrak{p}}$ of the differential $\pi_{*e} : \bar{\mathfrak{g}} \to T_o\bar{M}$ of π at e yields a linear isomorphism $\bar{\mathfrak{p}} \to T_o\bar{M}$. In the following, we will always identify $\bar{\mathfrak{p}}$ and $T_o\bar{M}$ via this isomorphism. From the

$\mathrm{Ad}(\bar{K})$-invariant inner product $\langle\cdot,\cdot\rangle$ on $\bar{\mathfrak{p}} \cong T_o\bar{M}$ we get a $\bar{G}$-invariant Riemannian metric on $\bar{M}$. Then $\bar{M} = \bar{G}/\bar{K}$ is the Riemannian symmetric space of noncompact type that is associated with $(\bar{\mathfrak{g}}, \sigma, \langle\cdot,\cdot\rangle)$.

We put

$$K'_+ = \{k \in \bar{K} | \mathrm{Ad}(k)\nu = \nu\}.$$

Then K'_+ is a closed Lie subgroup whose Lie algebra is $\mathfrak{k}_+$. The homogeneous space $M' = \bar{K}/K'_+$ is diffeomorphic to the orbits $\mathrm{Ad}(\bar{K}) \cdot \nu \subset \bar{\mathfrak{p}}$ and $\bar{K} \cdot \pi(\exp \nu) \subset \bar{M}$, where $\exp : \bar{\mathfrak{g}} \to \bar{G}$ denotes the Lie exponential map from $\bar{\mathfrak{g}}$ into $\bar{G}$. We equip M' with the induced Riemannian metric from $\bar{M}$. Then M' is a compact Riemannian symmetric space associated to the orthogonal symmetric Lie algebra $(\bar{\mathfrak{k}}, \tau|\bar{\mathfrak{k}})$, where $\tau|\bar{\mathfrak{k}}$ is the restriction of τ to $\bar{\mathfrak{k}}$. The symmetric spaces M' arising in this manner are precisely the symmetric R-spaces. If $\bar{\mathfrak{g}}$ is simple, then M' is an irreducible symmetric R-space.

The subspace $\mathfrak{p}_-$ is a Lie triple system in $\bar{\mathfrak{p}} = T_o\bar{M}$ and $[\mathfrak{p}_-, \mathfrak{p}_-] \subset \mathfrak{k}_+$. Thus, there exists a complete totally geodesic submanifold M of $\bar{M}$ with o and $T_oM = \mathfrak{p}_-$. Since M is the image of $\mathfrak{p}_-$ under the exponential map of $\bar{M}$ at o, we see that M is simply connected. We define a Lie subalgebra $\mathfrak{g}$ of $\bar{\mathfrak{g}}$ by $\mathfrak{g} = \mathfrak{k}_+ + \mathfrak{p}_-$ and denote by G the connected Lie subgroup of $\bar{G}$ that corresponds to $\mathfrak{g}$. Then, by construction, M is the G-orbit through o. We denote by K_+ the isotropy subgroup at o of the action of G on $\bar{M}$. The Lie algebra of K_+ is $\mathfrak{k}_+$. Since $M = G/K_+$ is simply connected, K_+ is connected. The restriction $\tau|\mathfrak{g}$ of τ to $\mathfrak{g}$ is an involutive automorphism of $\mathfrak{g}$ and $(\mathfrak{g}, \tau|\mathfrak{g})$ is an orthogonal symmetric Lie algebra dual to $(\bar{\mathfrak{k}}, \tau|\bar{\mathfrak{k}})$. Moreover, M is a Riemannian symmetric space of noncompact type associated with $(\mathfrak{g}, \tau|\mathfrak{g})$. Since both $\mathfrak{p}_-$ and $\mathfrak{p}_+$ are Lie triple systems, M is a reflective submanifold of $\bar{M}$. The corresponding Grassmann geometry $\mathfrak{G}(M, \bar{M})$ is a geometry according to Theorem 9.3.4 (5).

We will construct a one-parameter family of symmetric submanifolds in $\bar{M}$ consisting of submanifolds belonging to that Grassmann geometry that contains the totally geodesic submanifold M and the symmetric R-space M'. For each $c \in \mathbb{R}$ we define a subspace $\mathfrak{p}_c$ of $\mathfrak{p}_- + \mathfrak{k}_- = \bar{\mathfrak{g}}_{-1} + \bar{\mathfrak{g}}_1$ by

$$\mathfrak{p}_c = \{X + c\,\mathrm{ad}(\nu)X \mid \in \mathfrak{p}_-\}.$$

In particular, $\mathfrak{p}_1 = \bar{\mathfrak{g}}_1$ and $\mathfrak{p}_{-1} = \bar{\mathfrak{g}}_{-1}$ are abelian subalgebras of $\bar{\mathfrak{g}}$. Then $\mathfrak{g}_c = \mathfrak{k}_+ + \mathfrak{p}_c$ is a τ-invariant Lie subalgebra of $\bar{\mathfrak{g}}$ and $(\mathfrak{g}_c, \tau|\mathfrak{g}_c)$ is an orthogonal symmetric Lie algebra. We denote by G_c the connected Lie subgroup of $\bar{G}$ with Lie algebra $\mathfrak{g}_c$ and by M_c the orbit of the origin o by G_c in $\bar{M}$.

PROPOSITION 9.3.8

For each $c \in \mathbb{R}$, the orbit M_c is a symmetric submanifold of $\bar{M}$ belonging to the Grassmann geometry $\mathfrak{G}(M, \bar{M})$. The submanifolds M_c and M_{-c} are congruent via the geodesic symmetry s_o of $\bar{M}$ at o. The submanifolds M_c, $0 \leq c < 1$, form a family of noncompact symmetric submanifolds that are homothetic *to the the reflective submanifold M. The submanifolds M_c, $1 < c < \infty$, form a family of compact symmetric submanifolds that are homothetic to the symmetric R-space M'. The submanifold M_1 is a flat symmetric space that is isometric to a Euclidean space. The second fundamental form α_c of M_c is given by*

$$\alpha_c(X,Y) = c[\mathrm{ad}(\nu)X,Y] \in \mathfrak{p}_+ = \nu_o M_c \ , \ X,Y \in \mathfrak{p}_- = T_o M_c \ .$$

In particular, all submanifolds M_c, $0 \leq c < \infty$, are pairwise noncongruent.

It was proved in [18] that every symmetric submanifold of an irreducible Riemannian symmetric space of noncompact type and rank ≥ 2 arises in this way. The crucial point for the proof is a generalization of the fundamental theorem of submanifold geometry in space forms to certain Grassmannian geometries.

9.4 Submanifolds with parallel second fundamental form

a) ... in real space forms

With what we have achieved so far, the classification of submanifolds with parallel second fundamental form in spaces of constant curvature becomes very simple. When $\bar{M}$ has constant curvature, each subspace of any tangent space of $\bar{M}$ is curvature-invariant. From Proposition 9.3.1 we therefore get

COROLLARY 9.4.1
A complete submanifold of S^n, $\mathbb{R}^n$ or H^n has parallel second fundamental form if and only if it is a symmetric submanifold.

b) ... in complex space forms

When the ambient space has nonconstant curvature, one cannot expect that complete submanifolds with parallel second fundamental form are symmetric submanifolds. This can be seen most easily in complex projective space $\mathbb{C}P^n$. A totally geodesic real projective space $\mathbb{R}P^k$, $k = 1, \ldots, n-1$, is complete and obviously has parallel second fundamental form. But, at each point, the normal space is isomorphic to the vector space $\mathbb{R}^{n-k} \oplus \mathbb{C}^k$, and this cannot be the tangent space of a totally geodesic submanifold of $\mathbb{C}P^n$. Hence, the normal spaces are not curvature-invariant, and it follows that $\mathbb{R}P^k$ is not a symmetric submanifold of $\mathbb{C}P^n$.

The classification of submanifolds with parallel second fundamental form in complex projective space $\mathbb{C}P^n$ and complex hyperbolic space $\mathbb{C}H^n$ has been achieved by Naitoh [154, 155].

THEOREM 9.4.2 (Naitoh)
Let M be a complete submanifold of $\mathbb{C}P^n$ or $\mathbb{C}H^n$, $n \geq 2$, with parallel second fundamental form and that is not totally geodesic. Then M is

(i) a complex submanifold, or

(ii) a submanifold that is contained in a totally geodesic $\mathbb{R}P^k$ resp. $\mathbb{R}H^k$ for some $k \in \{1, \ldots, n\}$, or

(iii) a k-dimensional totally real submanifold that is contained in a totally geodesic $\mathbb{C}P^k$ resp. $\mathbb{C}H^k$ for some $k \in \{1, \ldots, n\}$.

Complex submanifolds always have their normal spaces curvature-invariant. Thus, the classification of complex submanifolds with parallel second fundamental form reduces to the one of symmetric complex submanifolds that has been discussed in Theorem 9.3.5 in the case of complex projective space. In the case of complex hyperbolic space [120] proved:

THEOREM 9.4.3 (Kon)
Let M be a complex submanifold of complex hyperbolic space $\mathbb{C}H^n$. If M has parallel second fundamental form then M is totally geodesic.

In case (ii), M also has parallel second fundamental form when considered as a submanifold in $\mathbb{R}P^k$ resp. $\mathbb{R}H^k$. So this case reduces to the corresponding problem in real space forms that has been discussed above.

In the last case (iii), M has parallel second fundamental form when considered as a submanifold in $\mathbb{C}P^k$ resp. $\mathbb{C}H^k$. Since M is totally real and has half the dimension of these smaller ambient spaces, this case reduces to the study of half-dimensional symmetric totally real submanifolds in $\mathbb{C}P^k$ resp. $\mathbb{C}H^k$. In the projective case, the classification has been given in Theorem 9.3.6. The classification in the hyperbolic case has been achieved by Naitoh in [155].

c) ...in quaternionic space forms

The classification of submanifolds with parallel second fundamental form in quaternionic projective space $\mathbb{H}P^n$ and quaternionic hyperbolic space $\mathbb{H}H^n$ is due to Tsukada [226].

THEOREM 9.4.4 (Tsukada)
Let M be a complete submanifold of $\mathbb{H}P^n$ or $\mathbb{H}H^n$, $n \geq 2$, with parallel second fundamental form and that is not totally geodesic. Then M is

(i) a submanifold with parallel second fundamental form in a totally geodesic $\mathbb{R}P^k$ resp. $\mathbb{R}H^k$ for some $k \in \{1, \ldots, n\}$, or

(ii) a totally real submanifold with parallel second fundamental form in a totally geodesic $\mathbb{C}P^k$ resp. $\mathbb{C}H^k$ for some $k \in \{1, \ldots, n\}$, or

(iii) *a complex submanifold with parallel second fundamental form in a totally geodesic* $\mathbb{C}P^k$ *resp.* $\mathbb{C}H^k$ *for some* $k \in \{1, \ldots, n\}$*, or*

(iv) *a totally complex submanifold with parallel second fundamental form and complex dimension* k *in a totally geodesic* $\mathbb{H}P^k$ *resp.* $\mathbb{H}H^k$ *for some* $k \in \{1, \ldots, n\}$*, or*

(v) *a submanifold with parallel second fundamental form in a totally geodesic* $\mathbb{H}P^1 = S^4$ *resp.* $\mathbb{H}H^1 = \mathbb{R}H^4$*.*

Cases (i)-(iii) and (v) have been discussed already above, so we are left with case (iv). In this situation, the normal spaces are curvature-invariant, so M is symmetric. In the projective case, the classification was given in Theorem 9.3.7. In the hyperbolic case, Tsukada proved in [226] that M is totally geodesic in case (iv).

d) ...in Cayley projective or hyperbolic plane

The classification of submanifolds with parallel second fundamental form in Cayley projective plane $\mathbb{O}P^2$ and Cayley hyperbolic plane $\mathbb{O}H^2$ was also derived by Tsukada [227].

THEOREM 9.4.5 (Tsukada)
Let M *be a complete submanifold of* $\mathbb{O}P^2$ *or* $\mathbb{O}H^2$ *having parallel second fundamental form and being not totally geodesic. Then* M *is*

(i) *a submanifold with parallel second fundamental form in a totally geodesic* $\mathbb{O}P^1 = S^8$ *resp.* $\mathbb{O}H^1 = \mathbb{R}H^8$*, or*

(ii) *a submanifold with parallel second fundamental form in a totally geodesic* $\mathbb{H}P^2$ *resp.* $\mathbb{H}H^2$*.*

This reduces the classification problem to several that have been discussed above.

e) ...in symmetric spaces of higher rank

The previous discussion shows that the classification problem for submanifolds with parallel second fundamental form in Riemannian symmetric spaces of rank one is completely solved. In contrast, apart from the classification of symmetric submanifolds, not much is known about submanifolds with parallel second fundamental form in Riemannian symmetric spaces of higher rank. One exception is a paper by Tsukada [228] in which he classifies the complex submanifolds with parallel second fundamental form in Hermitian symmetric spaces. Suppose $\bar{M}$ is a Hermitian symmetric space of compact type and $\mathbb{C}P^k$ is a complex and totally geodesically embedded complex projective space. Then each complex submanifold of $\mathbb{C}P^k$ with parallel second fundamental form in $\mathbb{C}P^k$ also has parallel second fundamental form in $\bar{M}$. The submanifolds given in Theorem 9.3.5 therefore provide examples via such totally geodesic embeddings. One can use these examples as building blocks of more general examples provided one has products of totally geodesic complex projective spaces embedded totally geodesically and holomorphically in $\bar{M}$. Tsukada proved that, in the compact case, all submanifolds with parallel second fundamental form arise in this manner. In the noncompact case, the situation is quite simple, because Tsukada obtained the following result as a generalization of Theorem 9.4.3:

THEOREM 9.4.6 (Tsukada)
Let M be a complex submanifold in a Hermitian symmetric space of non-compact type. If M has parallel second fundamental form then M is totally geodesic.

9.5 Homogeneous hypersurfaces

The classification of homogeneous hypersurfaces in spheres has already been discussed in 3.8.6. We now turn to the more general case of symmetric spaces and start with homogeneous hypersurfaces in projective spaces. Of course, for real projective spaces, the classification is the same as for spheres, modulo the two-fold covering map $S^n \to \mathbb{R}P^n$.

a) Homogeneous hypersurfaces in complex projective spaces

An interesting fact is that in complex projective spaces the theories of isoparametric hypersurfaces and hypersurfaces with constant principal curvatures are different. In fact, Wang [237] showed that certain nonhomogeneous isoparametric hypersurfaces in spheres project via the Hopf map $S^{2n+1} \to \mathbb{C}P^n$ to isoparametric hypersurfaces in complex projective spaces with nonconstant principal curvatures. It is still an open problem whether any hypersurface with constant principal curvatures in $\mathbb{C}P^n$ is isoparametric or homogeneous. The classification of homogeneous hypersurfaces in $\mathbb{C}P^n$ was achieved by Takagi [209]. It is easy to see that every homogeneous hypersurface in $\mathbb{C}P^n$ is the projection of a homogeneous hypersurface in S^{2n+1}. But not every homogeneous hypersurface in S^{2n+1} is invariant under the S^1-action and hence does not project to a homogeneous hypersurface in $\mathbb{C}P^n$. In fact, Takagi proved that those that do project are precisely those that arise from isotropy representations of *Hermitian* symmetric spaces of rank two. In detail, this gives the following classification:

THEOREM 9.5.1 (Takagi)
A hypersurface in $\mathbb{C}P^n$, $n \geq 2$, is homogeneous if and only if it is congruent to

(1) a tube around a k-dimensional totally geodesic subspace $\mathbb{C}P^k \subset \mathbb{C}P^n$ for some $k \in \{1, \ldots, n-1\}$, or

(2) a tube around the complex quadric $\{[z] \in \mathbb{C}P^n \mid z_0^2 + \ldots + z_n^2 = 0\}$ in $\mathbb{C}P^n$, or

(3) a tube around the Segre embedding of $\mathbb{C}P^1 \times \mathbb{C}P^k$ into $\mathbb{C}P^{2k+1}$, $k \geq 1$, or

(4) a tube around the Plücker embedding of the complex Grassmann manifold $G_2(\mathbb{C}^5)$ into $\mathbb{C}P^9$, or

(5) a tube around the half spin embedding of the symmetric space $SO(10)/U(5)$ into $\mathbb{C}P^{15}$.

The corresponding Hermitian symmetric spaces of rank two whose s-representations yield these embeddings via the Hopf map are (1) $\mathbb{C}P^{k+1} \times \mathbb{C}P^{n-k}$, (2) $G_2^+(\mathbb{R}^{n+3})$, (3) $G_2(\mathbb{C}^{k+3})$, (4) $SO(10)/U(5)$, (5) $E_6/T \cdot Spin(10)$. Takagi's result was improved by Uchida [230], who classified all connected closed subgroups of $SU(n+1)$ acting on $\mathbb{C}P^n$ with cohomogeneity one, that is, whose principal orbits have codimension one. Uchida's approach to the classification problem is completely different and uses cohomological methods. In fact, Uchida classified all connected compact Lie groups acting with an orbit of codimension one on a simply connected smooth manifold whose rational cohomology ring is isomorphic to the one of a complex projective space. This includes, for instance, all odd-dimensional complex quadrics (which are real Grassmannians) $G_2^+(\mathbb{R}^{2n+1}) = SO(2n+1)/SO(2) \times SO(2n-1)$.

b) Homogeneous hypersurfaces in quaternionic projective spaces

For the quaternionic projective space $\mathbb{H}P^n$, Iwata [106] used a method analogous to the one of Uchida and classified all connected compact Lie groups acting with an orbit of codimension one on a simply connected smooth manifold whose rational cohomology ring is isomorphic to the one of a quaternionic projective space. For instance, the symmetric space $G_2/SO(4)$ has the same rational cohomology as the quaternionic projective plane $\mathbb{H}P^2$. For the special case of $\mathbb{H}P^n$ Iwata's classification yields

THEOREM 9.5.2 (Iwata)
A hypersurface of $\mathbb{H}P^n$, $n \geq 2$, is homogeneous if and only if it is a tube around a totally geodesic $\mathbb{H}P^k \subset \mathbb{H}P^n$ for some $k \in \{0, \ldots, n-1\}$ or if it is a tube around a totally geodesic $\mathbb{C}P^n \subset \mathbb{H}P^n$.

The tubes around $\mathbb{H}P^k \subset \mathbb{H}P^n$ are the principal orbits of the action of $Sp(k+1) \times Sp(n-k) \subset Sp(n+1)$ on $\mathbb{H}P^n$. The two singular orbits of this action are totally geodesic $\mathbb{H}P^k$ and $\mathbb{H}P^{n-k-1}$. The tubes around $\mathbb{C}P^n \subset \mathbb{H}P^n$ are the principal orbits of the action of $U(n+1) \subset Sp(n+1)$ on $\mathbb{H}P^n$. A different proof, following the lines of Takagi, has been given by D'Atri [65].

c) Homogeneous hypersurfaces in Cayley projective plane

For the Cayley projective plane $\mathbb{O}P^2$ Iwata [107] could also apply his cohomological methods and obtain:

THEOREM 9.5.3 (Iwata)
A hypersurface in $\mathbb{O}P^2$ is homogeneous if and only if it is a geodesic hypersphere or a tube around a totally geodesic $\mathbb{H}P^2 \subset \mathbb{O}P^2$.

The geodesic hyperspheres are obviously the principal orbits of the isotropy group $Spin(9) \subset F_4$. The second singular orbit of this action is a totally geodesic $S^8 = \mathbb{O}P^1 \subset \mathbb{O}P^2$. The tubes around $\mathbb{H}P^2$ are the principal orbits of the action of maximal compact subgroup $Sp(3) \times Sp(1)$ of F_4. Here, the second singular orbit is an 11-dimensional sphere $S^{11} = Sp(3)/Sp(2)$, which is not totally geodesic but minimal in $\mathbb{O}P^2$.

d) Homogeneous hypersurfaces in Riemannian symmetric spaces of compact type

The classification of homogeneous hypersurfaces in irreducible simply connected Riemannian symmetric spaces of compact type is part of the more general classification of hyperpolar actions (up to orbit equivalence) on these spaces due to Kollross [119]. Hyperpolar actions on symmetric spaces are sometimes viewed as generalizations of s-representations, that is, of isotropy representations of semisimple Riemannian symmetric spaces. An isometric action of a closed Lie group on a semisimple Riemannian symmetric space M is said to be *hyperpolar* if there exists a closed, totally geodesic, flat submanifold of M meeting each orbit of the action and intersecting it perpendicularly. It is obvious that the cohomogeneity of a hyperpolar action must be less or equal than the rank of the symmetric space. In particular, the hyperpolar actions on Riemannian symmetric spaces of rank one are precisely the isometric actions of cohomogeneity one, whose classification up to orbit equivalence we described above for the compact case.

A large class of hyperpolar actions was discovered by Hermann [101]. Suppose (G, K) and (G, H) are two semisimple Riemannian symmetric pairs of compact type. Then the action of H on the Riemannian symmetric space G/K is hyperpolar. Also, the action of $H \times K$ on G given by $(h, k) \cdot g := hgk^{-1}$ is hyperpolar. Note that, in particular, the action of the isotropy group of a semisimple Riemannian symmetric space is hyperpolar.

We describe the idea for the classification by Kollross in the special case when the action is of cohomogeneity one and the symmetric space $M = G/K$ is of rank ≥ 2 and not of group type. Suppose H is a maximal closed subgroup of G. If H is not transitive on M, then its cohomogeneity is at least one. Since the cohomogeneity of the action of any closed subgroup of H is at least the cohomogeneity of the action of H, and we are interested only in classification up the orbit equivalence, it is sufficient to consider only maximal closed subgroups of G. But it may happen that H acts transitively on G/K. This happens in precisely four cases, where we write down $G/K = H/(H \cap K)$:

$$\begin{aligned} SO(2n)/U(n) &= SO(2n-1)/U(n-1)\ (n \geq 4)\ , \\ SU(2n)/Sp(n) &= SU(2n-1)/Sp(n-1)\ (n \geq 3)\ , \\ G_2^+(\mathbb{R}^7) = SO(7)/SO(2) \times SO(5) &= G_2/U(2)\ , \\ G_3^+(\mathbb{R}^8) = SO(8)/SO(3) \times SO(5) &= Spin(7)/SO(4)\ . \end{aligned}$$

In these cases, one has to go one step further and consider maximal closed subgroups of H that then never happen to act also transitively. Thus, it is sufficient to consider maximal closed subgroups of G, with the few exceptions just mentioned. In order to have a closed subgroup H act with cohomogeneity one, it obviously must satisfy $\dim H \geq \dim M - 1$. This already rules out a lot of possibilities. For the remaining maximal closed subgroups, one has to calculate the cohomogeneity case by case. One way to do this is to calculate the codimension of the slice representation; this is the action of the isotropy group $H \cap K$ on the normal space at the corresponding point of the orbit through that point. This procedure eventually leads to the classification of all cohomogeneity one actions up to orbit equivalence, and hence to the classification of homogeneous hypersurfaces on $M = G/K$. It turns out that, with five exceptions, all homogeneous hypersurfaces arise via the construction of Hermann. The exceptions come from the following actions:

1. The action of $G_2 \subset SO(7)$ on $SO(7)/U(3) = SO(8)/U(4) = G_2^+(\mathbb{R}^8)$.
2. The action of $G_2 \subset SO(7)$ on $SO(7)/SO(3) \times SO(4) = G_3^+(\mathbb{R}^7)$.
3. The action of $Spin(9) \subset SO(16)$ on $SO(16)/SO(2) \times SO(14) = G_2^+(\mathbb{R}^{16})$.
4. The action of $Sp(n)Sp(1) \subset SO(4n)$ on $SO(4n)/SO(2) \times SO(4n-2) = G_2^+(\mathbb{R}^{4n})$.
5. The action of $SU(3) \subset G_2$ on $G_2/SO(4)$.

All other homogeneous hypersurfaces can be obtained via the construction of Hermann. We refer to [119] for an explicit list of all Hermann actions of cohomogeneity one.

e) Homogeneous hypersurfaces in Riemannian symmetric spaces of noncompact type

Every homogeneous hypersurface in real hyperbolic space $\mathbb{R}H^n$ is obviously isoparametric. Conversely, as the classification of isoparametric hypersurfaces in $\mathbb{R}H^n$

by E. Cartan shows (see Section 3.8), any complete isoparametric hypersurface in $\mathbb{R}H^n$ is homogeneous. This gives

THEOREM 9.5.4 (E. Cartan)
A connected complete hypersurface in real hyperbolic space $\mathbb{R}H^n$ is homogeneous if and only if it is

(1) a geodesic hypersphere in $\mathbb{R}H^n$, or

(2) a tube around a totally geodesic $\mathbb{R}H^k \subset \mathbb{R}H^n$ for some $k \in \{1, \ldots, n-2\}$, or

(3) a real hyperbolic hyperplane $\mathbb{R}H^{n-1}$ or an equidistant hypersurface to it, or

(4) a horosphere in $\mathbb{R}H^n$.

As subgroups of $SO^o(1,n)$ giving these hypersurfaces as orbits one can choose (1) a maximal compact subgroup $SO(n)$; (2) $SO^o(1,k) \times SO(n-k)$; (3) $SO^o(1,n-1)$; (4) the nilpotent subgroup in an Iwasawa decomposition of $SO^o(1,n)$.

The method of Cartan does not work for the hyperbolic spaces $\mathbb{C}H^n$, $\mathbb{H}H^n$ and $\mathbb{O}H^2$. The reason is that the Gauss-Codazzi equations become too complicated. Nevertheless, we can apply the method of Cartan to the special class of curvature-adapted hypersurfaces. For these hypersurfaces the equations of Gauss and Codazzi simplify considerably. A hypersurface M of a Riemannian manifold $\bar{M}$ is called *curvature-adapted* if its shape operator and its normal Jacobi operator commute with each other. Recall that the normal Jacobi operator of M is the self-adjoint (local) tensor field on M defined by $\bar{R}(.,\xi)\xi$, where $\bar{R}$ is the Riemannian curvature tensor of $\bar{M}$ and ξ is a (local) unit normal vector field of M. If $\bar{M}$ is a space of constant curvature, then the normal Jacobi operator is a multiple of the identity at each point, and hence every hypersurface is curvature-adapted. But, for more general ambient spaces, this condition is quite restrictive. For instance, in a nonflat complex space form, say $\mathbb{C}P^n$ or $\mathbb{C}H^n$, a hypersurface M is curvature-adapted if and only if the structure vector field on M is a principal curvature vector everywhere. Recall that the structure vector field of M is the vector field obtained by rotating a local unit normal vector field to a tangent vector field using the ambient Kähler structure. In [13] the first author obtained the classification of all curvature-adapted hypersurfaces in $\mathbb{C}H^n$ with constant principal curvatures.

THEOREM 9.5.5 (Berndt)
A hypersurface of $\mathbb{C}H^n$, $n \geq 2$, is curvature-adapted and has constant principal curvatures if and only if it is

(1) a geodesic hypersphere in $\mathbb{C}H^n$, or

(2) a tube around a totally geodesic $\mathbb{C}H^k \subset \mathbb{C}H^n$ *for some* $k \in \{1, \ldots, n-1\}$, *or*

(3) a horosphere in $\mathbb{C}H^n$, *or*

(4) a tube around a totally geodesic $\mathbb{R}H^n \subset \mathbb{C}H^n$.

Note that all the hypersurfaces listed here are homogeneous. The geodesic hyperspheres are obviously the principal orbits of the isotropy group $S(U(1) \times U(n))$ of $SU(1,n)$ at a point. The tubes around a totally geodesic $\mathbb{C}H^k$ are the principal orbits of the action of $S(U(1,k) \times U(n-k)) \subset SU(1,n)$. The horospheres arise as the orbits (there are only principal orbits in this case) of the nilpotent part in an Iwasawa decomposition of $SU(1,n)$. Note that this nilpotent part is isomorphic to the $(2n-1)$-dimensional Heisenberg group. Eventually, the tubes around $\mathbb{R}H^n$ are the principal orbits of the action of $SO(1,n) \subset SU(1,n)$. A natural question is whether there are other homogeneous hypersurfaces in $\mathbb{C}H^n$. We will discuss this question in more detail below.

A hypersurface M in a quaternionic space form $\bar{M}$, say $\mathbb{H}P^n$ or $\mathbb{H}H^n$, is curvature-adapted if and only if the three-dimensional distribution that is obtained by rotating the normal bundle of M into the tangent bundle of M by the almost-Hermitian structures in the quaternionic Kähler structure of $\bar{M}$, is invariant under the shape operator of M. The curvature-adapted hypersurfaces in $\mathbb{H}H^n$ with constant principal curvatures were classified in [14].

THEOREM 9.5.6 (Berndt)

A hypersurface in $\mathbb{H}H^n$, $n \geq 2$, *is curvature-adapted and has constant principal curvatures if and only if it is*

(1) a geodesic hypersphere in $\mathbb{H}H^n$, *or*

(2) a tube around a totally geodesic $\mathbb{H}H^k \subset \mathbb{H}H^n$ *for some* $k \in \{1, \ldots, n-1\}$, *or*

(3) a horosphere in $\mathbb{H}H^n$, *or*

(4) a tube around a totally geodesic $\mathbb{C}H^n \subset \mathbb{H}H^n$.

The proof is based on the Gauss-Codazzi equations and uses focal set theory. It is an open problem whether there exist curvature-adapted hypersurfaces in $\mathbb{H}H^n$ with nonconstant principal curvatures. All the hypersurfaces listed in the previous theorem are homogeneous. The geodesic hyperspheres are the principal orbits of the isotropy group $Sp(1) \times Sp(n)$ of $Sp(1,n)$ at a point. The tubes around a totally geodesic $\mathbb{H}H^k$ are the principal orbits of the action of $Sp(1,k) \times Sp(n-k) \subset Sp(1,n)$. The horospheres arise as the orbits of the nilpotent part in an Iwasawa decomposition of $Sp(1,n)$. And the tubes around $\mathbb{C}H^n$ are the principal orbits of the action of $SU(1,n) \subset Sp(1,n)$.

Of course, now the question arises whether any homogeneous hypersurface in $\mathbb{C}H^n$ or $\mathbb{H}H^n$ is curvature-adapted. As the classifications by Takagi and Iwata show, the answer for the corresponding question in $\mathbb{C}P^n$ and $\mathbb{H}P^n$ is yes. But, in 1999, Lohnherr and Reckziegel [131] found an example of a homogeneous ruled hypersurface in $\mathbb{C}H^n$ that is not curvature-adapted. Consider a horocycle in a totally geodesic and totally real $\mathbb{R}H^2 \subset \mathbb{C}H^n$. At each point of the horocycle we attach a totally geodesic $\mathbb{C}H^{n-1}$ orthogonal to the complex hyperbolic line determined by the tangent vector of the horocycle at that point. By varying with the points on the horocycle, we get a homogeneous ruled hypersurface in $\mathbb{C}H^n$. In [16] the first author constructed this hypersurface by an algebraic method. Using this method, more examples of homogeneous hypersurfaces in $\mathbb{C}H^n$ were found. This method also generalizes to other Riemannian symmetric spaces of noncompact type and can be used to produce examples of homogeneous hypersurfaces. We will now describe this construction in more detail.

Let $M = G/K$ be a Riemannian symmetric space of noncompact type with $G = I^o(M)$ and K the isotropy group of G at a point $o \in M$. We denote by n the dimension of M and by r the rank of M. Any homogeneous hypersurface in M is an orbit of a connected closed subgroup of G acting on M with cohomogeneity one. We denote by $\mathfrak{M}$ the moduli space of all isometric cohomogeneity one actions on M modulo orbit equivalence. Clearly, to classify the homogeneous hypersurfaces in M, we just have to determine $\mathfrak{M}$.

The orbit space of an isometric cohomogeneity one action on a connected complete Riemannian manifold is homeomorphic to $\mathbb{R}$ or $[0, \infty)$. Geometrically, this means that either all orbits are principal and form a Riemannian foliation on M or there exists exactly one singular orbit with codimension ≥ 2 and the principal orbits are tubes around the singular orbit. This induces a disjoint union $\mathfrak{M} = \mathfrak{M}_F \cup \mathfrak{M}_S$, where $\mathfrak{M}_F$ is the set of all homogeneous codimension one foliations on M modulo isometric congruence and $\mathfrak{M}_S$ is the set of all connected normal homogeneous submanifolds with codimension ≥ 2 in M modulo isometric congruence. Here a submanifold of M is called *normal homogeneous* if it is an orbit of a connected closed subgroup of $I^o(M)$ and the slice representation at a point acts transitively on the unit sphere in the normal space at that point.

Let $\mathfrak{g}$ and $\mathfrak{k}$ be the Lie algebra of G and K, respectively, and B the Killing form of $\mathfrak{g}$. If $\mathfrak{p}$ is the orthogonal complement of $\mathfrak{k}$ in $\mathfrak{g}$ with respect to B then $\mathfrak{g} = \mathfrak{k} \oplus \mathfrak{p}$ is a Cartan decomposition of $\mathfrak{g}$. If $\theta : \mathfrak{g} \to \mathfrak{g}$ is the corresponding Cartan involution, we get a positive definite inner product on $\mathfrak{g}$ by $\langle X, Y \rangle = -B(X, \theta Y)$ for all $X, Y \in \mathfrak{g}$. We normalize the Riemannian metric on M such that its restriction to $T_oM \times T_oM$ coincides with $\langle \cdot, \cdot \rangle$, where we identify $\mathfrak{p}$ and T_oM in the usual manner.

Let $\mathfrak{a}$ be a maximal abelian subspace in $\mathfrak{p}$ and denote by $\mathfrak{a}^*$ the dual vector space of $\mathfrak{a}$. Moreover, let

$$\mathfrak{g} = \mathfrak{g}_0 \oplus \bigoplus_{\lambda \in \Sigma} \mathfrak{g}_\lambda$$

be the restricted root space decomposition of $\mathfrak{g}$ with respect to $\mathfrak{a}$. The root system Σ is either reduced and then of type $A_r, B_r, C_r, D_r, E_6, E_7, E_8, F_4, G_2$ or nonreduced and then of type BC_r. For each $\lambda \in \mathfrak{a}^*$ let $H_\lambda \in \mathfrak{a}$ be the dual vector in $\mathfrak{a}$ with respect to the Killing form, that is, $\lambda(H) = \langle H_\lambda, H\rangle$ for all $H \in \mathfrak{a}$. Then we get an inner product on $\mathfrak{a}^*$, which we also denote by $\langle \cdot, \cdot \rangle$, by means of $\langle \lambda, \mu\rangle = \langle H_\lambda, H_\mu\rangle$ for all $\lambda, \mu \in \mathfrak{a}^*$. We choose a set $\Lambda = \{\alpha_1, \ldots, \alpha_r\}$ of simple roots in Σ and denote the resulting set of positive restricted roots by Σ^+.

By $\mathrm{Aut}(DD)$ we denote the group of symmetries of the Dynkin diagram associated to Λ. There are just three possibilities, namely

$$\mathrm{Aut}(DD) = \begin{cases} \mathfrak{S}_3 & , \text{ if } \Sigma = D_4 \ , \\ \mathbb{Z}_2 & , \text{ if } \Sigma \in \{A_r \ (r \geq 2), \ D_r \ (r \geq 2, r \neq 4), \ E_6\} \ , \\ \mathrm{id} & , \text{ otherwise} \ . \end{cases}$$

where $\mathfrak{S}_3$ is the group of permutations of a set of three elements. The first two cases correspond to triality and duality principles on the symmetric space that were discovered by E. Cartan.

The symmetric spaces with a triality principle are $SO(8, \mathbb{C})/SO(8)$ and the hyperbolic Grassmannian $G_4^*(\mathbb{R}^8)$. Each symmetry $P \in \mathrm{Aut}(DD)$ can be linearly extended to a linear isometry of $\mathfrak{a}^*$, which we also denote by P. Denote by Φ the linear isometry from $\mathfrak{a}^*$ to $\mathfrak{a}$ defined by $\Phi(\lambda) = H_\lambda$ for all $\lambda \in \mathfrak{a}^*$. Then $\widetilde{P} = \Phi \circ P \circ \Phi^{-1}$ is a linear isometry of $\mathfrak{a}$ with $\widetilde{P}(H_\lambda) = H_\mu$ if and only if $P(\lambda) = \mu$, $\lambda, \mu \in \mathfrak{a}^*$. Since P is an orthogonal transformation, $\widetilde{P}$ is just the dual map of $P^{-1} : \mathfrak{a}^* \to \mathfrak{a}^*$. In this way, each symmetry $P \in \mathrm{Aut}(DD)$ induces linear isometries of $\mathfrak{a}^*$ and $\mathfrak{a}$, both of which we will denote by P, since it will always be clear from the context which of these two we are using.

We now define a nilpotent subalgebra $\mathfrak{n}$ of $\mathfrak{g}$ by

$$\mathfrak{n} = \bigoplus_{\lambda \in \Sigma^+} \mathfrak{g}_\lambda \ ,$$

which then induces an Iwasawa decomposition $\mathfrak{g} = \mathfrak{k} \oplus \mathfrak{a} \oplus \mathfrak{n}$ of $\mathfrak{g}$. Then $\mathfrak{a} + \mathfrak{n}$ is a solvable subalgebra of $\mathfrak{g}$ with $[\mathfrak{a} + \mathfrak{n}, \mathfrak{a} + \mathfrak{n}] = \mathfrak{n}$. The connected subgroups A, N, AN of G with Lie algebras $\mathfrak{a}, \mathfrak{n}, \mathfrak{a} + \mathfrak{n}$, respectively, are simply connected and AN acts simply transitively on M. The symmetric space M is isometric to the connected, simply connected, solvable Lie group AN equipped with the left-invariant Riemannian metric that is induced from the inner product $\langle \cdot, \cdot \rangle$.

Let ℓ be a linear line in $\mathfrak{a}$. Since ℓ lies in the orthogonal complement of the derived subalgebra of $\mathfrak{a} + \mathfrak{n}$, the orthogonal complement $\mathfrak{s}_\ell = (\mathfrak{a} + \mathfrak{n}) \ominus \ell$ of ℓ in $\mathfrak{a} + \mathfrak{n}$ is a subalgebra of $\mathfrak{a} + \mathfrak{n}$ of codimension one. Let S_ℓ be the connected Lie subgroup of AN with Lie algebra $\mathfrak{s}_\ell$. Then the orbits of the action of S_ℓ on M form a Riemannian foliation $\mathfrak{F}_\ell$ on M whose leaves are homogeneous hypersurfaces. If M has rank one then $\mathfrak{a}$ is one-dimensional and hence there exists only one such foliation, namely the one given by $S_\ell = S_\mathfrak{a} = N$. This is precisely the horosphere foliation on M, all of whose leaves are isometrically congruent to each other. One can show that, for higher rank also, all leaves of $\mathfrak{F}_\ell$ are isometrically congruent to each other. Using structure theory of semisimple and solvable Lie algebras, one can show that two foliations $\mathfrak{F}_\ell$ and $\mathfrak{F}_{\ell'}$ are isometrically congruent to each other if and only if there exists a symmetry $P \in \mathrm{Aut}(DD)$ with $P(\ell) = \ell'$. It follows that the set of all congruence classes of such foliations is parametrized my $\mathbb{R}P^{r-1}/\mathrm{Aut}(DD)$. Here, $\mathbb{R}P^{r-1}$ is the projective space of all linear lines ℓ in $\mathfrak{a}$, and the action of $\mathrm{Aut}(DD)$ on $\mathbb{R}P^{r-1}$ is the induced one from the linear action of $\mathrm{Aut}(DD)$ on $\mathfrak{a}$.

Let $\alpha_i \in \Lambda$, $i \in \{1,\ldots,r\}$, be a simple root. For each unit vector $\xi \in \mathfrak{g}_{\alpha_i}$ the subspace $\mathfrak{s}_\xi = \mathfrak{a} + (\mathfrak{n} \ominus \mathbb{R}\xi)$ is a subalgebra of $\mathfrak{a} + \mathfrak{n}$. Let S_ξ be the connected Lie subgroup of AN with Lie algebra $\mathfrak{s}_\xi$. Then the orbits of the action of S_ξ on M form a Riemannian foliation $\mathfrak{F}_\xi$ on M whose leaves are homogeneous hypersurfaces. If $\eta \in \mathfrak{g}_{\alpha_i}$ is another unit vector, the induced foliation $\mathfrak{F}_\eta$ is congruent to $\mathfrak{F}_\xi$ under an isometry in the centralizer of $\mathfrak{a}$ in K. Thus, for each simple root $\alpha_i \in \Lambda$, we obtain a congruence class of homogeneous foliations of codimension one on M. We denote a representative of this congruence class by $\mathfrak{F}_i$. By investigating the geometry of these foliations one can prove that $\mathfrak{F}_i$ and $\mathfrak{F}_j$ are isometrically congruent if and only if there exists a symmetry $P \in \mathrm{Aut}(DD)$ with $P(\alpha_i) = \alpha_j$. Thus, the set of all congruence classes of such foliations is parametrized by $\{1,\ldots,r\}/\mathrm{Aut}(DD)$, where the action of $\mathrm{Aut}(DD)$ on $\{1,\ldots,r\}$ is given by identifying $\{1,\ldots,r\}$ with the vertices of the Dynkin diagram. The geometry of these foliations is quite fascinating. Among all leaves there exists exactly one that is minimal. All leaves together form a *homogeneous* isoparametric system on M, and if the rank of M is ≥ 3, there exist among these systems some that are noncongruent but have the same principal curvatures with the same multiplicities. Such a feature had already been discovered by Ferus, Karcher and Münzner [87] for *inhomogeneous* isoparametric systems on spheres.

Using structure theory of semisimple and solvable Lie algebras Berndt and Tamaru proved in [19] that every homogeneous codimension one foliation on M is isometrically congruent to one of the above.

THEOREM 9.5.7 (Berndt-Tamaru)

Let M be a connected irreducible Riemannian symmetric space of noncompact type and with rank r. The moduli space $\mathfrak{M}_F$ of all noncongruent homogeneous codimension one foliations on M is isomorphic to the orbit space of the action of $\mathrm{Aut}(DD)$ on $\mathbb{R}P^{r-1} \cup \{1,\ldots,r\}$:

$$\mathfrak{M}_F \cong (\mathbb{R}P^{r-1} \cup \{1,\ldots,r\})/\mathrm{Aut}(DD) .$$

It is very surprising and remarkable that $\mathfrak{M}_F$ depends only on the rank and on possible duality or triality principles on the symmetric space. For instance, for the symmetric spaces $SO(17,\mathbb{C})/SO(17)$, $Sp(8,\mathbb{R})/U(8)$, $Sp(8,\mathbb{C})/Sp(8)$, $SO(16,\mathbb{H})/U(16)$, $SO(17,\mathbb{H})/U(17)$, $E_8^8/SO(16)$, $E_8^{\mathbb{C}}/E_8$ and for the hyperbolic Grassmannians $G_8^*(\mathbb{R}^{n+16})$ $(n \geq 1)$, $G_8^*(\mathbb{C}^{n+16})$ $(n \geq 0)$, $G_8^*(\mathbb{H}^{n+16})$ $(n \geq 0)$ the moduli space $\mathfrak{M}_F$ of all noncongruent homogeneous codimension one foliations is isomorphic to $\mathbb{R}P^7 \cup \{1,\ldots,8\}$.

We now discuss the case when the rank r is one, that is, M is a hyperbolic space over one of the normed real division algebras $\mathbb{R}$, $\mathbb{C}$, $\mathbb{H}$ or $\mathbb{O}$. From Theorem 9.5.7 we see that there are exactly two congruence classes of homogeneous codimension one foliations on M. The first one, coming from the 0-dimensional real projective space, is the well known horosphere foliation. The second foliation is not so well known except for the real hyperbolic case. In the case of $\mathbb{R}H^n$, we get the foliation whose leaves are a totally geodesic $\mathbb{R}H^{n-1} \subset \mathbb{R}H^n$ and its equidistant hypersurfaces. Comparing this with Cartan's classification of homogeneous hypersurfaces in Theorem 9.5.4 we see that we indeed got all homogeneous hypersurfaces of $\mathbb{R}H^n$ that are not tubes around a lower dimensional submanifold. In the case of $\mathbb{C}H^n$, the minimal orbit of the second foliation is precisely the minimal ruled real hypersurface of $\mathbb{C}H^n$ discovered by Lohnherr and Reckziegel, as mentioned above. The geometry of the second foliation has been investigated for all hyperbolic spaces in [16].

To complete the classification of homogeneous hypersurfaces in connected irreducible Riemannian symmetric spaces of noncompact type we must also determine the moduli space $\mathfrak{M}_S$. In the case of $\mathbb{R}H^n$ we already know from Theorem 9.5.4 that $\mathfrak{M}_S$ consists of $n-1$ elements given by the real hyperbolic subspace $\mathbb{R}H^k \subset \mathbb{R}H^n$, $k \in \{0, \ldots, n-2\}$, and the tubes around $\mathbb{R}H^k$. An obvious consequence from Cartan's classification is the nonobvious fact that a singular orbit of a cohomogeneity one action of $\mathbb{R}H^n$ is totally geodesic.

In [20], Berndt and Tamaru determined the subset $\mathfrak{M}_s^{tg}$ of $\mathfrak{M}_S$ consisting of all cohomogeneity one actions for which the singular orbit is totally geodesic. In this special situation, one can use duality between symmetric spaces of compact type and noncompact type to derive the classification. An explicit list of all totally geodesic singular orbits can be found in [20], which can be summarized as follows. The set $\mathfrak{M}_s^{tg}$ is empty for the exceptional symmetric spaces of $E_7^{\mathbb{C}}$ and $E_8^{\mathbb{C}}$ and all their noncompact real forms, and of $E_6^{\mathbb{C}}$ and its split real form. For all other symmetric spaces, and this includes all classical symmetric spaces, $\mathfrak{M}_S^{tg}$ is nonempty and finite. It is $\# \mathfrak{M}_S^{tg} = n > 3$ only for the hyperbolic spaces $\mathbb{R}H^{n+1}$, $\mathbb{C}H^{n-1}$ and $\mathbb{H}H^{n-1}$. For the symmetric spaces $\mathbb{R}H^4$, $\mathbb{C}H^2$, $\mathbb{H}H^2$, $\mathbb{O}H^2$, $G_3^*(\mathbb{R}^7)$, $G_2^*(\mathbb{R}^{2n})$ $(n \geq 3)$ and $G_2^*(\mathbb{C}^{2n})$ $(n \geq 3)$ we have $\# \mathfrak{M}_S^{tg} = 3$. For the symmetric spaces $\mathbb{R}H^3$, $G_k^*(\mathbb{R}^n)$ $(1 < k < n-k, (k,n) \neq (3,7), (2,2m), m > 2)$, $G_3^*(\mathbb{R}^6)$, $G_k^*(\mathbb{C}^n)$ $(1 < k < n-k, (k,n) \neq (2,2m), m > 2)$, $G_k^*(\mathbb{H}^n)$ $(1 < k < n-k)$, $SL(3,\mathbb{H})/Sp(3)$, $SL(3,\mathbb{C})/SU(3)$, $SL(4,\mathbb{C})/SU(4) = SO(6,\mathbb{C})/SO(6)$, $SO(7,\mathbb{C})/SO(7)$, $G_2^2/SO(4)$ and E_6^{-24}/F_4 we have $\# \mathfrak{M}_S^{tg} = 2$. In all remaining cases we have $\# \mathfrak{M}_S^{tg} = 1$.

Of course, the natural question now is whether a singular orbit of a cohomogeneity one action on M is totally geodesic. As we already know from Theorem 9.5.4, the answer is yes for $\mathbb{R}H^n$. In [17] Berndt and Brück investigated this question for the other hyperbolic spaces $\mathbb{C}H^n$, $\mathbb{H}H^n$ and $\mathbb{O}H^2$. The surprising outcome of their investigations is that in all these spaces there exist cohomogeneity one actions with non-totally geodesic singular orbits. In the following, we describe the construction of these actions. Let M be one of these hyperbolic spaces and consider an Iwasawa decomposition $\mathfrak{g} = \mathfrak{k} + \mathfrak{a} + \mathfrak{n}$ of the Lie algebra of the isometry group of M. The restricted root system Σ associated to M is of type BC_1 and hence nonreduced. The nilpotent Lie algebra $\mathfrak{n}$ decomposes into root spaces $\mathfrak{n} = \mathfrak{g}_\alpha + \mathfrak{g}_{2\alpha}$, where α is a simple root in Σ. The root space $\mathfrak{g}_{2\alpha}$ is the center of $\mathfrak{n}$. The Lie algebra $\mathfrak{n}$ is a Heisenberg algebra in case of $\mathbb{C}H^n$, a generalized Heisenberg algebra with 3-dimensional center in case of $\mathbb{H}H^n$, and a generalized Heisenberg algebra with 7-dimensional center in case of $\mathbb{O}H^2$.

We first consider the case of $\mathbb{C}H^n$, $n \geq 3$, in which case $\mathfrak{g}_\alpha$ is a complex vector space of complex dimension ≥ 2. Denote by J its complex structure. We choose a linear subspace $\mathfrak{v}$ of $\mathfrak{g}_\alpha$ such that its orthogonal complement $\mathfrak{v}^\perp$ in $\mathfrak{g}_\alpha$ has constant Kähler angle, that is, there exists a real number $\varphi \in [0, \pi/2]$ such that the angle between $J(\mathbb{R}v)$ and $\mathfrak{v}^\perp$ is φ for all nonzero vectors $v \in \mathfrak{v}^\perp$. If $\varphi = 0$ then $\mathfrak{v}$ is a complex subspace. It is easy to classify all subspaces with constant Kähler angle in a complex vector space. In particular, such subspaces exist for each given angle φ. It is clear that $\mathfrak{s} = \mathfrak{a} + \mathfrak{v} + \mathfrak{g}_{2\alpha}$ is a subalgebra of $\mathfrak{a} + \mathfrak{n}$. Let S be the connected closed subgroup of AN with Lie algebra S and $N_K^o(S)$ the identity component of the normalizer of S in $K = S(U(1) \times U(n))$. Then $N_K^o(S)S \subset KAN = G$ acts on $\mathbb{C}H^n$ with cohomogeneity one and singular orbit $S \subset AN = G/K = \mathbb{C}H^n$. If $\varphi \neq 0$ then S is not totally geodesic.

A similar construction works in the quaternionic hyperbolic space $\mathbb{H}H^n$, $n \geq 3$. In this case, the root space $\mathfrak{g}_\alpha$ is a quaternionic vector space of quaternionic dimension $n-1$ and for $\mathfrak{v}$ one has to choose linear subspaces for which the orthogonal complement $\mathfrak{v}^\perp$ of $\mathfrak{v}$ in $\mathfrak{g}_\alpha$ has constant quaternionic Kähler angle. If $n = 2$ we can choose any linear subspace $\mathfrak{v}$ of $\mathfrak{g}_\alpha$ of real dimension one or two.

Finally, in the case of the Cayley hyperbolic plane $\mathbb{O}H^2$, the root space $\mathfrak{g}_\alpha$ is isomorphic to the Cayley algebra $\mathbb{O}$. Let $\mathfrak{v}$ be a linear subspace of $\mathfrak{g}_\alpha$ of real dimension 1, 2, 4, 5 or 6. Let S be the connected closed subgroup of AN with Lie algebra $\mathfrak{s} = \mathfrak{a} + \mathfrak{v} + \mathfrak{g}_{2\alpha}$ and $N_K^o(S)$ the identity component of the normalizer of S in $K = Spin(9)$. For instance, if $\dim \mathfrak{v} = 1$ then $N_K^o(S)$ is isomorphic to the exceptional Lie group G_2. The action of G_2 on the 7-dimensional normal space $\mathfrak{v}^\perp$ is equivalent to the standard 7-dimensional representation of G_2. Since this is transitive on the 6-dimensional sphere, it follows that $G_2 S \subset KAN = G = F_4$ acts on $\mathbb{O}H^2$ with cohomogeneity one and with S as a nontotally geodesic singular orbit. For the dimensions 2, 4, 5 and 6, the corresponding normalizer is isomorphic to $U(4)$, $SO(4)$, $SO(3)$ and $SO(2)$ respectively, and one also gets cohomogeneity one actions on $\mathbb{O}H^2$ with a nontotally geodesic singular orbit. Surprisingly, if $\mathfrak{v}$ is 3-dimensional this method does not yield such a cohomogeneity one action.

It is an open problem whether for $\mathbb{C}H^n$, $\mathbb{H}H^n$ or $\mathbb{O}H^2$ the moduli space $\mathfrak{M}_S$ contains more elements than described above. Also, for higher rank, the explicit structure of $\mathfrak{M}_S$ is still unknown.

9.6 Exercises

Exercise 9.6.1 Prove that a polar in $\mathbb{C}P^n$ is a totally geodesic $\mathbb{C}P^{n-1}$.

Exercise 9.6.2 Show that $\mathbb{R}P^k$ is not a reflective submanifold of $\mathbb{C}P^n$ if $k < n$.

Exercise 9.6.3 Use duality between symmetric spaces of compact and noncompact type to deduce the classification of totally geodesic submanifolds in $\mathbb{C}H^n$ from the one in $\mathbb{C}P^n$.

Exercise 9.6.4 Let γ be a circle in $\mathbb{C}P^n$. Construct explicitly a one-parameter group of isometries of $\mathbb{C}P^n$ that has γ as an orbit.

Exercise 9.6.5 Calculate explicitly the second canonical embedding of $\mathbb{C}P^n$ into a complex projective space.

Exercise 9.6.6 Prove that the embedding of $SU(n)/SO(n)$ in CP^N as described in Section 9.3 e) is totally real, where $N = (n-1)(n+2)/2$.

Exercise 9.6.7 Prove that the focal set of the complex quadric in $\mathbb{C}P^n$ is a totally geodesic $\mathbb{R}P^n$.

Exercise 9.6.8 The action of $Sp(3) \times Sp(1) \subset F_4$ on the Cayley projective plane $\mathbb{O}P^2$ has a totally geodesic $\mathbb{H}P^2$ as a singular orbit. Prove that the second singular orbit is an 11-dimensional sphere S^{11}.

Exercise 9.6.9 Show that the action of the exceptional Lie group $G_2 \subset SO(7)$ on $SO(7)/U(3) = SO(8)/U(4) = G_2^+(\mathbb{R}^8)$ is of cohomogeneity one. Prove that the two singular orbits of this action are $G_2/U(2) = G_2^+(\mathbb{R}^7)$ and $G_2/SU(3) = S^6$.

Appendix Basic material

Our study of submanifolds is mainly carried out in the framework of Riemannian geometry. For the reader's convenience and the purpose of fixing notations, in this appendix, we summarize some basic concepts regarding Riemannian manifolds, Lie groups, homogeneous and symmetric spaces. This is not an attempt to introduce these topics. At the beginning of each section, we provide a list of textbooks where the interested reader can find further details.

A.1 Riemannian manifolds

Modern introductions to Riemannian geometry can be found in the books by Chavel [47], Gallot-Hulin-Lafontaine [89], Jost [108], Petersen [188] and Sakai [195].

Riemannian manifolds

Let M be an m-dimensional smooth manifold. By smooth we always mean C^∞, and, as manifolds are always assumed to satisfy the second countability axiom, they are paracompact. For each $p \in M$ we denote by T_pM (or $T_p(M)$) the tangent space of M at p. The tangent bundle of M is denoted by TM.

Suppose each tangent space T_pM is equipped with an inner product $\langle \cdot , \cdot \rangle_p$. If the function $p \mapsto \langle X_p, Y_p \rangle_p$ is smooth for any two smooth vector fields X, Y on M, then this family of inner products is called a *Riemannian metric*, or *Riemannian structure*, on M. We usually denote a Riemannian metric, and each inner product it consists of, by $\langle \cdot , \cdot \rangle$. Paracompactness implies that any smooth manifold admits a Riemannian structure. A smooth manifold equipped with a Riemannian metric is called a *Riemannian manifold*.

Length, distance and completeness

The presence of an inner product on each tangent space allows measurement of the length of tangent vectors, by which we can define the length of curves and a distance function. For the latter, we have to assume that M is connected. If $c : [a, b] \to M$ is any smooth curve into a Riemannian manifold M, the *length* $L(c)$ of c is defined by

$$L(c) := \int_a^b \sqrt{\langle \dot{c}(t), \dot{c}(t) \rangle} dt ,$$

where $\dot{c}$ denotes the tangent vector field of c. The length $L(c)$ of a piecewise smooth curve $c : [a, b] \to M$ is then defined in the usual way by means of a suitable subdivision of $[a, b]$. The *distance* $d(p, q)$ between two points $p, q \in M$ is defined as the infimum over all $L(c)$, where $c : [a, b] \to M$ is a piecewise smooth curve in M with $c(a) = p$ and $c(b) = q$. The distance function $d : M \times M \to \mathbb{R}$ turns M into a metric space. The topology on M induced by this metric coincides with the underlying manifold topology. A *complete Riemannian manifold* is a Riemannian manifold M that is complete when considered as a metric space, that is, every Cauchy sequence in M converges in M.

Isometries

Let M be a Riemannian manifold. A smooth diffeomorphism $f : M \to M$ is called an *isometry* if $\langle f_* X, f_* Y \rangle = \langle X, Y \rangle$ for all $X, Y \in T_p M$, $p \in M$, where f_* denotes the differential of f at p. If M is connected, a surjective continuous map $f : M \to M$ is an isometry if and only if it preserves the distance function d on M, that is, if $d(f(p), f(q)) = d(p, q)$ for all $p, q \in M$. An isometry of a connected Riemannian manifold is completely determined by both its value and its differential at some point. In particular, an isometry that fixes a point and whose differential at this point is the identity, is the identity map. If M is a connected, simply connected, complete, real analytic Riemannian manifold, then every local isometry of M can be extended to a global isometry of M.

The isometries of a Riemannian manifold form a group in an obvious manner. We will denote it by $I(M)$ and call it the *isometry group* of M. We always consider this group as a topological group equipped with the compact-open topology. With respect to this topology $I(M)$ carries the structure of a Lie group acting on M as a Lie transformation group. We usually denote by $I^o(M)$ the identity component of $I(M)$, that is, the connected component of $I(M)$ containing the identity transformation of M.

Covariant derivatives

While there is a natural way to differentiate smooth functions on a smooth manifold, there is no such natural way to differentiate smooth vector fields. The theory of studying the various possibilities for such a differentiation process is called theory of connections, or covariant derivatives. A *covariant derivative* (or *connection*) on a smooth manifold M is an operator ∇ assigning to two vector fields X, Y on M a third vector field $\nabla_X Y$ and satisfying the following axioms:

(i) ∇ is $\mathbb{R}$-bilinear,

(ii) $\nabla_{fX} Y = f \nabla_X Y$,

(iii) $\nabla_X fY = f \nabla_X Y + X(f) Y$,

where f is any smooth function on M and $X(f) = df(X)$ is the derivative of f in direction X.

If M is a Riemannian manifold, it is important to consider covariant derivatives that are compatible with the metric, that is to say, covariant derivatives satisfying

(iv) $Z\langle X,Y\rangle = \langle\nabla_Z X,Y\rangle + \langle X,\nabla_Z Y\rangle$.

A covariant derivative ∇ satisfying (iv) is called *metric*. A covariant derivative ∇ is called *torsion-free* if it satisfies

(v) $\nabla_X Y - \nabla_Y X = [X,Y]$.

On a Riemannian manifold there exists a unique torsion-free metric covariant derivative, i.e., a covariant derivative satisfying properties (iv) and (v). This covariant derivative is usually called the *Riemannian covariant derivative* or *Levi Civita covariant derivative* of the Riemannian manifold M. Unless otherwise stated, ∇ usually denotes the Levi Civita covariant derivative of a Riemannian manifold. Explicitly, from these properties, the Levi-Civita covariant derivative can be computed by means of the well-known Koszul formula:

$$\begin{aligned}2\langle\nabla_X Y,Z\rangle &= X\,\langle Y,Z\rangle + Y\,\langle X,Z\rangle - Z\,\langle X,Y\rangle + \\ &\quad + \langle[X,Y],Z\rangle - \langle[X,Z],Y\rangle - \langle[Y,Z],X\rangle\,.\end{aligned}$$

Riemannian curvature tensor, Ricci tensor, scalar curvature

The major concept of Riemannian geometry is curvature. There are various kinds of curvature of great interest. All of them can be deduced from the so-called *Riemannian curvature tensor*

$$R(X,Y)Z = \nabla_X\nabla_Y Z - \nabla_Y\nabla_X Z - \nabla_{[X,Y]}Z\ .$$

The Riemannian curvature tensor has the properties:

$$\begin{aligned}\langle R(X,Y)Z,W\rangle &= -\langle R(Y,X)Z,W\rangle\ ,\\ \langle R(X,Y)Z,W\rangle &= -\langle R(X,Y)W,Z\rangle\ ,\\ \langle R(X,Y)Z,W\rangle &= \langle R(Z,W)X,Y\rangle\ ,\end{aligned}$$

and

$$R(X,Y)Z + R(Y,Z)X + R(Z,X)Y = 0\ .$$

These equations are often called *algebraic curvature identities* of R, the latter, in particular, will be referred to as *algebraic Bianchi identity* or *first Bianchi identity*. Moreover, R satisfies

$$(\nabla_X R)(Y,Z)W + (\nabla_Y R)(Z,X)W + (\nabla_Z R)(X,Y)W = 0\ ,$$

known as the *differential Bianchi identity* or *second Bianchi identity* .

Let $p \in M$, $X, W \in T_pM$, and denote by $\mathrm{ric}_p(X, W)$ the real number which is obtained contracting the bilinear map

$$T_pM \times T_pM \to \mathbb{R} \;,\; (Y, Z) \mapsto \langle R(X, Y)Z, W\rangle \;.$$

The algebraic curvature identities show that ric_p is a symmetric bilinear map on T_pM. The tensor field ric is called the *Ricci tensor* of M. The corresponding self-adjoint tensor field of type (1,1) is denoted by Ric. A Riemannian manifold for which the Ricci tensor satisfies

$$\mathrm{ric} = f\langle \cdot, \cdot \rangle$$

with some smooth function f on M is called an *Einstein manifold.* For instance, each quaternionic Kähler manifold of dimension $4m$, $m \geq 2$, is an Einstein manifold.

The weakest notion of curvature on a Riemannian manifold is the *scalar curvature.* This is the smooth function on M that is obtained by contracting the Ricci tensor.

Sectional curvature

Perhaps the most geometric interpretation of the Riemannian curvature tensor arises via the sectional curvature. Consider a 2-dimensional linear subspace σ of T_pM, $p \in M$, and choose an orthonormal basis X, Y of σ. Since $\exp_p$ is a local diffeomorphism near 0 in T_pM, it maps an open neighborhood of 0 in σ onto some 2-dimensional surface S in M. Then the Gaussian curvature of S at p, which we denote by $K(\sigma)$, satisfies

$$K(\sigma) = \langle R(X, Y)Y, X\rangle \;.$$

Let $G_2(TM)$ be the Grassmann bundle over M consisting of all 2-dimensional linear subspaces $\sigma \subset T_pM, p \in M$. The map

$$K : G_2(TM) \to \mathbb{R} \;,\; \sigma \mapsto K(\sigma)$$

is called *sectional curvature function* of M, and $K(\sigma)$ is called *sectional curvature* of M with respect to σ. It is worthwhile to mention that one can reconstruct the Riemannian curvature tensor from the sectional curvature function by using the curvature identities.

A Riemannian manifold M is said to have *constant curvature* if the sectional curvature function is constant. If $\dim M \geq 3$, the second Bianchi identity and Schur's Lemma imply the following well-known result: *If the sectional curvature function depends only on the point p, then M has constant curvature.* A space of constant curvature is also called a *space form.* The Riemannian curvature tensor of a space form with constant curvature κ is given by

$$R(X, Y)Z = \kappa(\langle Y, Z\rangle X - \langle X, Z\rangle Y) \;.$$

Every connected three-dimensional Einstein manifold is a space form. It is an algebraic fact (i.e., does not involve the second Bianchi identity) that a Riemannian manifold M has constant sectional curvature equal to zero if and only if M is flat, i.e., the Riemannian curvature tensor of M vanishes.

A connected, simply connected, complete Riemannian manifold of nonpositive sectional curvature is called a *Hadamard manifold.* The *Hadamard Theorem* states that, for each point p in a Hadamard manifold M, the exponential map $\exp_p : T_pM \to M$ is a diffeomorphism. More generally, if M is a connected, complete Riemannian manifold of nonpositive sectional curvature, then the exponential map $\exp_p : T_pM \to M$ is a covering map for each $p \in M$.

Vector fields and flows

A vector field X on a Riemannian manifold M is called a *Killing vector field* if the local diffeomorphisms $\Phi_t^X : U \to M$ are isometries into M. This just means that the Lie derivative of the Riemannian metric of M with respect to X vanishes. A useful characterization of Killing vector fields is that a vector field X on a Riemannian manifold is a Killing vector field if and only if its covariant derivative ∇X is a skew-symmetric tensor field on M. A Killing vector field is completely determined by its value and its covariant derivative at any given point. In particular, a Killing vector field X for which $X_p = 0$ and $(\nabla X)_p = 0$ at some point $p \in M$ vanishes at each point of M. For a complete Killing vector field X on a Riemannian manifold M, the corresponding one-parameter group (Φ_t^X) consists of isometries of M. Conversely, suppose we have a one-parameter group Φ_t of isometries on a Riemannian manifold M. Then

$$X_p := \left.\frac{d}{dt}\right|_{t=0} \Phi_t(p)$$

defines a complete Killing vector field X on M with $\Phi_t^X = \Phi_t$ for all $t \in \mathbb{R}$. If X is a Killing vector field on M and $X_p = 0$, then

$$\left.\frac{d}{dt}\right|_{t=0} (\Phi_t^X)_{*p} = (\nabla X)_p$$

for all $t \in \mathbb{R}$.

Distributions and the Frobenius Theorem

A *distribution* on a Riemannian manifold M is a smooth vector subbundle $\mathcal{H}$ of the tangent bundle TM. A distribution $\mathcal{H}$ on M is called *integrable* if for any $p \in M$ there exists a connected submanifold L_p of M such that $T_qL_p = \mathcal{H}_q$ for all $q \in L_p$. Such a submanifold L_p is called an *integral manifold* of $\mathcal{H}$. The *Frobenius Theorem* states that $\mathcal{H}$ is integrable if and only if it is involutive, that is, if the Lie bracket of any two vector fields tangent to $\mathcal{H}$ is also a vector field tangent to $\mathcal{H}$. If $\mathcal{H}$ is integrable, there exists through each point $p \in M$ a maximal integral manifold of $\mathcal{H}$ containing p. Such a maximal integral manifold is called the *leaf of $\mathcal{H}$ through p.* A distribution $\mathcal{H}$ on M is called *autoparallel* if $\nabla_{\mathcal{H}}\mathcal{H} \subset \mathcal{H}$, that is, if for any two vector fields X, Y tangent to $\mathcal{H}$ the vector field $\nabla_X Y$ is also tangent to $\mathcal{H}$. By the Frobenius Theorem every autoparallel distribution is integrable. An integrable distribution is autoparallel if and only if its leaves are totally geodesic submanifolds of the ambient space. A distribution $\mathcal{H}$ on M is called *parallel* if $\nabla_X \mathcal{H} \subset \mathcal{H}$ for any vector field X on M. Obviously, any parallel distribution is autoparallel. Since ∇ is a metric connection, for each parallel distribution $\mathcal{H}$ on M, its orthogonal complement $\mathcal{H}^\perp$ in TM is also a parallel distribution on M.

Covariant derivatives along curves

Given a piecewise differentiable curve $c(t)$ on M, defined on an interval I, there is a covariant derivative operator $\frac{D^c}{dt}$ along $c(t)$ which maps (differentiable) tangent vector fields of M along c to (differentiable) tangent vector fields of M along c (see [71]). Frequently, when it is clear from the context, we will write $\frac{D}{dt}X$ or $X'(t)$ instead of $\frac{D^c}{dt}X$. The covariant derivative along c is completely determined by the following properties:

(i) $\frac{D}{dt}(Z_1(t) + Z_2(t)) = \frac{D}{dt}Z_1(t) + \frac{D}{dt}Z_2(t)$ for all vector fields $Z_1(t), Z_2(t)$ along $c(t)$;

(ii) $\frac{D}{dt}f(t)Z(t) = f'(t)Z(t) + f(t)\frac{D}{dt}Z(t)$ for all vector fields $Z(t)$ along $c(t)$ and all smooth functions $f(t)$ defined on I;

(iii) $\frac{D}{dt}Y(c(t)) = \nabla_{c'(t)}Y$, for all vector fields Y on M.

Since ∇ is metric, we have

$$\frac{d}{dt}\langle X(t), Y(t)\rangle = \langle \frac{D}{dt}X(t), Y(t)\rangle + \langle X(t), \frac{D}{dt}Y(t)\rangle$$

for all vector fields $X(t), Y(t)$ along $c(t)$.

REMARK A.1.1 If $c(t) \equiv p$ is a constant curve and $X(t)$ is a vector field along $c(t)$, i.e., for all t we have $X(t) \in T_pM$, then $\frac{D}{dt}X(t) = \frac{d}{dt}X(t)$, where the last derivative is the usual one in the vector space T_pM. □

A vector field $X(t)$ along $c(t)$ is called *parallel* if $\frac{D}{dt}X(t) \equiv 0$. The above equality implies that $\langle X(t), Y(t)\rangle$ is constant if both vector fields are parallel along c. From the theory of ordinary differential equations, one can easily see that, for each $v \in T_{c(t_o)}M$, $t_o \in I$, there exists a unique parallel vector field $X_v(t)$ along $c(t)$ such that $X_v(t_o) = v$. For each $t \in I$ there is then a well-defined linear isometry $\tau^c(t) : T_{c(t_o)} \to T_{c(t)}$, called the *parallel transport* along c, given by

$$\tau^c(t)(v) = X_v(t) .$$

The covariant derivative operator and parallel transport along $c(t)$ are related by

$$\frac{D}{dt}X(t) = \frac{d}{dh}\bigg|_{h=0} (\tau^c(t+h))^{-1}X(t+h) .$$

Note that the parallel transport does not depend on the parametrization of the curve.

A *parametrized surface* in M is a smooth map f from an open subset of $\mathbb{R}^2$ into M. We do not assume that the differential of such a map is injective (in this case, the surface is called regular). As in the case of curves, we will be considering smooth vector fields $X(s,t)$ along f. Then we have $X(s,t) \in T_{f(s,t)}M$. As usual, we will denote by $\frac{D}{\partial s}$ the covariant derivative along the curve $s \mapsto f(s,t)$ with t fixed. The corresponding tangent vector field of this curve is denoted by $\frac{\partial f}{\partial s}$. In the same way, we define $\frac{D}{\partial t}$ and $\frac{\partial f}{\partial t}$. From the fact that the Levi Civita covariant derivative is torsion-free, we deduce

$$\frac{D}{\partial s}\frac{\partial f}{\partial t} = \frac{D}{\partial t}\frac{\partial f}{\partial s}.$$

But the covariant derivatives with respect to s and t do not commute in general if the curvature tensor does not vanish. More precisely,

$$\frac{D}{\partial s}\frac{D}{\partial t}X(s,t) - \frac{D}{\partial t}\frac{D}{\partial s}X(s,t) = R\left(\frac{\partial f}{\partial s}, \frac{\partial f}{\partial t}\right)X(s,t).$$

We will often omit f in the partial derivatives and simply write $\frac{\partial}{\partial s}$ and $\frac{\partial}{\partial t}$.

Note that, when a smooth curve $c(t)$ is defined on a closed interval $[a,b]$, this means that $c(t)$ is the restriction to $[a,b]$ of a smooth curve that is defined on an open interval $I \supset [a,b]$. A similar remark applies to a surface that is defined on a closed subset of $\mathbb{R}^2$.

Holonomy

A Riemannian manifold M is said to be *flat* if its curvature tensor vanishes. This implies that, locally, the parallel transport does not depend on the curve used for joining two given points. If the curvature tensor does not vanish, the parallel transport depends on the curve. A way of measuring how far the space deviates from being flat is given by the *holonomy group*. Let $p \in M$ and $\Omega(p)$ the set of all piecewise smooth curves $c : [0,1] \to M$ with $c(0) = c(1) = p$. Then the parallel translation along any curve $c \in \Omega(p)$ from $c(0)$ to $c(1)$ is an orthogonal transformation of T_pM. In an obvious manner, the set of all these parallel translations forms a subgroup $\mathrm{Hol}_p(M)$ of the orthogonal group $O(T_pM)$, which is called the *holonomy group of M at p*. As a subset of $O(T_pM)$, it carries a natural topology. With respect to this topology, the identity component $\mathrm{Hol}_p^o(M)$ of $\mathrm{Hol}_p(M)$ is called the *restricted holonomy group* of M at p. The restricted holonomy group consists of all those transformations arising from null homotopic curves in $\Omega(p)$. If M is connected, then all (restricted) holonomy groups are congruent to each other, and, in this situation, one speaks of the (restricted) holonomy group of the manifold M, which we will then denote by $\mathrm{Hol}(M)$ resp. $\mathrm{Hol}^o(M)$. The connected Lie group $\mathrm{Hol}^o(M)$ is always compact, whereas $\mathrm{Hol}(M)$ is, in general, not closed in the orthogonal group. A reduction of the holonomy group corresponds to an additional geometric structure on M. For instance, $\mathrm{Hol}(M)$ is contained in $SO(T_pM)$ for some $p \in M$ if and only if M is orientable. An excellent introduction to holonomy groups can be found in the book by Salamon [196].

Geodesics

Of great importance in Riemannian geometry are the curves that minimize the distance between two given points. Of course, given two arbitrary points, such curves do not exist in general. But they do exist provided the manifold is connected and complete. Distance-minimizing curves γ are solutions of a variational problem. The corresponding first variation formula shows that any such curve γ satisfies $\frac{D}{dt}\dot{\gamma} = 0$. A smooth curve γ satisfying this equation is called a *geodesic*. Every geodesic is locally distance-minimizing, but not globally, as a great circle on a sphere illustrates. The basic theory of ordinary differential equations implies that, for each point $p \in M$ and each tangent vector $X \in T_pM$, there exists a unique geodesic $\gamma : I \to M$ with $0 \in I$, $\gamma(0) = p$, $\dot{\gamma}(0) = X$, and such that, for any other geodesic $\alpha : J \to M$ with $0 \in J$, $\alpha(0) = p$ and $\dot{\alpha}(0) = X$, we have $J \subset I$. This curve γ is often called the *maximal geodesic* in M through p tangent to X, and we denote it sometimes by γ_X. The *Hopf-Rinow Theorem* states that a Riemannian manifold is complete if and only if γ_X is defined on $\mathbb{R}$ for each $X \in TM$.

Kähler manifolds

An *almost complex structure* on a smooth manifold M is a tensor field J of type (1,1) on M satisfying $J^2 = -\mathrm{id}_{TM}$. An *almost complex manifold* is a smooth manifold equipped with an almost complex structure. Each tangent space of an almost complex manifold is isomorphic to a complex vector space, which implies that the dimension of an almost complex manifold is an even number. A *Hermitian metric* on an almost complex manifold M is a Riemannian metric $\langle \cdot, \cdot \rangle$ for which the almost complex structure J on M is orthogonal, that is,

$$\langle JX, JY \rangle = \langle X, Y \rangle$$

for all $X, Y \in T_pM$, $p \in M$. An orthogonal almost complex structure on a Riemannian manifold is called an *almost Hermitian structure*.

Every complex manifold M has a canonical almost complex structure. In fact, if $z = x + iy$ is a local coordinate on M, define

$$J\frac{\partial}{\partial x_\nu} = \frac{\partial}{\partial y_\nu}\ ,\ J\frac{\partial}{\partial y_\nu} = -\frac{\partial}{\partial x_\nu}\ .$$

These local almost complex structures are compatible on the intersection of any two coordinate neighborhoods and hence induce an almost complex structure, which is called the *induced complex structure* of M. An almost complex structure J on a smooth manifold M is *integrable* if M can be equipped with the structure of a complex manifold so that J is the induced complex structure. A famous result by Newlander-Nirenberg says that the almost complex structure J of an almost complex manifold M is integrable if and only if

$$[X, Y] + J[JX, Y] + J[X, JY] - [JX, JY] = 0$$

for all $X, Y \in T_pM$, $p \in M$. A *Hermitian manifold* is an almost Hermitian manifold with an integrable almost complex structure. The almost Hermitian structure of a Hermitian manifold is called a *Hermitian structure*.

The 2-form ω on a Hermitian manifold M defined by

$$\omega(X,Y) = \langle X, JY \rangle$$

is called the *Kähler form* of M. A *Kähler manifold* is a Hermitian manifold whose Kähler form is closed. A Hermitian manifold M is a Kähler manifold if and only if its Hermitian structure J is parallel with respect to the Levi Civita connection ∇ of M, that is, if $\nabla J = 0$. The latter condition characterizes the Kähler manifolds among all Hermitian manifolds by the geometric property that parallel translation along curves commutes with the Hermitian structure J. A $2m$-dimensional connected Riemannian manifold M can be equipped with the structure of a Kähler manifold if and only if its holonomy group $\mathrm{Hol}(M)$ is contained in the unitary group $U(m)$. The standard examples of Kähler manifolds are the complex vector space $\mathbb{C}^m$, the complex projective space $\mathbb{C}P^m$, and the complex hyperbolic space $\mathbb{C}H^m$.

The curvature and Ricci tensors of a Kähler manifold satisfy

$$\begin{aligned} &R(X,Y)JZ = J(R(X,Y)Z)\,, \\ &2\,\mathrm{ric}\,(X,JY) = \langle R(X,Y), J\rangle\,, \qquad X,Y,Z \in T_pM,\ p \in M\ , \end{aligned}$$

where $\langle .\,,.\rangle$ denotes the inner product on tensors induced by the Riemannian metric.

Quaternionic Kähler manifolds

A *quaternionic Kähler structure* on a Riemannian manifold M is a rank three vector subbundle $\mathfrak{J}$ of the endomorphism bundle $\mathrm{End}(TM)$ over M with the following properties: (1) For each p in M there exist an open neighborhood U of p in M and sections J_1, J_2, J_3 of $\mathfrak{J}$ over U so that J_ν is an almost Hermitian structure on U and

$$J_\nu J_{\nu+1} = J_{\nu+2} = -J_{\nu+1}J_\nu \quad \text{(index modulo three)}$$

for all $\nu = 1,2,3$; (2) $\mathfrak{J}$ is a parallel subbundle of $\mathrm{End}(TM)$, that is, if J is a section in $\mathfrak{J}$ and X a vector field on M, then $\nabla_X J$ is also a section in $\mathfrak{J}$. Each triple J_1, J_2, J_3 of the above kind is called a *canonical local basis* of $\mathfrak{J}$, or, if restricted to the tangent space T_pM of M at p, a *canonical basis* of $\mathfrak{J}_p$. A *quaternionic Kähler manifold* is a Riemannian manifold equipped with a quaternionic Kähler structure. The canonical bases of a quaternionic Kähler structure turn the tangent spaces of a quaternionic Kähler manifold into quaternionic vector spaces. Therefore, the dimension of a quaternionic Kähler manifold is $4m$ for some $m \in \mathbb{N}$. A $4m$-dimensional connected Riemannian manifold M can be equipped with a quaternionic Kähler structure if and only if its holonomy group $\mathrm{Hol}(M)$ is contained in $Sp(m)\cdot Sp(1)$. The standard examples of quaternionic Kähler manifolds are the quaternionic vector space $\mathbb{H}^m$, the quaternionic projective space $\mathbb{H}P^m$, and the quaternionic hyperbolic space $\mathbb{H}H^m$.

Riemannian products and covering spaces

Let M_1 and M_2 be Riemannian manifolds. At each point $(p_1, p_2) \in M_1 \times M_2$, the tangent space $T_{(p_1,p_2)}(M_1 \times M_2)$ is canonically isomorphic to the direct sum $T_{p_1}M_1 \oplus T_{p_2}M_2$. The inner products on $T_{p_1}M_1$ and $T_{p_2}M_2$ therefore induce an inner product on $T_{(p_1,p_2)}(M_1 \times M_2)$. In this way, we get a Riemannian metric on $M_1 \times M_2$. The product manifold $M_1 \times M_2$ equipped with this Riemannian metric is called the *Riemannian product* of M_1 and M_2. For each connected Riemannian manifold M there exists a connected, simply connected Riemannian manifold $\tilde{M}$ and an isometric covering map $\tilde{M} \to M$. Such a manifold $\tilde{M}$ is unique up to isometry and is called the *Riemannian universal covering space* of M. A Riemannian manifold M is called *reducible* if its Riemannian universal covering space $\tilde{M}$ is isometric to the Riemannian product of at least two Riemannian manifolds of dimension ≥ 1. Otherwise, M is called *irreducible.* A Riemannian manifold M is said to be *locally reducible* if, for each point $p \in M$, there exists an open neighborhood of p in M that is isometric to the Riemannian product of at least two Riemannian manifolds of dimension ≥ 1. Otherwise, M is said to be *locally irreducible.*

The de Rham Decomposition Theorem

The *de Rham Decomposition Theorem* states that a connected Riemannian manifold M is locally reducible if and only if T_pM is reducible as a $\mathrm{Hol}^o(M)$-module for some, and hence for every, point $p \in M$. Since $\mathrm{Hol}^o(M)$ is compact, there exists a decomposition

$$T_pM = V_0 \oplus V_1 \oplus \ldots \oplus V_k$$

of T_pM into $\mathrm{Hol}^o(M)$-invariant subspaces of T_pM, where $V_0 \subset T_pM$ is the fixed point set of the action of $\mathrm{Hol}^o(M)$ on T_pM and $V_1, \ldots, V_k$ are irreducible $Hol^o(M)$-modules. It might happen that $V_0 = T_pM$, for instance, when $M = \mathbb{R}^n$, or $V_0 = \{0\}$, or when M is the sphere S^n, $n > 1$. The above decomposition is unique up to order of the factors and determines integrable distributions $V_0, \ldots, V_k$ on M. Then there exists an open neighborhood of p in M that is isometric to the Riemannian product of sufficiently small integral manifolds of these distributions through p. The global version of the de Rham decomposition theorem states that a connected, simply connected, complete Riemannian manifold M is reducible if and only if T_pM is reducible as a $\mathrm{Hol}^o(M)$-module. If M is reducible and $T_pM = V_0 \oplus \ldots \oplus V_k$ is the decomposition of T_pM as described above, then M is isometric to the Riemannian product of the maximal integral manifolds $M_0, \ldots, M_k$ through p of the distributions $V_0, \ldots, V_k$. In this situation, $M = M_0 \times \ldots \times M_k$ is called the *de Rham decompositon* of M. The Riemannian manifold M_0 is isometric to a possibly zero-dimensional Euclidean space. If $\dim M_0 > 0$ then M_0 is called the *Euclidean factor* of M. A connected, complete Riemannian manifold M is said to have *no Euclidean factor* if the de Rham decomposition of the Riemannian universal covering space $\tilde{M}$ of M has no Euclidean factor.

Exponential map and normal coordinates

Of great importance is the exponential map exp of a Riemannian manifold. To define it, we denote by $\widetilde{TM} \subset TM$ the set of all tangent vectors for which $\gamma_X(1)$ is defined. This is an open subset of TM containing the zero section. A Riemannian manifold is complete if and only if $\widetilde{TM} = TM$. The map

$$\exp : \widetilde{TM} \to M \ , \ X \mapsto \gamma_X(1)$$

is called the *exponential map* of M. For each $p \in M$, we denote the restriction of exp to $T_pM \cap \widetilde{TM}$ by $\exp_p$. The map $\exp_p$ is a diffeomorphism from some open neighborhood of $0 \in T_pM$ onto some open neighborhood of $p \in M$. If we choose an orthonomal basis $e_1, \ldots, e_m$ of T_pM, then the map

$$(x_1, \ldots, x_m) \mapsto \exp_p\left(\sum_{i=1}^{m} x_i e_i\right)$$

defines local coordinates of M in some open neighborhood of p. Such coordinates are called *normal coordinates*.

Jacobi vector fields

Let $\gamma : I \to M$ be a geodesic parametrized by arc length. A vector field Y along γ is called a *Jacobi vector field* if it satisfies the second order differential equation

$$Y'' + R(Y, \dot{\gamma})\dot{\gamma} = 0 \ .$$

Standard theory of ordinary differential equations implies that the Jacobi vector fields along a geodesic form a $2n$-dimensional vector space. Every Jacobi vector field is uniquely determined by the initial values $Y(t_0)$ and $Y'(t_0)$ at a fixed number $t_0 \in I$. The Jacobi vector fields arise geometrically as infinitesimal variational vector fields of geodesic variations. Jacobi vector fields can be used to describe the differential of the exponential map. Indeed, let $p \in M$ and $\exp_p$ be the exponential map of M restricted to T_pM. For each $X \in T_pM$ we identify $T_X(T_pM)$ with T_pM in the canonical way. Then, for each $Z \in T_pM$, we have

$$exp_{p*X}Z = Y_Z(1) \ ,$$

where Y_Z is the Jacobi vector field along γ_X with initial values $Y_Z(0) = 0$ and $Y_Z'(0) = Z$.

A.2 Lie groups and Lie algebras

Lie groups were introduced by Sophus Lie in the framework of his studies on differential equations as local transformation groups. The global theory of Lie groups was developed by Hermann Weyl and Élie Cartan. Lie groups are both groups and manifolds, which allows us to use concepts from both algebra and analysis to study these objects. Some modern books on this topic are Adams [2], Carter-Segal-Macdonald [37], Knapp [113], Varadarajan [231]. Foundations on Lie theory can also be found in Onishchik [182], and the structure of Lie groups and Lie algebras is discussed in Onishchik-Vinberg [183]. A good introduction to the exceptional Lie groups may be found in Adams [3].

Lie groups

A *real Lie group*, or briefly *Lie group*, is an abstract group G that is equipped with a smooth manifold structure such that $G \times G \to G$, $(g_1, g_2) \mapsto g_1 g_2$ and $G \to G$, $g \mapsto g^{-1}$ are smooth maps. For a *complex Lie group* G, one requires that G is equipped with a complex analytic structure and that multiplication and inversion are holomorphic maps. A simple example of a real Lie group is $\mathbb{R}^n$ equipped with its additive group structure, turning it into an Abelian Lie group. Suppose Γ is a lattice in $\mathbb{R}^n$, that is, Γ is a discrete subgroup of rank n of the group of translations of $\mathbb{R}^n$. Then $T^n = \mathbb{R}^n/\Gamma$ is a compact Abelian Lie group, a so-called n-dimensional *torus*. Every Abelian Lie group is isomorphic to the product $\mathbb{R}^n \times T^k$ for some nonnegative integers $n, k \geq 0$. Another basic example of a Lie group is the isometry group $I(M)$ of a Riemannian manifold M. For any Lie group G the connected component of G containing the identity of G is called the identity component of G. We denote this component usually by G^o.

A subgroup H of a Lie group G is called a *Lie subgroup* if H is a Lie group and if the inclusion $H \to G$ is a smooth map. For instance the identity component G^o of a Lie group G is a Lie subgroup. Each closed subgroup of a Lie group is a Lie subgroup.

For each $g \in G$, the smooth diffeomorphisms

$$L_g : G \to G \ , \ g' \mapsto gg' \quad \text{and} \quad R_g : G \to G \ , \ g' \mapsto g'g$$

are called the *left translation* and *right translation* on G with respect to g, respectively. A vector field X on G is called *left-invariant* resp. *right-invariant* if it is invariant under any left translation resp. right translation, that is $L_{g*}X = X \circ L_g$ resp. $R_{g*}X = X \circ R_g$ for all $g \in G$. The smooth diffeomorphism

$$I_g = L_g \circ R_{g^{-1}} : G \to G \ , \ g' \mapsto gg'g^{-1}$$

is called an *inner automorphism* of G.

Lie algebras

A (real or complex) *Lie algebra* is a finite-dimensional (real or complex) vector space $\mathfrak{g}$ equipped with a skew-symmetric bilinear map $[\cdot,\cdot] : \mathfrak{g} \times \mathfrak{g} \to \mathfrak{g}$ satisfying

$$[[X,Y],Z] + [[Y,Z],X] + [[Z,X],Y] = 0$$

for all $X, Y, Z \in \mathfrak{g}$. The latter identity is called the *Jacobi identity*. To every Lie group G, there is associated a Lie algebra $\mathfrak{g}$, namely the vector space of all left-invariant vector fields equipped with the bilinear map arising from the Lie bracket. Since each left-invariant vector field is uniquely determined by its value at the identity $e \in G$, $\mathfrak{g}$ is isomorphic as a vector space to T_eG. In particular, we have $\dim \mathfrak{g} = \dim G$.

Let $\mathfrak{g}$ be a real Lie algebra and $\mathfrak{g}^{\mathbb{C}} = \mathfrak{g} \oplus i\mathfrak{g}$ be the complexification of $\mathfrak{g}$ considered as a vector space. By extending the Lie algebra structure on $\mathfrak{g}$ complex linearly to $\mathfrak{g}^{\mathbb{C}}$ we turn $\mathfrak{g}^{\mathbb{C}}$ into a complex Lie algebra, the *complexification* of $\mathfrak{g}$. Any complex Lie algebra $\mathfrak{h}$ can be considered canonically as a real Lie algebra $\mathfrak{h}^{\mathbb{R}}$ by restricting the scalar multiplication to $\mathbb{R} \subset \mathbb{C}$. If $\mathfrak{g}$ is a real Lie algebra and $\mathfrak{h}$ is a complex Lie algebra so that $\mathfrak{h}$ is isomorphic to $\mathfrak{g}^{\mathbb{C}}$ then $\mathfrak{g}$ is a *real form* of $\mathfrak{h}$.

Lie exponential map

Let G be a Lie group with Lie algebra $\mathfrak{g}$. Any $X \in \mathfrak{g}$ is a left-invariant vector field on G and hence determines a flow $\Phi^X : \mathbb{R} \times G \to G$. The smooth map

$$\mathrm{Exp} : \mathfrak{g} \to G \, , \; X \mapsto \Phi^X(1, e)$$

is called the *Lie exponential map* of $\mathfrak{g}$ or G. For each $X \in \mathfrak{g}$, the curve $t \mapsto \mathrm{Exp}(tX)$ is a one-parameter subgroup of G and we have $\Phi^X(t, g) = R_{\mathrm{Exp}(tX)}(g)$ for all $g \in G$ and $t \in \mathbb{R}$. The Lie exponential map is crucial when studying the interplay between Lie groups and Lie algebras. It is a diffeomorphism of some open neighborhood of $0 \in \mathfrak{g}$ onto some open neighborhood of $e \in G$.

The Lie algebra of the isometry group

Let M be a connected Riemannian manifold. The Lie algebra $\mathfrak{g}$ of the isometry group $G = I(M)$ can be identified with the Lie algebra $\mathcal{K}(M)$ of Killing vector fields on M in the following way. The Lie bracket on $\mathcal{K}(M)$ is the usual Lie bracket for vector fields. For $X \in \mathfrak{g}$ we define a vector field X^* on M by

$$X_p^* = \left.\frac{d}{dt}\right|_{t=0} \mathrm{Exp}(tX)(p)$$

for all $p \in M$. Then the map

$$\mathfrak{g} \to \mathcal{K}(M) \, , \; X \mapsto X^*$$

is a vector space isomorphism satisfying

$$[X, Y]^* = -[X^*, Y^*] \, .$$

In other words, if one should define the Lie algebra $\mathfrak{g}$ of G by using right-invariant vector fields instead of left-invariant vector fields, then the map $\mathfrak{g} \to \mathcal{K}(M)$, $X \mapsto X^*$ would be a Lie algebra isomorphism.

Adjoint representation

The inner automorphisms I_g of G determine the so-called *adjoint representation* of G by

$$\mathrm{Ad} : G \to \mathrm{GL}(\mathfrak{g}) \ , \ g \mapsto I_{g*e} \ ,$$

where I_{g*e} denotes the differential of I_g at e and we identify T_eG with $\mathfrak{g}$ by means of the vector space isomorphism

$$\mathfrak{g} \to T_eG \ , \ X \mapsto X_e \ .$$

The adjoint representation of $\mathfrak{g}$ is the homomorphism

$$\mathrm{ad} : \mathfrak{g} \to \mathrm{End}(\mathfrak{g}) \ , \ X \mapsto (\mathfrak{g} \to \mathfrak{g} \ , \ Y \mapsto [X,Y]) \ .$$

It can be obtained from Ad by means of

$$\mathrm{ad}(X)Y = \frac{d}{dt}|_{t=0}(t \mapsto \mathrm{Ad}(\mathrm{Exp}(tX))Y) \ .$$

The relationship between Ad and ad is described by

$$\mathrm{Ad}(\mathrm{Exp}(X)) = \exp(\mathrm{ad}(X)) \ ,$$

where $\exp$ denotes here the exponential map for endomorphisms of the vector space $\mathfrak{g}$.

Killing form

The symmetric bilinear form B on $\mathfrak{g}$ defined by

$$B(X,Y) = \mathrm{tr}(\mathrm{ad}(X) \circ \mathrm{ad}(Y))$$

for all $X, Y \in \mathfrak{g}$ is called the *Killing form*, or *Cartan-Killing form*, of $\mathfrak{g}$. Every automorphism σ of $\mathfrak{g}$ has the property

$$B(\sigma X, \sigma Y) = B(X,Y)$$

for all $X, Y \in \mathfrak{g}$. This implies that

$$B(\mathrm{ad}(Z)X, Y) + B(X, \mathrm{ad}(Z)Y) = 0$$

for all $X, Y, Z \in \mathfrak{g}$.

Solvable and nilpotent Lie algebras and Lie groups

Let $\mathfrak{g}$ be a Lie algebra. The *commutator ideal* $[\mathfrak{g},\mathfrak{g}]$ of $\mathfrak{g}$ is the ideal in $\mathfrak{g}$ generated by all vectors in $\mathfrak{g}$ of the form $[X,Y]$, $X, Y \in \mathfrak{g}$. The commutator series of $\mathfrak{g}$ is the decreasing sequence

$$\mathfrak{g}^0 = \mathfrak{g} \ , \ \mathfrak{g}^1 = [\mathfrak{g}^0, \mathfrak{g}^0] \ , \ \mathfrak{g}^2 = [\mathfrak{g}^1, \mathfrak{g}^1] \ , \ \ldots$$

of ideals of $\mathfrak{g}$. The Lie algebra $\mathfrak{g}$ is *solvable* if this sequence is finite, that is, if $\mathfrak{g}^k = 0$ for some $k \in \mathbb{N}$. The lower central series of $\mathfrak{g}$ is the decreasing sequence

$$\mathfrak{g}_0 = \mathfrak{g} \ , \ \mathfrak{g}_1 = [\mathfrak{g}, \mathfrak{g}_0] \ , \ \mathfrak{g}_2 = [\mathfrak{g}, \mathfrak{g}_1] \ , \ \ldots$$

of ideals in $\mathfrak{g}$. The Lie algebra $\mathfrak{g}$ is *nilpotent* if this sequence is finite, that is, if $\mathfrak{g}_k = 0$ for some $k \in \mathbb{N}$. Each nilpotent Lie algebra is solvable. A Lie group G is solvable or nilpotent if and only if its Lie algebra $\mathfrak{g}$ is solvable or nilpotent, respectively.

Simple and semisimple Lie algebras and Lie groups

Let $\mathfrak{g}$ be a Lie algebra. There is a unique solvable ideal in $\mathfrak{g}$ that contains all solvable ideals in $\mathfrak{g}$, the so-called *radical* of $\mathfrak{g}$. If this radical is trivial, the Lie algebra is called *semisimple*. A criterion by Cartan says that a Lie algebra is semisimple if and only if its Killing form is nondegenerate. A semisimple Lie algebra $\mathfrak{g}$ is called *simple* if it contains no ideals different from $\{0\}$ and $\mathfrak{g}$. A Lie group is semisimple or simple if and only if its Lie algebra is semisimple or simple, respectively.

Structure theory of semisimple complex Lie algebras

Let $\mathfrak{g}$ be a semisimple complex Lie algebra and B its Killing form. A *Cartan subalgebra* of $\mathfrak{g}$ is a maximal Abelian subalgebra $\mathfrak{h}$ of $\mathfrak{g}$ so that all endomorphisms $\mathrm{ad}(H)$, $H \in \mathfrak{h}$, are simultaneously diagonalizable. There always exists a Cartan subalgebra in $\mathfrak{g}$, and any two of them are conjugate by an inner automorphism of $\mathfrak{g}$. The common value of the dimension of these Cartan subalgebras is called the *rank* of $\mathfrak{g}$.

Any semisimple complex Lie algebra can be decomposed into the direct sum of simple complex Lie algebras, which were classified by Elie Cartan. The simple complex Lie algebras are

$$A_n = \mathfrak{sl}(n+1, \mathbb{C}) \ , B_n = \mathfrak{so}(2n+1, \mathbb{C}) \ , C_n = \mathfrak{sp}(n, \mathbb{C}) \ , D_n = \mathfrak{so}(2n, \mathbb{C})(n \geq 3) \ ,$$

which are the simple complex Lie algebras of classical type, and

$$G_2 \ , \ F_4 \ , \ E_6 \ , \ E_7 \ , \ E_8 \ ,$$

which are the simple complex Lie algebras of exceptional type. Here, the index refers to the rank of the Lie algebra. Note that there are isomorphisms $A_1 = B_1 = C_1$, $B_2 = C_2$ and $A_3 = D_3$. The Lie algebra $D_2 = \mathfrak{so}(4, \mathbb{C})$ is not simple since $D_2 = A_1 \oplus A_1$.

Let $\mathfrak{h}$ be a Cartan subalgebra of a semisimple complex Lie algebra $\mathfrak{g}$. For each one-form α in the dual vector space $\mathfrak{g}^*$ of $\mathfrak{g}$ we define

$$\mathfrak{g}_\alpha = \{X \in \mathfrak{g} \mid \mathrm{ad}(H)X = \alpha(H)X \text{ for all } H \in \mathfrak{h}\} \ .$$

If $\mathfrak{g}_\alpha$ is nontrivial and α is nonzero, α is called a *root* of $\mathfrak{g}$ with respect to $\mathfrak{h}$ and $\mathfrak{g}_\alpha$ is called the *root space* of $\mathfrak{g}$ with respect to α. The complex dimension of $\mathfrak{g}_\alpha$ is always one. We denote by Δ the set of all roots of $\mathfrak{g}$ with respect to $\mathfrak{h}$. The direct sum decomposition

$$\mathfrak{g} = \mathfrak{h} \oplus \bigoplus_{\alpha \in \Delta} \mathfrak{g}_\alpha$$

is called the *root space decomposition* of $\mathfrak{g}$ with respect to the Cartan subalgebra $\mathfrak{h}$.

Structure theory of compact real Lie groups

Let G be a connected, compact, real Lie group. The Lie algebra $\mathfrak{g}$ of G admits an inner product so that each $\mathrm{Ad}(g)$, $g \in G$, acts as an orthogonal transformation on $\mathfrak{g}$ and each $\mathrm{ad}(X)$, $X \in \mathfrak{g}$, acts as a skew-symmetric transformation on $\mathfrak{g}$. This yields the direct sum decomposition

$$\mathfrak{g} = \mathfrak{z}(\mathfrak{g}) \oplus [\mathfrak{g}, \mathfrak{g}] ,$$

where $\mathfrak{z}(\mathfrak{g})$ is the center of $\mathfrak{g}$ and $[\mathfrak{g}, \mathfrak{g}]$ is the commutator ideal in $\mathfrak{g}$, which is always semisimple. The Killing form of $\mathfrak{g}$ is negative semidefinite. If, in addition, $\mathfrak{g}$ is semisimple, or equivalently, if $\mathfrak{z}(\mathfrak{g}) = 0$, then its Killing form B is negative definite and hence $-B$ induces an $\mathrm{Ad}(G)$-invariant Riemannian metric on G. This metric is biinvariant, that is, both left and right translations are isometries of G. Let $Z(G)^o$ be the identity component of the center $Z(G)$ of G and G^s the connected Lie subgroup of G with Lie algebra $[\mathfrak{g}, \mathfrak{g}]$. Both $Z(G)^o$ and G^s are closed subgroups of G, G^s is semisimple and has finite centre, and G is isomorphic to the direct product $Z(G)^o \times G^s$.

A *torus* in G is a connected Abelian Lie subgroup T of G. The Lie algebra $\mathfrak{t}$ of a torus T in G is an Abelian Lie subalgebra of $\mathfrak{g}$. A torus T in G that is not properly contained in any other torus in G is called a *maximal torus*. Analogously, an Abelian Lie subalgebra $\mathfrak{t}$ of $\mathfrak{g}$ which is not properly contained in any other Abelian Lie subalgebra of $\mathfrak{g}$ is called a *maximal Abelian subalgebra*. There is a natural correspondence between the maximal tori in G and the maximal Abelian subalgebras of $\mathfrak{g}$. Any maximal Abelian subalgebra $\mathfrak{t}$ of $\mathfrak{g}$ is of the form

$$\mathfrak{t} = \mathfrak{z}(\mathfrak{g}) \oplus \mathfrak{t}^s ,$$

where $\mathfrak{t}^s$ is some maximal Abelian subalgebra of the semisimple Lie algebra $[\mathfrak{g}, \mathfrak{g}]$.

Any two maximal Abelian subalgebras of $\mathfrak{g}$ are conjugate via $\mathrm{Ad}(g)$ for some $g \in G$. This readily implies that any two maximal tori in G are conjugate. Furthermore, if T is a maximal torus in G, then any $g \in G$ is conjugate to some $t \in T$. Any two elements in T are conjugate in G if and only if they are conjugate via the Weyl group $W(G, T)$ of G with respect to T. The *Weyl group* of G with respect to T is defined by

$$W(G, T) = N_G(T)/Z_G(T) ,$$

where $N_G(T)$ is the normalizer of T in G and $Z_G(T) = T$ is the centralizer of T in G. In particular, the conjugacy classes in G are parametrized by $T/W(G, T)$. The common dimension of the maximal tori of G (resp. of the maximal Abelian subalgebras of $\mathfrak{g}$) is called the *rank* of G (resp. the *rank* of $\mathfrak{g}$). Let $\mathfrak{t}$ be a maximal Abelian subalgebra of $\mathfrak{g}$. Then $\mathfrak{t}^{\mathbb{C}}$ is a Cartan subalgebra of $\mathfrak{g}^{\mathbb{C}}$. For this reason, $\mathfrak{t}$ is also called a *Cartan subalgebra* of $\mathfrak{g}$ and the rank of $\mathfrak{g}$ coincides with the rank of $\mathfrak{g}^{\mathbb{C}}$.

We assume from now on that $\mathfrak{g}$ is semisimple, that is, the centre of $\mathfrak{g}$ is trivial. Then $\mathfrak{g}$ is called a *compact real form* of $\mathfrak{g}^{\mathbb{C}}$. Each semisimple complex Lie algebra has a compact real form that is unique up to conjugation by an element in the connected Lie subgroup of the group of real automorphisms of $\mathfrak{g}^{\mathbb{C}}$ with Lie algebra $\mathrm{ad}(\mathfrak{g})$. The compact real forms of the simple complex Lie algebras are for the classical complex Lie algebras

$$\mathfrak{su}(n+1) \subset A_n\ ,\ \mathfrak{so}(2n+1) \subset B_n\ ,\ \mathfrak{sp}(n) \subset C_n\ ,\ \mathfrak{so}(2n) \subset D_n\ ,$$

and, for the exceptional complex Lie algebras

$$\mathfrak{g}_2 \subset G_2\ ,\ \mathfrak{f}_4 \subset F_4\ ,\ \mathfrak{e}_6 \subset E_6\ ,\ \mathfrak{e}_7 \subset E_7\ ,\ \mathfrak{e}_8 \subset E_8\ .$$

Let

$$\mathfrak{g}^{\mathbb{C}} = \mathfrak{t}^{\mathbb{C}} \oplus \bigoplus_{\alpha \in \Delta} (\mathfrak{g}^{\mathbb{C}})_\alpha$$

be the root space decomposition of $\mathfrak{g}^{\mathbb{C}}$ with respect to $\mathfrak{t}^{\mathbb{C}}$. Each root $\alpha \in \Delta$ is imaginary-valued on $\mathfrak{t}$ and real-valued on $i\mathfrak{t}$. The subalgebra $i\mathfrak{t}$ of $\mathfrak{t}^{\mathbb{C}}$ is a real form of $\mathfrak{t}^{\mathbb{C}}$ and we may view each root $\alpha \in \Delta$ as a one-form on the dual space $(i\mathfrak{t})^*$. Since the Killing form B of $\mathfrak{g}$ is negative definite, it leads via complexification to a positive definite inner product on $i\mathfrak{t}$, which we also denote by B. For each $\lambda \in (i\mathfrak{t})^*$ there exists a vector $H_\lambda \in i\mathfrak{t}$ such that

$$\lambda(H) = B(H, H_\lambda)$$

for all $H \in i\mathfrak{t}$. The inner product on $i\mathfrak{t}$ induces an inner product $\langle \cdot, \cdot \rangle$ on $(i\mathfrak{t})^*$. For each $\lambda, \mu \in \Delta$, we then have

$$\langle \lambda, \mu \rangle = B(H_\lambda, H_\mu)\ .$$

For each $\alpha \in \Delta$ we define the *root reflection*

$$s_\alpha(\lambda) = \lambda - \frac{2\langle \lambda, \alpha \rangle}{\langle \alpha, \alpha \rangle}\alpha \quad (\lambda \in (i\mathfrak{t})^*)\ ,$$

which is a transformation on $(i\mathfrak{t})^*$. The Weyl group of G with respect to T is isomorphic to the group generated by all s_α, $\alpha \in \Delta$. Equivalently, one might view $W(G,T)$ as the group of transformations on $\mathfrak{t}$ generated by the reflections in the hyperplanes perpendicular to iH_λ, $\lambda \in \Delta$.

Structure theory of semisimple real Lie algebras

Let G be a connected, semisimple, real Lie group, $\mathfrak{g}$ its Lie algebra and B its Killing form. A *Cartan involution* on $\mathfrak{g}$ is an involution θ on $\mathfrak{g}$ so that

$$B_\theta(X,Y) = -B(X, \theta Y)$$

is a positive definite inner product on $\mathfrak{g}$. Each semisimple real Lie algebra has a Cartan involution, and any two of them are conjugate via $\mathrm{Ad}(g)$ for some $g \in G$. Let θ be a Cartan involution on $\mathfrak{g}$. Denoting by $\mathfrak{k}$ the $(+1)$-eigenspace of θ and by $\mathfrak{p}$ the (-1)-eigenspace of θ we get the *Cartan decomposition*

$$\mathfrak{g} = \mathfrak{k} \oplus \mathfrak{p} .$$

This decomposition is orthogonal with respect to B and B_θ, B is negative definite on $\mathfrak{k}$ and positive definite on $\mathfrak{p}$, and

$$[\mathfrak{k}, \mathfrak{k}] \subset \mathfrak{k} \, , \; [\mathfrak{k}, \mathfrak{p}] \subset \mathfrak{p} \, , \; [\mathfrak{p}, \mathfrak{p}] \subset \mathfrak{k} \, .$$

The Lie algebra $\mathfrak{k} \oplus i\mathfrak{p}$ is a compact real form of $\mathfrak{g}^{\mathbb{C}}$.

Let K be the connected Lie subgroup of G with Lie algebra $\mathfrak{k}$. Then there exists a unique involutive automorphism Θ of G whose differential at the identity of G coincides with θ. Then K is the fixed point set of Θ, is closed, and contains the center $Z(G)$ of G. If K is compact, then $Z(G)$ is finite, and if $Z(G)$ is finite, then K is a maximal compact subgroup of G. Moreover, the map

$$K \times \mathfrak{p} \to G \, , \; (k, X) \mapsto k\mathrm{Exp}(X)$$

is a diffeomorphism onto G. This is known as a *polar decomposition* of G.

Let $\mathfrak{a}$ be a maximal Abelian subspace of $\mathfrak{p}$. Then all $\mathrm{ad}(H)$, $H \in \mathfrak{a}$, form a commuting family of self-adjoint endomorphisms of $\mathfrak{g}$ with respect to the inner product B_θ. For each $\alpha \in \mathfrak{a}^*$ we define

$$\mathfrak{g}_\alpha = \{X \in \mathfrak{g} \mid \mathrm{ad}(H)X = \alpha(H)X \text{ for all } H \in \mathfrak{a}\} .$$

If $\lambda \neq 0$ and $\mathfrak{g}_\lambda \neq 0$, then λ is called a *restricted root* and $\mathfrak{g}_\lambda$ a *restricted root space* of $\mathfrak{g}$ with respect to $\mathfrak{a}$. We denote by Σ the set of all restricted roots of $\mathfrak{g}$ with respect to $\mathfrak{a}$. The *restricted root space decomposition* of $\mathfrak{g}$ is the direct sum decomposition

$$\mathfrak{g} = \mathfrak{g}_0 \oplus \bigoplus_{\lambda \in \Sigma} \mathfrak{g}_\lambda \, .$$

We always have

$$[\mathfrak{g}_\lambda, \mathfrak{g}_\mu] \subset \mathfrak{g}_{\lambda+\mu}$$

and

$$\theta(\mathfrak{g}_\lambda) = \mathfrak{g}_{-\lambda}$$

for all $\lambda, \mu \in \Sigma$. Moreover,

$$\mathfrak{g}_0 = \mathfrak{a} \oplus \mathfrak{m} \, ,$$

where $\mathfrak{m}$ is the centralizer of $\mathfrak{a}$ in $\mathfrak{k}$. We now choose a notion of positivity for $\mathfrak{a}^*$, which leads to a subset Σ^+ of positive restricted roots. Then

$$\mathfrak{n} = \bigoplus_{\lambda \in \Sigma^+} \mathfrak{g}_\lambda$$

is a nilpotent Lie subalgebra of $\mathfrak{g}$. Any two such nilpotent Lie subalgebras are conjugate via $\mathrm{Ad}(k)$ for some k in the normalizer of $\mathfrak{a}$ in K. The vector space direct sum

$$\mathfrak{g} = \mathfrak{k} \oplus \mathfrak{a} \oplus \mathfrak{n}$$

is called an *Iwasawa decomposition of* $\mathfrak{g}$. The vector space $\mathfrak{s} = \mathfrak{a} \oplus \mathfrak{n}$ is, in fact, a solvable Lie subalgebra of $\mathfrak{g}$ with $[\mathfrak{s}, \mathfrak{s}] = \mathfrak{n}$. Let A, N be the Lie subgroups of G with Lie algebra $\mathfrak{a}, \mathfrak{n}$ respectively. Then A and N are simply connected and the map

$$K \times A \times N \to G\ ,\ (k, a, n) \mapsto kan$$

is a diffeomorphism onto G, a so-called *Iwasawa decomposition of* G.

If $\mathfrak{t}$ is a maximal Abelian subalgebra of $\mathfrak{m}$, then $\mathfrak{h} = \mathfrak{a} \oplus \mathfrak{t}$ is a Cartan subalgebra of $\mathfrak{g}$, that is, $\mathfrak{h}^{\mathbb{C}}$ is a Cartan subalgebra of $\mathfrak{g}^{\mathbb{C}}$. Consider the root space decomposition of $\mathfrak{g}^{\mathbb{C}}$ with respect to $\mathfrak{h}^{\mathbb{C}}$,

$$\mathfrak{g}^{\mathbb{C}} = \mathfrak{h}^{\mathbb{C}} \oplus \bigoplus_{\alpha \in \Delta} (\mathfrak{g}^{\mathbb{C}})_\alpha\ .$$

Then we have

$$\mathfrak{g}_\lambda = \mathfrak{g} \cap \bigoplus_{\alpha \in \Delta,\ \alpha|\mathfrak{a} = \lambda} (\mathfrak{g}^{\mathbb{C}})_\alpha$$

for all $\lambda \in \Sigma$ and

$$\mathfrak{m}^{\mathbb{C}} = \mathfrak{t}^{\mathbb{C}} \oplus \bigoplus_{\alpha \in \Delta,\ \alpha|\mathfrak{a} = 0} (\mathfrak{g}^{\mathbb{C}})_\alpha\ .$$

In particular, all roots are real on $\mathfrak{a} \oplus i\mathfrak{t}$. Of particular interest are those real forms of $\mathfrak{g}^{\mathbb{C}}$ for which $\mathfrak{a}$ is a Cartan subalgebra of $\mathfrak{g}$. In this case, $\mathfrak{g}$ is called a *split real form* of $\mathfrak{g}^{\mathbb{C}}$. Note that $\mathfrak{g}$ is a split real form if and only if $\mathfrak{m}$, the centralizer of $\mathfrak{a}$ in $\mathfrak{k}$, is trivial.

A.3 Homogeneous spaces

A homogeneous space is a manifold with a transitive group of transformations. Homogeneous spaces provide excellent examples for studying the interplay of analysis, geometry, algebra and topology. A modern introduction to homogeneous spaces can be found in Kawakubo [111]. Further results on Lie transformation groups can be found in [182].

The quotient space G/K

Let G be a Lie group and K a closed subgroup of G. By G/K we denote the set of left cosets of K in G,

$$G/K = \{gK \mid g \in G\}\ ,$$

and by π the canonical projection

$$\pi : G \to G/K \ , \ g \mapsto gK \ .$$

We equip G/K with the quotient topology relative to π. Then π is a continuous map and, since K is closed in G, a Hausdorff space. There is exactly one smooth manifold structure on G/K (which is even real analytic) so that π becomes a smooth map and local smooth sections of G/K in G exist. If K is a normal subgroup of G, then G/K becomes a Lie group with respect to the multiplication $g_1K \cdot g_2K = (g_1g_2)K$.

If K is a closed subgroup of a Lie group G, then

$$G \times G/K \to G/K \ , \ (g_1, g_2K) \mapsto (g_1g_2)K$$

is a transitive smooth action of G on G/K. In fact, the smooth structure on G/K can be characterized by the property that this action is smooth. Conversely, suppose we have a transitive smooth action

$$G \times M \to M \ , \ (g, p) \mapsto gp$$

of a Lie group G on a smooth manifold M. Let p be a point in M and

$$G_p = \{g \in G \mid gp = p\}$$

the isotropy subgroup of G at p. If q is another point in M and $g \in G$ with $gp = q$, then $G_q = gG_pg^{-1}$. Thus, the isotropy subgroups of G are all conjugate to each other. The isotropy group G_p is obviously closed in G. Thus, we can equip G/G_p with a smooth manifold structure as described above. With respect to this structure, the map

$$G/G_p \to M \ , \ gG_p \mapsto gp$$

is a smooth diffeomorphism. In this way, we will always identify the smooth manifold M with the coset space G/K. Moreover, $\pi : G \to G/K$ is a principal fibre bundle with fibre and structure group K, where K acts on G by multiplication from the right.

Homogeneous spaces

If M is a smooth manifold and G is a Lie group acting transitively on M, one says that M is a *homogeneous space*, or, more precisely, a *homogeneous G-space*. If M is a connected homogeneous G-space, then the identity component G^o of G also acts transitively on M. This allows us to reduce many problems on connected homogeneous spaces to connected Lie groups and thereby to Lie algebras. Another important fact, proved by Montgomery, is that, if $M = G/K$ is a compact homogeneous G-space with G and K connected, then there exists a compact subgroup of G acting transitively on M. This makes it possible to use the many useful features of compact Lie groups for studying compact homogeneous spaces.

Effective actions

Let M be a homogeneous G-space and $\phi : G \to \mathrm{Diff}(M)$ be the homomorphism from G into the diffeomorphism group of M assigning to each $g \in G$ the diffeomorphism

$$\varphi_g : M \to M \ , \ p \mapsto gp \ .$$

One says that the action of G on M is *effective* if $\ker \phi = \{e\}$, where e denotes the identity in G. In other words, an action is effective if just the identity of G acts as the identity transformation on M. Writing $M = G/K$, we can characterize $\ker \phi$ as the largest normal subgroup of G that is contained in K. Thus, $G/\ker \phi$ is a Lie group with an effective transitive action on M.

Reductive decompositions

Let $M = G/K$ be a homogeneous G-space. We denote by e the identity of G and put $o = eK \in M$. Let $\mathfrak{g}$ and $\mathfrak{k}$ be the Lie algebras of G and K, respectively. As usual, we identify the tangent space of a Lie group at the identity with the corresponding Lie algebra. We choose any linear subspace $\mathfrak{m}$ of $\mathfrak{g}$ complementary to $\mathfrak{k}$, so that $\mathfrak{g} = \mathfrak{k} + \mathfrak{m}$. Then the differential π_{*e} at e of the projection $\pi : G \to G/K$ gives rise to an isomorphism

$$\pi_{*e}|\mathfrak{m} : \mathfrak{m} \to T_oM \ .$$

One of the basic tools in studying homogeneous spaces is to use this isomorphism to identify tangent vectors of M at o with elements in the Lie algebra $\mathfrak{g}$. But there are many choices of complementary subspaces $\mathfrak{m}$, and certain ones turn out to be quite useful. We will describe this now.

Let $\mathrm{Ad} : G \to \mathrm{GL}(\mathfrak{g})$ be the adjoint representation of G. The subspace $\mathfrak{m}$ is said to be $\mathrm{Ad}(K)$-invariant if $\mathrm{Ad}(k)\mathfrak{m} \subset \mathfrak{m}$ for all $k \in K$. If $\mathfrak{m}$ is $\mathrm{Ad}(K)$-invariant and $k \in K$, the differential φ_{k*o} at o of the diffeomorphism $\varphi_k : M \to M \ , \ p \mapsto kp$ has the simple expression

$$\varphi_{k*o} = \mathrm{Ad}(k)|\mathfrak{m} \ .$$

For this reason, one is interested in finding $\mathrm{Ad}(K)$-invariant linear subspaces $\mathfrak{m}$ of $\mathfrak{g}$. Unfortunately, not every homogeneous space admits such subspaces. A homogeneous space G/K is called *reductive* if there is an $\mathrm{Ad}(K)$-invariant linear subspace $\mathfrak{m}$ of $\mathfrak{g}$ so that $\mathfrak{g} = \mathfrak{k} + \mathfrak{m}$ and $\mathfrak{k} \cap \mathfrak{m} = \{0\}$. In this situation, $\mathfrak{g} = \mathfrak{k} + \mathfrak{m}$ is called a *reductive decomposition* of $\mathfrak{g}$.

Isotropy representations and invariant metrics

The homomorphism

$$\chi : K \to \mathrm{GL}(T_oM) \ , \ k \mapsto \varphi_{k*o}$$

is called the *isotropy representation* of the homogeneous space G/K, and the image $\chi(K) \subset \mathrm{GL}(T_oM)$ is called the *linear isotropy group* of G/K. In case G/K is reductive and $\mathfrak{g} = \mathfrak{k} + \mathfrak{m}$ is a reductive decomposition, the isotropy representation of G/K coincides with the adjoint representation $\mathrm{Ad}|K : K \to \mathrm{GL}(\mathfrak{m})$ (via the identification $\mathfrak{m} = T_oM$).

The linear isotropy group contains the information that decides whether a homogeneous space G/K can be equipped with a G-invariant Riemannian structure. A *G-invariant Riemannian metric* $\langle \; , \; \rangle$ on $M = G/K$ is a Riemannian metric so that φ_g is an isometry of M for each $g \in G$, that is, if G acts on M by isometries. A homogeneous space $M = G/K$ can be equipped with a G-invariant Riemannian metric if and only if the linear isotropy group $\chi(K)$ is a relative compact subset of the topological space $L(T_oM, T_oM)$ of all linear maps $T_oM \to T_oM$. It follows that every homogeneous space G/K with K compact admits a G-invariant Riemannian metric. Each Riemannian homogeneous space is reductive. If G/K is reductive and $\mathfrak{g} = \mathfrak{k} + \mathfrak{m}$ is a reductive decomposition, then there is a one-to-one correspondence between the G-invariant Riemannian metrics on G/K and the positive definite $\mathrm{Ad}(K)$-invariant symmetric bilinear forms on $\mathfrak{m}$. Any such bilinear form defines a Riemannian metric on M by requiring that each φ_g be an isometry. The $\mathrm{Ad}(K)$-invariance of the bilinear form ensures that the inner product on each tangent space is well-defined. In particular, if $K = \{e\}$, that is, $M = G$ is a Lie group, then the G-invariant Riemannian metrics on M are exactly the left-invariant Riemannian metrics on G. We finally remark that a G-invariant Riemannian metric on a homogeneous space G/K is unique up to homothety in case the isotropy representation is irreducible. This is a consequence of the so-called Lemma of Schur.

Naturally reductive Riemannian homogeneous spaces

A homogeneous Riemannian manifold M is said to be a *naturally reductive Riemannian homogeneous space* if there exists a connected Lie subgroup G of the isometry group $I(M)$ of M that acts transitively and effectively on M and a reductive decomposition $\mathfrak{g} = \mathfrak{k} + \mathfrak{m}$ of the Lie algebra $\mathfrak{g}$ of G, where $\mathfrak{k}$ is the Lie algebra of the isotropy subgroup K of G at some point $o \in M$, such that

$$g([X,Z]_{\mathfrak{m}}, Y) + g(Z, [X,Y]_{\mathfrak{m}}) = 0$$

for all $X, Y, Z \in \mathfrak{m}$, where g denotes the inner product on $\mathfrak{m}$ that is induced by the Riemannian metric on M and $[\cdot,\cdot]_{\mathfrak{m}}$ denotes the canonical projection onto $\mathfrak{m}$ with respect to the decomposition $\mathfrak{g} = \mathfrak{k} + \mathfrak{m}$. Any such decomposition is called a *naturally reductive decomposition* of $\mathfrak{g}$. The above algebraic condition is equivalent to saying that every geodesic in M through o is the orbit through o of the one-parameter subgroup of G that is generated by some $X \in \mathfrak{m}$.

A.4 Symmetric spaces and flag manifolds

Symmetric spaces form a subclass of the homogeneous spaces and were studied intensely and also classified by Elie Cartan. The fundamental books on this topic are

Helgason [99] and Loos [132]. Another nice introduction can be found in [214]. Flag manifolds are homogeneous spaces that are intimately related to symmetric spaces.

(Locally) symmetric spaces

Let M be a Riemannian manifold, $p \in M$, and $r \in \mathbb{R}_+$ sufficiently small so that normal coordinates are defined on the open ball $B_r(p)$ consisting of all points in M with distance less than r to p. Denote by $\exp_p : T_pM \to M$ the exponential map of M at p. The map

$$s_p : B_r(p) \to B_r(p) \ , \ \exp(tv) \mapsto \exp(-tv)$$

reflects in p the geodesics of M through p and is called a *local geodesic symmetry* at p. A connected Riemannian manifold is called a *locally symmetric space* if, at each point p in M, there exists an open ball $B_r(p)$ such that the corresponding local geodesic symmetry s_p is an isometry. A connected Riemannian manifold is called a *symmetric space* if at each point $p \in M$ such a local geodesic symmetry extends to a global isometry $s_p : M \to M$. This is equivalent to saying that there exists an involutive isometry s_p of M such that p is an isolated fixed point of s_p. In such a case, one calls s_p the *symmetry* of M in p.

Let M be a Riemannian homogeneous space and suppose there exists a symmetry of M at some point $p \in M$. Let q be any point in M and g an isometry of M with $g(p) = q$. Then $s_q := gs_pg^{-1}$ is a symmetry of M at q. In order to show that a Riemannian homogeneous space is symmetric, it therefore suffices to construct a symmetry at one point. Using this, we can easily describe some examples of symmetric spaces. The Euclidean space $\mathbb{R}^n$ is symmetric with $s_0 : \mathbb{R}^n \to \mathbb{R}^n \ , \ p \mapsto -p$. The map

$$S^n \to S^n \ , \ (p_1, \ldots, p_n, p_{n+1}) \mapsto (-p_1, \ldots, -p_n, p_{n+1})$$

is a symmetry of the sphere S^n at $(0, \ldots, 0, 1)$. Let G be a connected compact Lie group. Any $\mathrm{Ad}(G)$-invariant inner product on $\mathfrak{g}$ extends to a biinvariant Riemannian metric on G. With respect to such a Riemannian metric, the inverse map $s_e : G \to G \ , \ g \mapsto g^{-1}$ is a symmetry of G at e. Thus, any connected compact Lie group is a symmetric space.

We recall some basic features of (locally) symmetric spaces. A Riemannian manifold is locally symmetric if and only if its Riemannian curvature tensor is parallel, that is, $\nabla R = 0$. If M is a connected, complete, locally symmetric space, then its Riemannian universal covering is a symmetric space. Using the symmetries, one can show easily that any symmetric space is homogeneous. Note that there are complete locally symmetric spaces that are not symmetric, even not homogeneous. For instance, let M be a compact Riemann surface with genus ≥ 2 and equipped with a Riemannian metric of constant curvature -1. It is known that the isometry group of M is finite, so M is not homogeneous and therefore also not symmetric. On the other hand, M is locally isometric to the real hyperbolic plane $\mathbb{R}H^2$ and hence locally symmetric.

Cartan decomposition and Riemannian symmetric pairs

One can associate a Riemannian symmetric pair to each symmetric space. We first recall the definition of a Riemannian symmetric pair. Let G be a connected Lie group and s a nontrivial involutive automorphism of G. We denote by $G_s \subset G$ the set of fixed points of s and by G_s^o the connected component of G_s containing the identity e of G. Let K be a closed subgroup of G with $G_s^o \subset K \subset G_s$. Then $\sigma := s_{*e}$ is an involutive automorphism of $\mathfrak{g}$ and

$$\mathfrak{k} = \{X \in \mathfrak{g} \mid \sigma X = X\} .$$

The linear subspace

$$\mathfrak{p} = \{X \in \mathfrak{g} \mid \sigma X = -X\}$$

of $\mathfrak{g}$ is called the *standard complement* of $\mathfrak{k}$ in $\mathfrak{g}$. Then we have $\mathfrak{g} = \mathfrak{k} \oplus \mathfrak{p}$ (direct sum of vector spaces) and

$$[\mathfrak{k}, \mathfrak{p}] \subset \mathfrak{p} \ , \ [\mathfrak{p}, \mathfrak{p}] \subset \mathfrak{k} \ .$$

This particular decomposition of $\mathfrak{g}$ is called the *Cartan decomposition* or standard decomposition of $\mathfrak{g}$ with respect to σ. In this situation, the pair (G, K) is called a *Riemannian symmetric pair* if $\mathrm{Ad}_G(K)$ is a compact subgroup of $\mathrm{GL}(\mathfrak{g})$ and $\mathfrak{p}$ is equipped with some $\mathrm{Ad}_G(K)$-invariant inner product.

Suppose (G, K) is a Riemannian symmetric pair. The inner product on $\mathfrak{p}$ determines a G-invariant Riemannian metric on the homogeneous space $M = G/K$, and the map

$$M \to M \ , \ gK \mapsto s(g)K \ ,$$

where s, the involutive automorphism on G, is a symmetry of M at $o = eK \in M$. Thus, M is a symmetric space. Conversely, suppose M is a symmetric space. Let G be the identity component of the full isometry group M, o any point in M, s_o the symmetry of M at o, and K the isotropy subgroup of G at o. Then

$$s : G \to G \ , \ g \mapsto s_o g s_o$$

is an involutive automorphism of G with $G_s^o \subset K \subset G_s$, and the inner product on the standard complement $\mathfrak{p}$ of $\mathfrak{k}$ in $\mathfrak{g}$ is $\mathrm{Ad}_G(K)$-invariant (using our usual identification $\mathfrak{p} = T_o M$). In this way, the symmetric space M determines a Riemannian symmetric pair (G, K). This Riemannian symmetric pair is effective, that is, each normal subgroup of G contained in K is trivial. As described here, there is a one-to-one correspondence between symmetric spaces and effective Riemannian symmetric pairs.

Riemannian geometry of symmetric spaces

Let M be a symmetric space, $o \in M$, $G = I^o(M)$, K the isotropy group at o and $\mathfrak{g} = \mathfrak{k} \oplus \mathfrak{p}$ the corresponding Cartan decomposition of $\mathfrak{g}$. For each $X \in \mathfrak{g}$ we have a one-parameter group $\mathrm{Exp}(tX)$ of isometries of M. We denote the corresponding complete Killing vector field on M by X^*. Note that

$$[X, Y]^* = -[X^*, Y^*]$$

for all $X \in \mathfrak{g}$, where the bracket on the left-hand side is in $\mathfrak{g}$ and the one on the right-hand side is the one for vector fields on manifolds. As usual, we identify $\mathfrak{p}$ and T_oM by means of the isomorphism $\mathfrak{p} \to T_oM$, $X \mapsto X^*_o$. Since X^* is a Killing vector field, its covariant derivative ∇X^* is a skew-symmetric tensor field on M. Its value at o is given by

$$(\nabla X^*)_o = \mathrm{ad}_{\mathfrak{p}}(X)$$

if $X \in \mathfrak{k}$ and

$$(\nabla X^*)_o = 0$$

if $X \in \mathfrak{p}$, where $\mathrm{ad}_{\mathfrak{p}}(X)Y = [X, Y]_{\mathfrak{p}}$ is the projection of $[X, Y]$ onto $\mathfrak{p}$ for all $Y \in \mathfrak{p}$. For each $X \in \mathfrak{p}$, the geodesic $\gamma_X : \mathbb{R} \to M$ with $\gamma_X(0) = o$ and $\dot{\gamma}_X(0) = X$ is the curve $t \mapsto \mathrm{Exp}(tX)o$. Let Φ^{X^*} be the flow of X^*. Then the parallel translation along γ_X from $o = \gamma_X(0)$ to $\gamma_X(t)$ is given by

$$(\Phi_t^{X^*})_{*o} : T_oM \to T_{\gamma_X(t)}M \ .$$

The Riemannian curvature tensor R_o of M at o is given by the simple formula

$$R_o(X, Y)Z = -[[X, Y], Z]$$

for all $X, Y, Z \in \mathfrak{p} = T_oM$.

Semisimple symmetric spaces, rank, and duality

Let M be a symmetric space and $\tilde{M}$ its Riemannian universal covering space. Let $\tilde{M}_0 \times \ldots \times \tilde{M}_k$ be the de Rham decomposition of $\tilde{M}$, where the Euclidean factor $\tilde{M}_0$ is isometric to some Euclidean space of dimension ≥ 0. Each $\tilde{M}_i$, $i > 0$, is a simply connected, irreducible, symmetric space. A *semisimple symmetric space* is a symmetric space for which $\tilde{M}_0$ has dimension zero. This notion is because, if $\tilde{M}_0$ is trivial, then $I^o(M)$ is a semisimple Lie group. A symmetric space M is said to be of *compact type* if M is semisimple and compact, and it is said to be of *noncompact type* if M is semisimple and noncompact. Symmetric spaces of noncompact type are always simply connected. An *s-representation* is the isotropy representation of a simply connected, semisimple, symmetric space $M = G/K$ with $G = I^o(M)$.

The *rank* of a semisimple symmetric space $M = G/K$ is the dimension of a maximal Abelian subspace of $\mathfrak{p}$ in some Cartan decomposition $\mathfrak{g} = \mathfrak{k} \oplus \mathfrak{p}$ of the Lie algebra $\mathfrak{g}$ of $G = I^o(M)$.

Let (G, K) be a Riemannian symmetric pair so that G/K is a simply connected Riemannian symmetric space of compact type or of noncompact type, respectively. Consider the complexification $\mathfrak{g}^{\mathbb{C}} = \mathfrak{g} + i\mathfrak{g}$ of $\mathfrak{g}$ and the Cartan decomposition $\mathfrak{g} = \mathfrak{k} \oplus \mathfrak{p}$ of $\mathfrak{g}$. Then $\mathfrak{g}^* = \mathfrak{k} \oplus i\mathfrak{p}$ is a real Lie subalgebra of $\mathfrak{g}^{\mathbb{C}}$ with respect to the induced Lie algebra structure. Let G^* be the real Lie subgroup of $G^{\mathbb{C}}$ with Lie algebra $\mathfrak{g}^*$. Then G^*/K is a simply connected Riemannian symmetric space of noncompact type or of compact type, respectively, with Cartan decomposition $\mathfrak{g}^* = \mathfrak{k} \oplus i\mathfrak{p}$. This feature is known as *duality* between symmetric spaces of compact type and of noncompact type and describes explicitly a one-to-one correspondence between these two types of simply connected symmetric spaces.

Classification of symmetric spaces

Any simply connected symmetric space decomposes into the Riemannian product of a Euclidean space and some simply connected, irreducible, symmetric spaces. Thus, the classification problem for simply connected symmetric spaces reduces to the classification of simply connected, irreducible symmetric spaces. Any such space is either of compact type or of noncompact type. The concept of duality enables one to reduce the classification problem to those of noncompact type. The crucial step for deriving the latter classification is to show that every noncompact irreducible symmetric space is of the form $M = G/K$ with some simple noncompact real Lie group G with trivial center and K a maximal compact subgroup of G. If the complexification of $\mathfrak{g}$ is simple as a complex Lie algebra, then M is said to be of type III, otherwise M is said to be of type IV. The corresponding compact, irreducible, symmetric spaces are said to be of types I and II, respectively. The complete list of simply connected, irreducible, symmetric spaces is as follows:

TABLE A.1: Classical symmetric spaces of types I and III

Type I (compact)	Type III (noncompact)	Dimension	Rank
$SU(n)/SO(n)$	$SL(n,\mathbb{R})/SO(n)$	$(n-1)(n+2)/2$	$n-1$
$SU(2n)/Sp(n)$	$SL(n,\mathbb{H})/Sp(n)$	$(n-1)(2n+1)$	$n-1$
$SU(p+q)/S(U(p)\times U(q))$	$SU(p,q)/S(U(p)\times U(q))$	$2pq$	$\min\{p,q\}$
$SO(p+q)/SO(p)\times SO(q)$	$SO^o(p,q)/SO(p)\times SO(q)$	pq	$\min\{p,q\}$
$SO(2n)/U(n)$	$SO^*(2n)/U(n)$	$n(n-1)$	$[n/2]$
$Sp(n)/U(n)$	$Sp(n,\mathbb{R})/U(n)$	$n(n+1)$	n
$Sp(p+q)/Sp(p)\times Sp(q)$	$Sp(p,q)/Sp(p)\times Sp(q)$	$4pq$	$\min\{p,q\}$

The symmetric space $SO(p+q)/SO(p)\times SO(q)$ is the Grassmann manifold of all p-dimensional oriented linear subspaces of $\mathbb{R}^{p+q}$ and will often be denoted by $G_p^+(\mathbb{R}^{p+q})$. The Grassmann manifold $G_2^+(\mathbb{R}^4)$ is isometric to the Riemannian product $S^2\times S^2$ and hence reducible. So, strictly speaking, this special case has to be excluded from the above table. Disregarding the orientation of the p-planes, we have a natural 2-fold covering map $G_p^+(\mathbb{R}^{p+q})\to G_p(\mathbb{R}^{p+q})$ onto the Grassmann manifold $G_p(\mathbb{R}^{p+q})$ of all p-dimensional linear subspaces of $\mathbb{R}^{p+q}$, which can be written as the homogeneous space $SO(p+q)/S(O(p)\times O(q))$. Similarily, the symmetric space $SU(p+q)/S(U(p)\times U(q))$ is the Grassmann manifold of all p-dimensional complex linear subspaces of $\mathbb{C}^{p+q}$ and will be denoted by $G_p(\mathbb{C}^{p+q})$. Eventually, the symmetric space $Sp(p+q)/Sp(p)\times Sp(q)$ is the Grassmann manifold of all p-dimensional quaternionic linear subspaces of $\mathbb{H}^{p+q}$ and will be denoted by $G_p(\mathbb{H}^{p+q})$. The Grassmann manifold $G_1^+(\mathbb{R}^{1+q})$ is the q-dimensional sphere S^q. And the Grassmann manifold $G_1(\mathbb{R}^{1+q})$ (resp. $G_1(\mathbb{C}^{1+q})$ or $G_1(\mathbb{H}^{1+q})$) is the q-dimensional real (resp. complex or quaternionic) projective space $\mathbb{R}P^q$ (resp. $\mathbb{C}P^q$ or $\mathbb{H}P^q$). The dual space of the sphere S^q is the real hyperbolic space $\mathbb{R}H^q$. And the dual space of the complex projective space $\mathbb{C}P^q$ (resp. the quaternionic projective space $\mathbb{H}P^q$) is the complex hyperbolic space $\mathbb{C}H^q$ (resp. the quaternionic hyperbolic space $\mathbb{H}H^q$).

In small dimensions, certain symmetric spaces are isometric to each other (with a suitable normalization of the Riemannian metric):

$$\begin{aligned} S^2 &= \mathbb{C}P^1 = SU(2)/SO(2) = SO(4)/U(2) = Sp(1)/U(1)\,, \\ S^4 &= \mathbb{H}P^1\,, \\ S^5 &= SU(4)/Sp(2)\,, \\ \mathbb{C}P^3 &= SO(6)/U(3)\,, \\ G_2^+(\mathbb{R}^5) &= Sp(2)/U(2)\,, \\ G_2^+(\mathbb{R}^6) &= G_2(\mathbb{C}^4)\,, \\ G_2^+(\mathbb{R}^8) &= SO(8)/U(4)\,, \\ G_3^+(\mathbb{R}^6) &= SU(4)/SO(4)\,. \end{aligned}$$

In the noncompact case, one has isometries between the corresponding dual symmetric spaces.

TABLE A.2: Exceptional symmetric spaces of types I and III

Type I (compact)	Type III (noncompact)	Dimension	Rank
$E_6/Sp(4)$	$E_6^6/Sp(4)$	42	6
$E_6/SU(6)\times SU(2)$	$E_6^2/SU(6)\times SU(2)$	40	4
$E_6/T\cdot Spin(10)$	$E_6^{-14}/T\cdot Spin(10)$	32	2
E_6/F_4	E_6^{-26}/F_4	26	2
$E_7/SU(8)$	$E_7^7/SU(8)$	70	7
$E_7/SO(12)\times SU(2)$	$E_7^{-5}/SO(12)\times SU(2)$	64	4
$E_7/T\cdot E_6$	$E_7^{-25}/T\cdot E_6$	54	3
$E_8/SO(16)$	$E_8^8/SO(16)$	128	8
$E_8/E_7\times SU(2)$	$E_8^{-24}/E_7\times SU(2)$	112	4
$F_4/Sp(3)\times SU(2)$	$F_4^4/Sp(3)\times SU(2)$	28	4
$F_4/Spin(9)$	$F_4^{-20}/Spin(9)$	16	1
$G_2/SO(4)$	$G_2^2/SO(4)$	8	2

Here we denote by E_6, E_7, E_8, F_4, G_2 the connected, simply connected, compact, real Lie group with Lie algebra $\mathfrak{e}_6, \mathfrak{e}_7, \mathfrak{e}_8, \mathfrak{f}_4, \mathfrak{g}_2$, respectively. This is the same notation as was used for the corresponding simple complex Lie algebras, but it should always be clear from the context what these symbols represent. The symmetric space $F_4/Spin(9)$ is the Cayley projective plane $\mathbb{O}P^2$ and the dual space $F_4^{-20}/Spin(9)$ is the Cayley hyperbolic plane $\mathbb{O}H^2$.

Since $Spin(2)$ is isomorphic to $U(1)$ and $Spin(4)$ is isomorphic to $SU(2)\times SU(2)$ we have to assume $n \geq 3$ for the spaces in the last row this table. In small

TABLE A.3: Classical symmetric spaces of types II and IV

Type II (compact)	Type IV (noncompact)	Dimension	Rank
$SU(n+1)$	$SL(n+1,\mathbb{C})/SU(n+1)$	$n(n+2)$	n
$Spin(2n+1)$	$SO(2n+1,\mathbb{C})/SO(2n+1)$	$n(2n+1)$	n
$Sp(n)$	$Sp(n,\mathbb{C})/Sp(n)$	$n(2n+1)$	n
$Spin(2n)$	$SO(2n,\mathbb{C})/SO(2n)$	$n(2n-1)$	n

dimensions there are the following additional isomorphisms:

$$
\begin{aligned}
Spin(3) &= SU(2) = Sp(1)\ , \\
Spin(5) &= Sp(2)\ , \\
Spin(6) &= SU(4)\ .
\end{aligned}
$$

In the noncompact case, there are isomorphisms between the corresponding dual spaces.

TABLE A.4: Exceptional symmetric spaces of types II and IV

Type II (compact)	Type IV (noncompact)	Dimension	Rank
E_6	$E_6^{\mathbb{C}}/E_6$	78	6
E_7	$E_7^{\mathbb{C}}/E_7$	133	7
E_8	$E_8^{\mathbb{C}}/E_8$	248	8
F_4	$F_4^{\mathbb{C}}/F_4$	52	4
G_2	$G_2^{\mathbb{C}}/G_2$	14	2

Hermitian symmetric spaces

A *Hermitian symmetric space* is a symmetric space that is equipped with some Kähler structure so that the geodesic symmetries are holomorphic maps. The simplest example of a Hermitian symmetric space is the complex vector space $\mathbb{C}^n$. For semisimple symmetric spaces, one can easily decide whether it is Hermitian or not. In fact, let (G, K) be the Riemannian symmetric pair of some irreducible semisimple symmetric space M. Then the center of K is either discrete or one-dimensional. The irreducible semisimple Hermitian symmetric spaces are precisely those for which the center of K is one-dimensional. This gives the list in Table A.5

Note that $SO(4)/SO(2)\times SO(2)$ is isometric to the Riemannian product $S^2 \times S^2$, therefore, we have to exclude the case $q = 2$ in the second row of the above table. Every semisimple Hermitian symmetric space is simply connected and hence decomposes into the Riemannian product of irreducible semisimple Hermitian symmetric spaces.

TABLE A.5: Irreducible semisimple Hermitian symmetric spaces

compact type	noncompact type
$SU(p+q)/S(U(p)\times U(q))$	$SU(p,q)/S(U(p)\times U(q))$
$SO(2+q)/SO(2)\times SO(q)$	$SO^o(2,q)/SO(2)\times SO(q)$
$SO(2n)/U(n)$	$SO^*(2n)/U(n)$
$Sp(n)/U(n)$	$Sp(n,\mathbb{R})/U(n)$
$E_6/T\cdot Spin(10)$	$E_6^{-14}/T\cdot Spin(10)$
$E_7/T\cdot E_6$	$E_7^{-25}/T\cdot E_6$

Complex flag manifolds

Let G be a connected, compact, semisimple, real Lie group with trivial center and $\mathfrak{g}$ its Lie algebra. Consider the action of G on $\mathfrak{g}$ by the adjoint representation $\mathrm{Ad}: G \to \mathrm{End}(\mathfrak{g})$. For each $0 \neq X \in \mathfrak{g}$ the orbit

$$G \cdot X = \{\mathrm{Ad}(g)X \mid g \in G\}$$

is a homogeneous G-space. Let $\mathfrak{t}_X$ be the intersection of all maximal Abelian subalgebras of $\mathfrak{g}$ containing X and T_X the torus in G with Lie algebra $\mathfrak{t}_X$. Then the isotropy subgroup of G at X is $Z_G(T_X)$, the centralizer of T_X in G, and therefore

$$G \cdot X = G/Z_G(T_X) .$$

In particular, if X is a regular element of $\mathfrak{g}$, that is, if there is a unique maximal Abelian subalgebra $\mathfrak{t}$ of $\mathfrak{g}$ that contains X, then $G \cdot X = G/T$, where T is the maximal torus in G with Lie algebra $\mathfrak{t}$. Any orbit $G \cdot X$ of the adjoint representation of G is called a *complex flag manifold* or *C-space*. The latter notion is used more frequently in earlier papers on this topic. In the special case of $G = SU(n)$, one obtains the flag manifolds of all possible flags in $\mathbb{C}^n$ in this way. In particular, when T is some maximal torus of $SU(n)$, then $SU(n)/T$ is the flag manifold of all full flags in $\mathbb{C}^n$, that is, of all possible arrangements $\{0\} \subset V^1 \subset \ldots \subset V^{n-1} \subset \mathbb{C}^n$, where V^k is a k-dimensional complex linear subspace of $\mathbb{C}^n$.

The importance of complex flag manifolds becomes clear from the following facts. Each orbit $G \cdot X$ admits a canonical complex structure that is also integrable. If G is simple, there exists a unique (up to homothety) G-invariant Kähler-Einstein metric on $G \cdot X$ with positive scalar curvature and compatible with the canonical complex structure on $G \cdot X$. Moreover, any Kähler-Einstein metric on $G \cdot X$ is homogeneous under its own group of isometries and is obtained from a G-invariant Kähler-Einstein metric via some automorphism of the complex structure. Conversely, any simply connected, compact, homogeneous Kähler manifold is isomorphic as a complex homogeneous manifold to some orbit $G \cdot X$ of the adjoint representation of G, where $G = I^o(M)$ and $X \in \mathfrak{g}$. Note that each compact homogeneous Kähler manifold is the Riemannian product of some flat complex torus and some simply connected, compact, homogeneous Kähler manifold.

Real flag manifolds

A *real flag manifold* is an orbit of an s-representation. Real flag manifolds are also known as *R-spaces*, an idea that is used more frequently in earlier papers on this topic. Note that the s-representation of a symmetric space of noncompact type is the same as the one of the corresponding dual symmetric space. Thus, in order to classify and study real flag manifolds, it is sufficient to consider just one type of symmetric spaces.

Let $M = G/K$ be a simply connected semisimple symmetric space of noncompact type with $G = I^o(M)$, $o \in M$ and K the isotropy subgroup of G at o. Note that K is connected as M is assumed to be simply connected and G is connected. We consider the corresponding Cartan decomposition $\mathfrak{g} = \mathfrak{k} \oplus \mathfrak{p}$ of the semisimple real Lie algebra $\mathfrak{g}$ of G. Let $0 \neq X \in \mathfrak{p}$ and $K \cdot X$ the orbit of K through X via the s-representation. For each $k \in K$ we have $k \cdot X = k_{*o}X = Ad(k)X$ and, therefore, $K \cdot X = K/K_X$ with $K_X = \{k \in K \mid Ad(k)X = X\}$. Let $\mathfrak{a}_X$ be the intersection of all maximal Abelian subspaces $\mathfrak{a}$ of $\mathfrak{p}$ with $X \in \mathfrak{a}$. We say that X is *regular* if $\mathfrak{a}_X$ is a maximal Abelian subspace of $\mathfrak{p}$, or equivalently, if there exists a unique maximal Abelian subspace of $\mathfrak{p}$ that contains X. Otherwise, we call X *singular*. The isotropy subgroup K_X is the centralizer of $\mathfrak{a}_X$ in K. If, in particular, $\mathfrak{g}$ is a split real form of $\mathfrak{g}^{\mathbb{C}}$ and X is regular, then $K \cdot X = K$.

In general, a real flag manifold is not a symmetric space. Consider the semisimple real Lie algebra $\mathfrak{g}$ equipped with the positive definite inner product $B_\sigma(X,Y) = -B(X,\sigma Y)$, where σ is the Cartan involution on $\mathfrak{g}$ coming from the symmetric space structure of G/K. For $0 \neq X \in \mathfrak{p}$, the endomorphism $\mathrm{ad}(X) : \mathfrak{g} \to \mathfrak{g}$ is self-adjoint and hence has real eigenvalues. The real flag manifold $K \cdot X$ is a symmetric space if and only if the eigenvalues of $\mathrm{ad}(X)$ are $-1, 0, +1$. Note that not every semisimple real Lie algebra $\mathfrak{g}$ admits such an element X. A real flag manifold that is a symmetric space is called a *symmetric R-space*. If, in addition, $\mathfrak{g}$ is simple, then it is called an *irreducible symmetric R-space*. Decomposing $\mathfrak{g}$ into its simple parts, one easily sees that every symmetric R-space is the Riemannian product of irreducible symmetric R-spaces.

The classification of the symmetric R-spaces was established by S. Kobayashi and T. Nagano [116]. It follows from their classification and a result by M. Takeuchi [213] that the symmetric R-spaces consist of the Hermitian symmetric spaces of compact type and their real forms. A *real form* M of a Hermitian symmetric space $\bar{M}$ is a connected, complete, totally real, totally geodesic submanifold of $\bar{M}$ whose real dimension equals the complex dimension of $\bar{M}$. These real forms were classified by M. Takeuchi [213] and independently by D.S.P. Leung [129].

Among the irreducible symmetric R-spaces, the Hermitian symmetric spaces are precisely those arising from simple complex Lie groups modulo some compact real form. This means that an irreducible symmetric R-space is a Hermitian symmetric space or a real form precisely if the symmetric space G/K is of type IV or III, respectively. The isotropy representation of a symmetric space G/K of noncompact type is the same as the isotropy representation of its dual simply connected compact symmetric space. Thus, we can also characterize the Hermitian symmetric spaces among the irreducible symmetric R-spaces as those spaces that arise as an orbit of the adjoint representation of a simply connected, compact, real Lie group G, or equivalently, that is a complex flag manifold. This leads to the following table:

TABLE A.6: Irreducible symmetric R-spaces of Hermitian type

G	$K \cdot X = \mathrm{Ad}(G) \cdot X$	Remarks
$Spin(n)$	$SO(n)/SO(2) \times SO(n-2)$	$n \geq 5$
$Spin(2n)$	$SO(2n)/U(n)$	$n \geq 3$
$SU(n)$	$SU(n)/S(U(p) \times U(n-p))$	$n \geq 2,\ 1 \leq p \leq [\frac{n}{2}]$
$Sp(n)$	$Sp(n)/U(n)$	$n \geq 2$
E_6	$E_6/T \cdot Spin(10)$	
E_7	$E_7/T \cdot E_6$	

The real forms are always non-Hermitian and, among the irreducible symmetric R-spaces, they are precisely those spaces arising from the isotropy representation of a symmetric space G/K of type I.

TABLE A.7: Irreducible symmetric R-spaces of non-Hermitian type

G/K	$K \cdot X$	Remarks
$SU(n)/SO(n)$	$G_p(\mathbb{R}^n)$	$n \geq 3,\ 1 \leq p \leq [\frac{n}{2}]$
$SU(2n)/Sp(n)$	$G_p(\mathbb{H}^n)$	$n \geq 2,\ 1 \leq p \leq [\frac{n}{2}]$
$SU(2n)/S(U(n) \times U(n))$	$U(n)$	$n \geq 2$
$SO(n)/SO(p) \times SO(n-p)$	$(S^{p-1} \times S^{n-p-1})/\mathbb{Z}_2$	$n \geq 3,\ 1 \leq p \leq [\frac{n}{2}]$
$SO(2n)/SO(n) \times SO(n)$	$SO(n)$	$n \geq 5$
$SO(4n)/U(2n)$	$U(2n)/Sp(n)$	$n \geq 3$
$Sp(n)/U(n)$	$U(n)/SO(n)$	$n \geq 3$
$Sp(2n)/Sp(n) \times Sp(n)$	$Sp(n)$	$n \geq 2$
$E_6/Sp(4)$	$G_2(\mathbb{H}^4)/\mathbb{Z}_2$	
E_6/F_4	$\mathbb{O}P^2$	
$E_7/SU(8)$	$(SU(8)/Sp(4))/\mathbb{Z}_2$	
$E_7/T \cdot E_6$	$T \cdot E_6/F_4$	

References

[1] Abresch, U., Isoparametric hypersurfaces with four or six distinct principal curvatures. Necessary conditions on the multiplicities, *Math. Ann.* 264, 283-302 (1983).

[2] Adams, J.F., *Lectures on Lie Groups*, W.A. Benjamin Inc., 1969.

[3] Adams, J.F., *Lectures on Exceptional Lie Groups*, University of Chicago Press, 1996.

[4] Alekseevskii, D.V., Kimelfeld, B.N., Structure of homogeneous Riemann spaces with zero Ricci curvature, *Funct. Anal. Appl.* 9, 97-102 (1975).

[5] Backes, E., Geometric applications of Euclidean Jordan triple systems, *Manuscr. Math.* 42, 265-272 (1983).

[6] Backes, E., Reckziegel, H., On symmetric submanifolds of spaces of constant curvature, *Math. Ann.* 263, 419-433 (1983).

[7] Bérard Bergery, L., Sur la courbure des metriques riemanniennes invariantes des groupes de Lie et des espaces homogènes, *Ann. Sci. Éc. Norm. Supér. IV. Sér.* 11, 543-576 (1978).

[8] Bérard Bergery, L., Sur de nouvelles variétés riemanniennes d'Einstein, *Inst. Elie Cartan Univ. Nancy I* 6, 1-60 (1983).

[9] Bérard Bergery, L., Ikemakhen, A., On the holonomy of Lorentzian manifolds, *Proc. Symp. Pure Math.* 54 Part 2, 27-40 (1993).

[10] Berger, M., Sur les groupes d'holonomie homogènes de variétés à connexion affine et des variétés riemanniennes, *Bull. Soc. Math. Fr.* 83, 279-330 (1955).

[11] Berger, M., Les espaces symétriques non compacts, *Ann. Sci. Éc. Norm. Supér. III. Sér.* 74, 85-177 (1959).

[12] Berger, M., Gostiaux, B., *Differential geometry: manifolds, curves and surfaces*, Springer, 1988.

[13] Berndt, J., Real hypersurfaces with constant principal curvatures in complex hyperbolic space, *J. Reine Angew. Math.* 395, 132-141 (1989).

[14] Berndt, J., Real hypersurfaces in quaternionic space forms, *J. Reine Angew. Math.* 419, 9-26 (1991).

[15] Berndt, J., Riemannian geometry of complex two-plane Grassmannians, *Rend. Sem. Mat. Torino* 55, 19-83 (1997).

[16] Berndt, J., Homogeneous hypersurfaces in hyperbolic spaces, *Math. Z.* 229, 589-600 (1998).

[17] Berndt, J., Brück, M., Cohomogeneity one actions on hyperbolic spaces, *J. Reine Angew. Math.* 541, 209-235 (2001).

[18] Berndt, J., Eschenburg, J.H., Naitoh, H., Tsukada, K., Symmetric submanifolds associated with the irreducible symmetric R-spaces, preprint.

[19] Berndt, J., Tamaru, H., Homogeneous codimension one foliations on noncompact symmetric spaces, preprint.

[20] Berndt, J., Tamaru, H., Cohomogeneity one actions on noncompact symmetric spaces with a totally geodesic singular orbit, preprint.

[21] Berndt, J., Tricerri, F., Vanhecke, L., *Generalized Heisenberg Groups and Damek-Ricci Harmonic Spaces*, Springer, 1995.

[22] Besse, A.L., *Einstein Manifolds*, Springer, 1987.

[23] Bieberbach, L., Eine singularitätenfreie Fläche konstanter negativer Krümmung im Hilbertschen Raum, *Comment. Math. Helv.* 4, 248-255 (1932).

[24] Bishop, R.L., Crittenden, R.J., *Geometry of Manifolds*, Academic Press, 1964.

[25] Bitossi, M., On the effective calculation of holonomy groups, *Rend. Ist. Mat. Univ. Trieste* 28, 293-302 (1996).

[26] Bott, R., The geometry and representation theory of compact Lie groups, in: *Representation Theory of Lie Groups*, Proc. SRC/LMS Res. Symp., Oxford 1977, Lond. Math. Soc. Lect. Note Ser. 34, 65-90 (1979).

[27] Bott, R., Samelson, H., Applications of the theory of Morse to symmetric spaces, *Am. J. Math.* 80, 964-1029 (1958). Correction: ibid. 83, 207-208 (1961).

[28] Bredon, G.E., *Introduction to Compact Transformation Groups*, Academic Press, 1972.

[29] Bröcker, T., tom Dieck, T., *Representations of Compact Lie Groups*, Corrected reprint of the 1985 original, Springer, 1995.

[30] Carfagna D'Andrea, A., Console, S., Immersions into the hyperbolic space invariant by reflections, *Beitr. Algebra Geom.* 40, 67-78 (1999).

[31] Carfagna D'Andrea, A., Mazzocco, R., Romani, G., Some characterizations of 2-symmetric submanifolds in spaces of constant curvature, *Czech. Math. J.* 44, 691-711 (1994).

[32] Cartan, E., Familles de surfaces isoparamétriques dans les espaces à courbure constante, *Ann. Mat. Pura Appl. IV. Ser.* 17, 177-191 (1938).

[33] Cartan, E., Sur des familles remarquables d'hypersurfaces isoparamétriques dans les espaces sphériques, *Math. Z.* 45, 335-367 (1939).

[34] Cartan, E., Sur quelques familles remarquables d'hypersurfaces, *C. R. Congr. Sci. Math.* 30-41 (1939).

[35] Cartan, E., Sur des familles d'hypersurfaces isoparamétriques des espaces spheriques a 5 et a 9 dimensions, *Rev. Univ. Nac. Tucuman Ser. A* 1, 5-22 (1940).

[36] Cartan, E., *Leçons sur la Géométrie des Espaces de Riemann,* 2e *éd.*, Gauthier-Villars, 1951.

[37] Carter, R., Segal, G., Macdonald, I., *Lectures on Lie Groups and Lie Algebras*, Cambridge University Press, 1995.

[38] Carter, S., West, A., Isoparametric systems and transnormality, *Proc. Lond. Math. Soc. III. Ser.* 51, 520-542 (1985).

[39] Carter, S., West, A., Generalized Cartan polynomials, *J. Lond. Math. Soc. II. Ser.* 32, 305-316 (1985).

[40] Carter, S., West, A., Partial tubes about immersed manifolds, *Geom. Dedicata* 54, 145-169 (1995).

[41] Cecil, T.E., *Lie Sphere Geometry. With Applications to Submanifolds*, Springer, 1992.

[42] Cecil, T.E., Taut and Dupin submanifolds, in: *Tight and Taut Submanifolds*, based on the workshop on differential systems, submanifolds and control theory, Berkeley, CA, USA, March 1-4, 1994, Cambridge University Press, 135-180 (1997).

[43] Cecil, T.E., Chi, Q.S., Jensen, G.R., Isoparametric hypersurfaces with four principal curvatures, preprint.

[44] Cecil, T.E., Ryan, P.J., Focal sets of submanifolds, *Pac. J. Math.* 78, 27-39 (1978).

[45] Cecil, T.E., Ryan, P.J., *Tight and Taut Immersions of Manifolds*, Pitman, 1985.

[46] Cecil, T.E., Ryan. P.J., The principal curvatures of the monkey saddle, *Am. Math. Mon.* 93, 380-382 (1986).

[47] Chavel, I., *Riemannian Geometry: A Modern Introduction*, Cambridge University Press, 1993.

[48] Chen, B.Y., *Geometry of Submanifolds*, Marcel Dekker, 1973.

[49] Chen, B.Y., Extrinsic spheres in Riemannian manifolds, *Houston J. Math.* 5, 319-324 (1979).

[50] Chen, B.Y., Totally umbilical Submanifolds, *Soochow J. Math.* 5, 9-37 (1980).

[51] Chen, B.Y., Classification of totally umbilical submanifolds in symmetric spaces, *J. Austral. Math. Soc. Ser. A* 30, 129-136 (1980).

[52] Chen, B.Y., Nagano, T., Totally geodesic submanifolds of symmetric spaces, I, *Duke Math. J.* 44, 745-755 (1977).

[53] Chen, B.Y., Nagano, T., Totally geodesic submanifolds of symmetric spaces, II, *Duke Math. J.* 45, 405-425 (1978).

[54] Chen, B.Y., Yano, K., Pseudo-umbilical submanifolds in a Riemannian manifold of constant curvature, in: *Differential Geometry, in Honor of Kentaro Yano*, Kinokuniya, 61-71 (1972).

[55] Chern, S.S., do Carmo, M.P., Kobayashi, S., Minimal submanifolds of a sphere with second fundamental form of constant length, in: *Functional Analysis and Related Fields*, Conf. Chicago 1968, 59-75 (1970).

[56] Conlon, L., Variational completeness and K-transversal domains, *J. Differ. Geom.* 5, 135-147 (1971).

[57] Conlon, L., A class of variationally complete representations, *J. Differ. Geom.* 7, 149-160 (1972).

[58] Console, S., Infinitesimally homogeneous submanifolds of Euclidean spaces, *Ann. Global Anal. Geom.* 12, 313-334 (1994).

[59] Console, S., Algebraic characterization of homogeneous submanifolds of space forms, *Boll. Unione Mat. Ital. VII. Ser. B* 10, 129-148 (1996).

[60] Console, S., Di Scala, A. J., Olmos, C., Holonomy and submanifold geometry, *Enseign. Math.* 48, 23-50 (2002).

[61] Console, S., Olmos, C., Submanifolds of higher rank, *Q. J. Math. Oxf. II. Ser.* 48, 309-321 (1997).

[62] Console, S., Olmos, C., Clifford systems, algebraically constant second fundamental form and isoparametric hypersurfaces, *Manuscr. Math.* 97, 335-342 (1998).

[63] Dadok, J., Polar coordinates induced by actions of compact Lie groups, *Trans. Am. Math. Soc.* 288, 125-137 (1985).

[64] Dajczer, M., Antonucci, M., Oliveira, G., Lima-Filho, P., Tojeiro, R., *Submanifolds and isometric immersions*, Publish or Perish, 1990.

[65] D'Atri, J.E., Certain isoparametric families of hypersurfaces in symmetric spaces, *J. Differ. Geom.* 14, 21-40 (1979).

[66] Dierkes, U., Hildebrandt, S., Küster, A., Wohlrab, O., *Minimal Surfaces I. Boundary Value Problems* and *Minimal Surfaces II. Boundary Regularity*, Springer, 1992.

[67] Di Scala, A.J., Reducibility of complex submanifolds of the complex Euclidean space, *Math. Z.* 235, 251-257 (2000).

[68] Di Scala, A.J., Minimal homogeneous submanifolds in Euclidean spaces, *Ann. Global Anal. Geom.* 21, 15-18 (2002).

[69] Di Scala, A.J., Olmos, C., The geometry of homogeneous submanifolds of hyperbolic space, *Math. Z.* 237, 199-209 (2001).

[70] Di Scala, A.J., Olmos, C., Submanifolds with curvature normals of constant length and the Gauss map, preprint.

[71] do Carmo, M.P., *Riemannian Geometry*, Birkhäuser, 1992.

[72] do Carmo, M.P., Wallach, N.R., Minimal immersions of spheres into spheres, *Ann. Math. (2)* 93, 43-62 (1971).

[73] Dorfmeister, J., Neher, E., Isoparametric hypersurfaces, case $g = 6$, $m = 1$, *Commun. Algebra* 13, 2299-2368 (1985).

[74] Duistermaat, J.J., Kolk, J.A.C., *Lie Groups*, Springer, 2000.

[75] Eberlein, P.B., *Geometry of Nonpositively Curved Manifolds*, University of Chicago Press, 1996.

[76] Eells, J., On equivariant harmonic maps, in: *Differential Geometry and Differential Equations*, Proc. Symp., Shanghai/China 1981, Science Press, 56-73 (1984).

[77] Ejiri, N., Minimal immersions of Riemannian products into real space forms, *Tokyo J. Math.* 2, 63–70 (1979).

[78] Erbacher, J., Reduction of the codimension of an isometric immersion, *J. Differ. Geom.* 5, 333-340 (1971).

[79] Eschenburg, J.H., Parallelity and extrinsic homogeneity, *Math. Z.* 229, 339-347 (1998).

[80] Eschenburg, J.H., Heintze, E., Extrinsic symmetric spaces and orbits of s-representations, *Manuscr. Math.* 88, 517-524 (1995). Erratum: ibid. 92, 408 (1997).

[81] Eschenburg, J.H., Heintze, E., Polar representations and symmetric spaces, *J. Reine Angew. Math.* 507, 93-106 (1999).

[82] Eschenburg, J.H., Olmos, C., Rank and symmetry of Riemannian manifolds, *Comment. Math. Helv.* 69, 483-499 (1994).

[83] Ferus, D., Produkt-Zerlegung von Immersionen mit paraller zweiter Fundamentalform, *Math. Ann.* 211, 1-5 (1974).

[84] Ferus, D., Immersions with parallel second fundamental form, *Math. Z.* 140, 87-93 (1974).

[85] Ferus, D., Immersionen mit paralleler zweiter Fundamentalform: Beispiele und Nicht-Beispiele, *Manuscr. Math.* 12, 153-162 (1974).

[86] Ferus, D., Symmetric submanifolds of Euclidean space, *Math. Ann.* 247, 81-93 (1980).

[87] Ferus, D., Karcher, H., Münzner, H.F., Cliffordalgebren und neue isoparametrische Hyperflächen, *Math. Z.* 177, 479-502 (1981).

[88] Fulton, W., Harris, J., *Representation Theory. A First Course*, Springer, 1991.

[89] Gallot, S., Hulin, D., Lafontaine, J., *Riemannian Geometry, 2nd ed.*, Springer, 1990.

[90] Gamkrelidze, R.V. (Ed.), *Geometry I: Basic Ideas and Concepts of Differential Geometry*, Springer, 1991.

[91] Gorodski C., Olmos C., Tojeiro R., Copolarity of isometric actions, preprint, math.DG/0208105.

[92] Gromoll, D., Grove, K., One-dimensional metric foliations in constant curvature spaces, in: *Differential Geometry and Complex Analysis*, Springer, 165-168 (1985).

[93] Harle, C.E., Isoparametric families of submanifolds, *Bol. Soc. Bras. Mat.* 13, 35-48 (1982).

[94] Heintze, E., Liu, X., Homogeneity of infinite-dimensional isoparametric submanifolds. *Ann. of Math.* (2) 149, 149–181 (1999).

[95] Heintze, E., Olmos, C., Normal holonomy groups and s-representations, *Indiana Univ. Math. J.* 41, 869-874 (1992).

[96] Heintze, E., Olmos, C., Thorbergsson, G., Submanifolds with constant principal curvatures and normal holonomy groups, *Int. J. Math.* 2, 167-175 (1991).

[97] Heintze, E., Palais, R., Terng, C.L., Thorbergsson, G., Hyperpolar actions and k-flat homogeneous spaces, *J. Reine Angew. Math.* 454, 163-179 (1994).

[98] Helgason, S., Totally geodesic spheres in compact symmetric spaces, *Math. Ann.* 165, 309-317 (1966).

[99] Helgason, S., *Differential Geometry, Lie Groups, and Symmetric Spaces*, Academic Press, 1978.

[100] Hermann, R., Existence in the large of totally geodesic submanifolds of Riemannian spaces, *Bull. Am. Math. Soc.* 66, 59-61 (1960).

[101] Hermann, R., Variational completeness for compact symmetric spaces, *Proc. Am. Math. Soc.* 11, 544-546 (1960).

[102] Hsiang, W.Y., Lawson, H.B.Jr., Minimal submanifolds of low cohomogeneity, *J. Differ. Geom.* 5, 1-38 (1971).

[103] Hsiang, W.Y., Palais, R., Terng, C.L., The topology of isoparametric submanifolds, *J. Differ. Geom.* 27, 423-460 (1988).

[104] Humphreys, J.E., *Reflection Groups and Coxeter Groups*, Cambridge University Press, 1990.

[105] Iwahori, N., Some remarks on tensor invariants of $\mathrm{O}(n), \mathrm{U}(n), \mathrm{Sp}(n)$, *J. Math. Soc. Japan* 10, 145-160 (1958).

[106] Iwata, K., Classification of compact transformation groups on cohomology quaternion projective spaces with codimension one orbits, *Osaka J. Math.* 15, 475-508 (1978).

[107] Iwata, K., Compact transformation groups on rational cohomology Cayley projective planes, *Tohoku Math. J. II. Ser.* 33, 429-442 (1981).

[108] Jost, J., *Riemannian Geometry and Geometric Analysis, 3rd ed.*, Springer, 2002.

[109] Karcher, H., A geometric classification of positively curved symmetric spaces and the isoparametric construction of the Cayley plane, *Astérisque* 163/164, 111-135 (1988).

[110] Kato, T., *Perturbation Theory for Linear Operators*, corr. printing of the 2nd ed., Springer, 1980.

[111] Kawakubo, K., *The Theory of Transformation Groups*, Oxford University Press, 1991.

[112] Kelly, E., Tight equivariant imbeddings of symmetric spaces, *J. Differ. Geom.* 7, 535-548 (1972).

[113] Knapp, A.W., *Lie Groups Beyond an Introduction*, Birkhäuser, 1996.

[114] Knarr, N., Kramer, L., Projective planes and isoparametric hypersurfaces, *Geom. Dedicata* 58, 193-202 (1995).

[115] Kobayashi, S., Isometric imbeddings of compact symmetric spaces, *Tohoku Math. J. II. Ser.* 20, 21-25 (1968).

[116] Kobayashi, S., Nagano, T., On filtered Lie algebras and geometric structures, I, *J. Math. Mech.* 13, 875-907 (1964).

[117] Kobayashi, S., Nomizu, K., *Foundations of Differential Geometry, I, II*, Interscience Publishers, 1963, 1969.

[118] Kobayashi, S., Takeuchi, M., Minimal imbeddings of R-spaces, *J. Differ. Geom.* 2, 203-215 (1968).

[119] Kollross, A., A classification of hyperpolar and cohomogeneity one actions, *Trans. Am. Math. Soc.* 354, 571-612 (2002).

[120] Kon, M., On some complex submanifolds in Kaehler manifolds, *Can. J. Math.* 26, 1442-1449 (1974).

[121] Kostant, B., Holonomy and the Lie algebra of infinitesimal motions of a Riemannian manifold, *Trans. Am. Math. Soc.* 80, 528-542 (1955).

[122] Kowalski, O., *Generalized Symmetric Spaces*, Springer, 1980.

[123] Kowalski, O., Counter-example to the "second Singer's theorem", *Ann. Global Anal. Geom.* 8, 211-214 (1990).

[124] Kowalski, O., Kulich, I., Generalized symmetric submanifolds of Euclidean spaces, *Math. Ann.* 277, 67-78 (1987).

[125] Leschke, K., Homogeneity and canonical connections of isoparametric manifolds, *Ann. Global Anal. Geom.* 15, 51-69 (1997).

[126] Leung, D.S.P., The reflection principle for minimal submanifolds of Riemannian symmetric spaces, *J. Differ. Geom.* 8, 153-160 (1973).

[127] Leung, D.S.P., On the classification of reflective submanifolds of Riemannian symmetric spaces, *Indiana Univ. Math. J.* 24, 327-339 (1974). Errata: ibid. 24, 1199 (1975).

[128] Leung, D.S.P., Reflective submanifolds. III. Congruency of isometric reflective submanifolds and corrigenda to the classification of reflective submanifolds, *J. Differ. Geom.* 14, 167-177 (1979).

[129] Leung, D.S.P., Reflective submanifolds. IV. Classification of real forms of Hermitian symmetric spaces, *J. Differ. Geom.* 14, 179-185 (1979).

[130] Levi-Civita, T., Famiglie di superficie isoparametriche nell'ordinario spazio euclideo, *Atti Accad. Naz. Lincei Rend. VI. Ser.* 26, 355-362 (1937).

[131] Lohnherr, M., Reckziegel, H., On ruled real hypersurfaces in complex space forms, *Geom. Dedicata* 74, 267-286 (1999).

[132] Loos, O., *Symmetric Spaces. I: General Theory. II: Compact Spaces and Classification*, W.A. Benjamin, 1969.

[133] Maeda, S., Ohnita, Y., Helical geodesic immersions into complex space forms, *Geom. Dedicata* 30, 93-114 (1989).

[134] Mashimo, K., Degree of the standard isometric minimal immersions of complex projective spaces into spheres, *Tsukuba J. Math.* 4, 133-145 (1980).

[135] Mashimo, K., Degree of the standard isometric minimal immersions of the symmetric spaces of rank one into spheres, *Tsukuba J. Math.* 5, 291-297 (1981).

[136] Mashimo, K., Tojo, K., Circles in Riemannian symmetric spaces, *Kodai Math. J.* 22, 1-14 (1999).

[137] Mercuri, F., Parallel and semi-parallel immersions into space forms, *Riv. Mat. Univ. Parma IV. Ser.* 17*, 91-108 (1991).

[138] Miyaoka, R., The linear isotropy group of $G_2/SO(4)$, the Hopf fibering and isoparametric hypersurfaces, *Osaka J. Math.* 30, 179-202 (1993).

[139] Moore, J.D., Isometric immersions of Riemannian products, *J. Differ. Geom.* 5, 159-168 (1971).

[140] Moore, J.D., Equivariant embeddings of Riemannian homogeneous spaces, *Indiana Univ. Math. J.* 25, 271-279 (1976).

[141] Montgomery, D., Samelson, H., Yang, C.T., Exceptional orbits of highest dimension, *Ann. of Math. II. Ser.* 64, 131-141 (1956).

[142] Montgomery, D., Yang, C.T., The existence of a slice, *Ann. of Math. II. Ser.* 65, 108-116 (1957).

[143] Mostert, P.S., On a compact Lie group acting on a manifold, *Ann. of Math. II. Ser.* 65, 447-455 (1957). Errata: ibid. 66, 589 (1957).

[144] Münzner, H.F., Isoparametrische Hyperflächen in Sphären, I, *Math. Ann.* 251, 57-71 (1980).

[145] Münzner, H.F., Isoparametrische Hyperflächen in Sphären, II, *Math. Ann.* 256, 215-232 (1981).

[146] Nagano, T., Transformation groups on compact symmetric spaces, *Trans. Am. Math. Soc.* 118, 428-453 (1965).

[147] Nagano, T., The involutions of compact symmetric spaces, *Tokyo J. Math.* 11, 57-79 (1988).

[148] Nagano, T., The involutions of compact symmetric spaces, II, *Tokyo J. Math.* 15, 39-82 (1992).

[149] Nagano, T., Tanaka, M.S., The involutions of compact symmetric spaces, III, *Tokyo J. Math.* 18, 193-212 (1995).

[150] Nagano, T., Tanaka, M.S., The involutions of compact symmetric spaces, IV, *Tokyo J. Math.* 22, 193-211 (1999).

[151] Nagano, T., Tanaka, M.S., The involutions of compact symmetric spaces, V, *Tokyo J. Math.* 23, 403-416 (2000).

[152] Nagano, T., Sumi, M., The spheres in symmetric spaces, *Hokkaido Math. J.* 20, 331-352 (1991).

[153] Naitoh, H., Totally real parallel submanifolds in $P^n(c)$, *Tokyo J. Math.* 4, 279-306 (1981).

[154] Naitoh, H., Parallel submanifolds of complex space forms, I, *Nagoya Math. J.* 90, 85-117 (1983).

[155] Naitoh, H., Parallel submanifolds of complex space forms, II, *Nagoya Math. J.* 91, 119-149 (1983).

[156] Naitoh, H., Symmetric submanifolds of compact symmetric spaces, *Tsukuba J. Math.* 10, 215-242 (1986).

[157] Naitoh, H., Compact simple Lie algebras with two involutions and submanifolds of compact symmetric spaces, I, *Osaka J. Math.* 30, 653-690 (1993).

[158] Naitoh, H., Compact simple Lie algebras with two involutions and submanifolds of compact symmetric spaces, II, *Osaka J. Math.* 30, 691-732 (1993).

[159] Naitoh, H., Grassmann geometries on compact symmetric spaces of general type, *J. Math. Soc. Japan* 50, 557-592 (1998).

[160] Naitoh, H., Grassmann geometries on compact symmetric spaces of exceptional type, *Japanese J. Math.* 26, 157-206 (2000).

[161] Naitoh, H., Grassmann geometries on compact symmetric spaces of classical type, *Japanese J. Math.* 26, 219-319 (2000).

[162] Naitoh, H., Takeuchi, M., Totally real submanifolds and symmetric bounded domains, *Osaka J. Math.* 19, 717-731 (1982).

[163] Naitoh, H., Takeuchi, M., Symmetric submanifolds of symmetric spaces, *Sugaku Exp.* 2, 157-188 (1989).

[164] Nash, J., The imbedding problem for Riemannian manifolds, *Ann. of Math. II. Ser.* 63, 20-63 (1956).

[165] Nakagawa, H., Takagi, R., On locally symmetric Kaehler submanifolds in a complex projective space. *J. Math. Soc. Japan* 28, 638-667 (1976).

[166] Nikolaevskij, Y.A., Totally umbilical submanifolds of symmetric spaces, *Mat. Fiz. Anal. Geom.* 1, 314-357 (1994).

[167] Nikolayevsky, I., Osserman conjecture in dimension $n \neq 8, 16$, preprint.

[168] Nölker, S., Isometric immersions with homothetical Gauss map, *Geom. Dedicata* 34, (1990), 271-280.

[169] Nomizu, K., Some results in E. Cartan's theory of isoparametric families of hypersurfaces, *Bull. Am. Math. Soc.* 79, 1184-1188 (1974).

[170] Nomizu, K., Elie Cartan's work on isoparametric families of hypersurfaces, *Proc. Symp. Pure Math.* 27 Part 1, 191-200 (1975).

[171] Nomizu, K., Yano, K., On circles and spheres in Riemannian geometry, *Math. Ann.* 210, 163-170 (1974).

[172] Ohnita, Y., The first standard minimal immersions of compact irreducible symmetric spaces, in: *Differential Geometry of Submanifolds*, Proc. Conf., Kyoto/Japan 1984, Springer, 37-49 (1984).

[173] Olmos, C., The normal holonomy group, *Proc. Am. Math. Soc.* 110, 813-818 (1990).

[174] Olmos, C., Isoparametric submanifolds and their homogeneous structures, *J. Differ. Geom.* 38, 225-234 (1993).

[175] Olmos, C., Homogeneous submanifolds of higher rank and parallel mean curvature, *J. Differ. Geom.* 39, 605-627 (1994).

[176] Olmos, C., Orbits of rank one and parallel mean curvature, *Trans. Am. Math. Soc.* 347, 2927-2939 (1995).

[177] Olmos, C., Salvai, M., Holonomy of homogeneous vector bundles and polar representations, *Indiana Univ. Math. J.* 44, 1007-1015 (1995).

[178] Olmos, C., Sánchez, C., A geometric characterization of the orbits of s-representations, *J. Reine Angew. Math.* 420, 195-202 (1991).

[179] Olmos, C., Will, A., Normal holonomy in Lorentzian space and submanifold geometry, *Indiana Univ. Math. J.* 50, 1777-1788 (2001).

[180] O'Neill, B., *Semi-Riemannian Geometry. With Applications to Relativity*, Academic Press, 1983.

[181] Onishchik, A.L., On totally geodesic submanifolds of symmetric spaces, *Geom. Metody Zadachakh Algebry Anal.* 2, 64-85 (1980).

[182] Onishchik, A.L., Gamkrelidze, R.V. (Eds.), *Lie Groups and Lie Algebras I. Foundations of Lie Theory. Lie Transformation Groups*, Springer, 1993.

[183] Onishchik, A.L., Vinberg, E.B., Gorbatsevich, V.V. (Eds.), *Lie Groups and Lie Algebras III. Structure of Lie Groups and Lie Algebras*, Springer, 1994.

[184] Ozeki, H., Takeuchi, M., On some types of isoparametric hypersurfaces in spheres, I, *Tohoku Math. J. II. Ser.* 27, 515-559 (1975).

[185] Ozeki, H., Takeuchi, M., On some types of isoparametric hypersurfaces in spheres, II, *Tohoku Math. J. II. Ser.* 28, 7-55 (1976).

[186] Palais, R.S., Terng, C.L., A general theory of canonical forms, *Trans. Am. Math. Soc.* 300, 771-789 (1987).

[187] Palais, R.S., Terng, C.L., *Critical Point Theory and Submanifold Geometry*, Springer, 1988.

[188] Petersen, P., *Riemannian Geometry*, Springer, 1998.

[189] Poor, W.A., *Differential Geometric Structures*, McGraw-Hill, 1981.

[190] Reckziegel, H., Krümmungsflächen von isometriscnen Immersionen in Räume konstanter Krümmung, *Math. Ann.* 223, 169-181 (1976).

[191] Reckziegel, H., On the eigenvalues of the shape operator of an isometric immersion into a space of constant curvature, *Math. Ann.* 243, 71-82 (1979).

[192] Reckziegel, H., On the problem whether the image of a given differentiable map into a Riemannian manifold is contained in a submanifold with parallel second fundamental form, *J. Reine Angew. Math.* 325, 87-104 (1981).

[193] Reinhart, B.L., Foliated manifolds with bundle-like metrics, *Ann. Math. (2)* 69, 119-132 (1959).

[194] Sagle, A.A., A note on triple systems and totally geodesic submanifolds in a homogeneous space, *Nagoya Math. J.* 32, 5-20 (1968).

[195] Sakai, T., *Riemannian Geometry*, American Mathematical Society, 1996.

[196] Salamon, S., *Riemannian Geometry and Holonomy Groups*, Longman, 1989.

[197] Sánchez, C., k-symmetric submanifolds of $\mathbb{R}^N$, *Math. Ann.* 270, 297–316 (1985).

[198] Sánchez, C., A characterization of extrinsic k-symmetric submanifolds of $\mathbb{R}^N$, *Rev. Un. Mat. Argentina* 38, No.1-2, 1-15 (1992).

[199] Segre, B., Famiglie di ipersuperficie isoparametriche negli spazi euclidei ad un qualunque numero di dimensioni, *Atti Accad. Naz. Lincei Rend. VI. Ser.* 27, 203-207 (1938).

[200] Simons, J., On the transitivity of holonomy systems, *Ann. Math. (2)* 76, 213-234 (1962).

[201] Singer, I.M., Infinitesimally homogeneous spaces, *Commun. Pure Appl. Math.* 13, 685-697 (1960).

[202] Singley, D.H., Smoothness theorems for the principal curvatures and principal vectors of a hypersurface, *Rocky Mt. J. Math.* 5, 135-144 (1975).

[203] Spivak, M., *A Comprehensive Introduction to Differential Geometry, Vols. I-V, 2nd ed.*, Publish or Perish, 1979.

[204] Strübing, W., Symmetric submanifolds of Riemannian manifolds, *Math. Ann.* 245, 37-44 (1979).

[205] Strübing, W., Isoparametric submanifolds, *Geom. Dedicata* 20, 367-387 (1986).

[206] Szenthe, J., A generalization of the Weyl group, *Acta Math. Hung.* 41, 347-357 (1983).

[207] Szenthe, J., Orthogonally transversal submanifolds and the generalizations of the Weyl group, *Period. Math. Hung.* 15, 281-299 (1984).

[208] Szenthe, J., Isometric actions having orthogonally transversal submanifolds, *Colloq. Math. Soc. János Bolyai* 46, 1155-1164 (1988).

[209] Takagi, R., On homogeneous real hypersurfaces in a complex projective space, *Osaka J. Math.* 10, 495-506 (1973).

[210] Takahashi, T., Minimal immersions of Riemannian manifolds, *J. Math. Soc. Japan* 18, 380-385 (1966).

[211] Takeuchi, M., Cell decompositions and Morse equalities on certain symmetric spaces, *J. Fac. Sci. Univ. Tokyo Sect. IA Math.* 12, 81-192 (1965).

[212] Takeuchi, M., Parallel submanifolds of space forms, in: *Manifolds and Lie Groups*, Papers in Honor of Y. Matsushima, Prog. Math. 14, 429-447 (1981)

[213] Takeuchi, M., Stability of certain minimal submanifolds of compact Hermitian symmetric spaces, *Tohoku Math. J. II. Ser.* 36, 293-314 (1984).

[214] Takeuchi, M., *Lie Groups I, II*, American Mathematical Society, 1991.

[215] Takeuchi, M., Kobayashi, S., Minimal imbeddings of R-spaces, *J. Differ. Geom.* 2, 203-215 (1968).

[216] Terng, C.L., Isoparametric submanifolds and their Coxeter groups, *J. Differ. Geom.* 21, 79-107 (1985).

[217] Terng, C.L., Submanifolds with flat normal bundle, *Math. Ann.* 277, 95-111 (1987).

[218] Tetzlaff, K., *Satz von Dadok und normale Holonomie*, Univ. Augsburg, 1993.

[219] Thorbergsson, G., Isoparametric foliations and their buildings, *Ann. Math. (2)* 133, 429-446 (1991).

[220] Thorbergsson, G., A survey on isoparametric hypersurfaces and their generalizations, in: *Handbook of Differential Geometry, Vol. I*, North-Holland, 963-995 (2000).

[221] Toth, G., *Harmonic Maps and Minimal Immersions Through Representation Theory*, Academic Press, 1990.

[222] Tricerri, F., Locally homogeneous Riemannian manifolds, *Rend. Semin. Mat. Torino* 50, 411-426 (1992).

[223] Tricerri, F., Vanhecke, L., *Homogeneous Structures on Riemannian Manifolds*, Cambridge University Press, 1983.

[224] Tricerri, F., Vanhecke, L., Special homogeneous structures on Riemannian manifolds, *Colloq. Math. Soc. János Bolyai* 46, 1211-1246 (1988).

[225] Tricerri, F., Vanhecke, L., Curvature homogeneous Riemannian manifolds, *Ann. Sci. Éc. Norm. Supér. IV. Sér.* 22, 535-554 (1989).

[226] Tsukada, K., Parallel submanifolds in a quaternion projective space, *Osaka J. Math.* 22, 187-241 (1985).

[227] Tsukada, K., Parallel submanifolds of Cayley plane, *Sci. Rep. Niigata Univ. Ser. A* 21, 19-32 (1985).

[228] Tsukada, K., Parallel Kähler submanifolds of Hermitian symmetric spaces, *Math. Z.* 190, 129-150 (1985).

[229] Tsukada, K., Totally geodesic submanifolds of Riemannian manifolds and curvature-invariant subspaces, *Kodai Math. J.* 19, 395-437 (1996).

[230] Uchida, F., Classification of compact transformation groups on cohomology complex projective spaces with codimension one orbits, *Japan. J. Math.* 3, 141-189 (1977).

[231] Varadarajan, V.S., *Lie Groups, Lie Algebras, and Their Representations*, Springer, 1984.

[232] Vargas, J., A symmetric space of noncompact type has no equivariant isometric immersions into the Euclidean space, *Proc. Am. Math. Soc.* 81, 149-151 (1981).

[233] Vilms, J., Submanifolds of Euclidean space with parallel second fundamental form, *Proc. Am. Math. Soc.* 32, 263-267 (1972).

[234] Walden, R., Untermannigfaltigkeiten mit paralleler zweiter Fundamentalform in euklidischen Räumen und Spären, *Manuscr. Math.* 10, 91-102 (1973).

[235] Wallach, N., Minimal immersions of symmetric spaces into spheres, in: *Symmetric spaces* (Short Courses, Washington Univ., St. Louis, Mo., 1969–1970), Pure and Appl. Math., Vol. 8, Dekker, 1-40 (1972).

[236] Wang, H.C., Closed manifolds with homogeneous complex structures, *Amer. J. Math.* 76, 1-32 (1954).

[237] Wang, Q.M., Isoparametric hypersurfaces in complex projective spaces, in: *Differential Geometry and Differential Equations*, Proc. 1980 Beijing Sympos., Vol. 3, 1509-1523 (1982).

[238] Wang, Q.M., Isoparametric maps of Riemannian manifolds and their applications, *Adv. Sci. China Math.* 2, 79-103 (1986).

[239] Wang, Q.M., Isoparametric functions on Riemannian manifolds, I, *Math. Ann.* 277, 639-646 (1987).

[240] Will, A., Homogeneous submanifolds of the hyperbolic space, *Rend. Sem. Mat. Torino* 56, 1-4 (1998)

[241] Will, A., Isoparametric Riemannian submanifolds of $\mathbb{R}^{n,k}$, *Geom. Dedicata* 76, 155-164 (1999).

[242] Wolf, J.A., Elliptic spaces in Grassmann manifolds, *Illinois J. Math.* 7, 447-462 (1963).

[243] Wolf, J.A., *Spaces of constant curvature, 3rd ed.*, Publish Perish, 1974.

[244] Wu, B., Isoparametric submanifolds of hyperbolic spaces, *Trans. Am. Math. Soc.* 331, 609-626 (1992).

[245] Yamabe, H., On an arcwise connected subgroup of a Lie group, *Osaka Math. J.* 2, 13-14 (1950).

List of Figures

List of Tables

Index